'Finding a reader-friendly balance between humour, irreverence and anally retentive scholarship, Pinker unpacks a remarkable variety of facts associated with the distinction between regular and irregular English words and their structure . . . the book provides a scholarly, persuasive, enjoyable and eminently readable account of important language phenomena'   David Poeppel, *Nature*

'Not only does Pinker breathe life into the topic, he makes the reading breathtakingly exciting'
                                   Howard Richler, *Montreal Gazette*

'[An] excellent work of popular science. Steven Pinker is a masterful explainer, a great collector of amusing examples, and smart'   Thomas Nagel, *The New Republic*

'Pinker has rightly been called the only linguist who can write readable prose, and if there is anybody who can make irregular verbs gripping it is he . . . a fascinating voyage of discovery'
                                   Matt Ridley, *Sunday Telegraph*

'Like Pinker's earlier best-selling books, *The Language Instinct* and *How the Mind Works*, *Words and Rules* is written in Pinkerese, witty, knowing, and full of apt allusions drawn from all over the place – literature, cartoons, popular songs, baseball, and recent American political history . . . Pinker has been called psychology's first rock star, but this is not pop psychology of the sort that lurks in airport bookstores or claims that men and women come from different planets. It is perhaps Pinker's most serious book to date, but still has the capacity to enlighten and inform at several levels of engagement'
                                   Michael Corballis, *New Zealand Listener*

'[Pinker] explains technical matters in day-to-day language . . . He does not oversimplify the issues, and his summarizing of academic articles is exemplary . . . [Readers will] have a good idea of how a thorough researcher approaches a task. Like a hungry dog, Pinker chews just about every possible piece of meat off this academic bone, and licks it clean'
                                   Jean Aitchison, *Literary Review*

Steven Pinker, a native of Montreal, studied experimental psychology at McGill University and Harvard University. He is Professor and a MacVicar Faculty Fellow in the Department of Brain and Cognitive Sciences at the Massachusetts Institute of Technology. Pinker has studied many aspects of language and visual cognition, with a focus on language acquisition in children. He is a fellow of several scientific societies, and has been awarded research prizes from the National Academy of Sciences and the American Psychological Association, two teaching prizes from MIT and book prizes from the American Psychological Association, the Linguistics Society of America and the *Los Angeles Times*. Steven Pinker is the author of the landmark bestsellers *The Language Instinct* (1994) and *How the Mind Works* (1998).

Words and Rules

The Ingredients of Language

STEVEN PINKER

PHŒNIX

A PHOENIX PAPERBACK

First published in Great Britain by Weidenfeld & Nicolson in 1999
This paperback edition published in 2000 by Phoenix,
an imprint of Orion Books Ltd,
Orion House, 5 Upper St Martin's Lane,
London WC2H 9EA

SCIENCE MASTERS are published in association with Basic Books,
A Member of Perseus Books, L.L.C.

A CIP catalogue record for this book
is available from the British Library.

ISBN 0 75381 025 5

Typeset at The Spartan Press Ltd,
Lymington, Hants
Printed and bound in Great Britain by
Clays Ltd, St Ives plc

TO THE PSYMORGS

CONTENTS

PREFACE

This book tries to illuminate the nature of language and mind by choosing a single phenomenon and examining it from every angle imaginable. That phenomenon is regular and irregular verbs, the bane of every language student.

At first glance that approach might seem to lie in the great academic tradition of knowing more and more about less and less until you know everything about nothing. But please don't put the book down just yet. Seeing the world in a grain of sand is often the way of science, as when geneticists agreed to study the lowly fruit fly so that their findings might cumulate into a deep understanding that would have been impossible had each scientist started from scratch with a different organism. Like fruit flies, regular and irregular verbs are small and easy to breed, and they contain, in an easily visible form, the machinery that powers larger phenomena in all their glorious complexity.

Since the dawn of the modern study of the mind in the late 1950s, children's language errors such as *breaked* and *holded*, which could not have been parroted from their parents' speech, have served as a vivid reminder that the mind of the child is not a sponge, but actively assembles words and concepts into new combinations guided by rules and regularities. Every new theory of the mind has tried to account for this feat of childhood creativity, and perhaps the most heated debate in contemporary cognitive science – on whether the mind is more like an artificial neural network or a symbol-manipulating computer – has used it as a benchmark.

The exploration of regular and irregular verbs will take us from the prehistoric tribes that originated our language to the brain-imaging and gene-sequencing technologies of the new millennium. Perhaps best of all, this case study immerses us in that mixture of mathematical beauty and human quirkiness called language. Discovering the rationale of a curious word or expression can bring the same blissful intellectual 'click' as completing a crossword puzzle or appreciating a witticism.

For the past dozen years my research has concentrated on regular and irregular verbs, and the pleasure of coming to understand one thing really well has been surpassed only by the pleasure of working with extraordinary people who were just as consumed by the topic: the members of the Psychology of Morphology Group at MIT, the Psymorgs. Many of the big ideas in this book originated with my friend and collaborator Alan Prince of Rutgers University, and others were thought up or brought to life by former graduate students, post-doctoral fellows, and research assistants: Chris Collins, Marie Coppola, Jenny Ganger, Greg Hickok, Michelle Hollander, John J. Kim, Gary Marcus, Sandeep Prasada, Jaemin Rhee, Annie Senghas, William Snyder, Karin Stromswold, Michael Ullman, and Fei Xu. Marcus and Ullman in particular had their own big ideas that I could not have dreamed of. This book is dedicated to all of them, with gratitude and affection.

It also has been a pleasure to work with Harald Clahsen, Richard Wiese, and Iris Berent on their ingenious studies of German and Hebrew. Hilary Bromberg and Cyrus Shaoul, former undergraduates at MIT, made important contributions in their senior research projects. Thanks go as well to other collaborators, especially Ursula Brinkmann, Suzanne Corkin, John Growdon, Walter Koroshetz, T. John Rosen, and Joseph Shimron.

I am happy to acknowledge the expert help of Patricia Claffey, the librarian of the Teuber Library in the Department of Brain and Cognitive Sciences at MIT, and of my assistants, Allison Baker, Sonia Chawla, and Marie Lamb.

I thank my editors, John Donatich of Basic Books, and Toby Mundy of Weidenfeld and Nicolson, for their invaluable encouragement and advice on every aspect of the book, and Michael Wilde for his excellent copyediting. I am particularly indebted to the friends and colleagues who generously provided detailed comments on an earlier draft: Iris Berent, Alfonso Caramazza,

Judith Rich Harris, David Kemmerer, Samuel Jay Keyser, Beth Levin, Gary Marcus, Sandeep Prasada, and Michael Ullman. My agent, John Brockman, conceived the Science Masters series in which this book is published, and I thank him for his advice and his efforts on my behalf.

I am grateful for the constant encouragement and support of my extended family, the Pinkers, Boodmans, and Subbiah-Adams. I am especially grateful to my wife, Ilavenil Subbiah, who designed the illustrations, commented on the manuscript, offered advice on every aspect, and was always there with love and support.

This research was funded by the National Institutes of Health (grant HD–18381), the National Science Foundation (BNS 91–09766), the McDonnell-Pew Center for Cognitive Neuroscience at MIT, the German–American Collaborative Research program of the American Council of Learned Societies and the Deutscher Akademischer Austauschdienst, and MIT's Undergraduate Research Opportunities Program.

Words and Rules

1

THE INFINITE LIBRARY

Language comes so naturally to us that it is easy to forget what a strange and miraculous gift it is. All over the world members of our species fashion their breath into hisses and hums and squeaks and pops and listen to others do the same. We do this, of course, not only because we like the sounds but because details of the sounds contain information about the intentions of the person making them. We humans are fitted with a means of sharing our ideas, in all their unfathomable vastness. When we listen to speech, we can be led to think thoughts that have never been thought before and that never would have occurred to us on our own. Behold, the bush burned with fire, and the bush was not consumed. Man is born free, and everywhere he is in chains. Emma Woodhouse, handsome, clever, and rich, with a comfortable home and happy disposition, seemed to unite some of the best blessings of existence. Energy equals mass times the speed of light squared. I have found it impossible to carry the heavy burden of responsibility and to discharge my duties as King without the help and support of the woman I love.

Language has fascinated people for thousands of years, and linguists have studied every detail, from the number of languages spoken in New Guinea to why we say *razzle-dazzle* instead of *dazzle-razzle*. Yet to me the first and deepest challenge in understanding language is accounting for its boundless expressive power. What is the trick behind our ability to fill one another's heads with so many different ideas?

The premise of this book is that there are two tricks, words and rules. They work by different principles, are learned and used in different ways, and may even reside in different parts of the brain. Their border disputes shape and reshape languages over centuries, and make language not only a tool for communication but also a medium for wordplay and poetry and an heirloom of endless fascination.

~

The first trick, the word, is based on a memorized arbitrary pairing between a sound and a meaning. 'What's in a name?' asks Juliet. 'That which we call a rose by any other name would smell as sweet.' What's in a name is that everyone in a language community tacitly agrees to use a particular sound to convey a particular idea. Although the word *rose* does not smell sweet or have thorns, we can use it to convey the idea of a rose because all of us have learned, at our mother's knee or in the playground, the same link between a noise and a thought. Now any of us can convey the thought by making the noise.

The theory that words work by a conventional pairing of sound and meaning is not banal or uncontroversial. In the earliest surviving debate on linguistics, Plato has Hermogenes say, 'Nothing has its name by nature, but only by usage and custom.' Cratylus disagrees: 'There is a correctness of name existing by nature for everything: a name is not simply that which a number of people jointly agree to call a thing.' Cratylus is a creationist, and suggests that 'a power greater than man assigned the first names to things.' Today, those who see a correctness of names might attribute it instead to onomatopoeia (words such as *crash* and *oink* that sound like what they mean) or to sound symbolism (words such as *sneer, cantankerous,* and *mellifluous* that naturally call to mind the things they mean).

Today this debate has been resolved in favor of Hermogenes' conventional pairing. Early in this century Ferdinand de Saussure, a founder of modern linguistics, called such pairing the *arbitrary sign* and made it a cornerstone of the study of language.[1] Onomatopoeia and sound symbolism certainly exist, but they are asterisks to the far more important principle of the arbitrary sign – or else we would understand the words in every foreign language instinctively, and never need a dictionary for our own! Even the most obviously onomatopoeic words – those for animal sounds – are notoriously unpredictable, with pigs oinking *boo-boo* in Japan and dogs barking *gong-gong* in Indonesia. Sound

symbolism, for its part, was no friend of the American woman in the throes of labor who overheard what struck her as the most beautiful word in the English language and named her newborn daughter *Meconium*, the medical term for fetal excrement.[2]

Though simple, the principle of the arbitrary sign is a powerful tool for getting thoughts from head to head. Children begin to learn words before their first birthday, and by their second they hoover them up at a rate of one every two hours. By the time they enter school children command 13,000 words, and then the pace picks up, because new words rain down on them from both speech and print. A typical high-school graduate knows about 60,000 words; a literate adult, perhaps twice that number.[3] People recognize words swiftly. The meaning of a spoken word is accessed by a listener's brain in about a fifth of a second, before the speaker has finished pronouncing it.[4] The meaning of a printed word is registered even more quickly, in about an eighth of a second.[5] People produce words almost as rapidly: It takes the brain about a quarter of a second to find a word to name an object, and about another quarter of a second to program the mouth and tongue to pronounce it.[6]

The arbitrary sign works because a speaker and a listener can call on identical entries in their mental dictionaries. The speaker has a thought, makes a sound, and counts on the listener to hear the sound and recover that thought. To depict an entry in the mental dictionary we need a way of showing the entry itself, as well as its sound and meaning. The entry for a word is simply its address in one's memory, like the location of the boldfaced entry for a word in a real dictionary. It's convenient to use an English letter sequence such as *r-o-s-e* to stand for the entry, as long as we remember this is just a mnemonic tag that allows us to remember which word the entry corresponds to; any symbol, such as *42759*, would do just as well. To depict the word's sound, we can use a phonetic notation, such as [rōz].* The meaning of a word is a link to an entry in the

---

*This book uses a simplified phonetic notation similar to that found in dictionaries, in which the long vowels *ā* in *bait*, *ē* in *beet*, *ī* in *bite*, *ō* in *boat* and *ū* in *boot* are distinguished from the short vowels *ă* in *bat*, *ĕ* in *bet*, *ĭ* in *bit*, *ŏ* in *pot*, and *ŭ* in *but*. An unadorned *a* stands for the first vowel in *father* or *papa*. The symbol *i* is used for the neutral vowel in the suffix of *melted* and *Rose's* (e.g., *mĕltid, rōziz*), a version of the vowel sometimes called *schwa*.

'Long vowel' and other technical terms in linguistics, psycholinguistics, and neuroscience are defined in the Glossary.

person's mental *encyclopedia*, which captures the person's concept of a rose. For convenience we can symbolize it with a picture, such as ❀. So a mental dictionary entry looks something like this:

> rose
>   sound: *rōz*
>   meaning: ❀

A final component is the word's part of speech, or grammatical category, which for *rose* is noun (N):

> rose
>   sound: *rōz*
>   meaning: ❀
>   part of speech: N

And that brings us to the second trick behind the vast expressive power of language.

~

People do not just blurt out isolated words but rather *combine* them into phrases and sentences, in which the meaning of the combination can be inferred from the meanings of the words and the way they are arranged. We talk not merely of roses, but of the red rose, proud rose, sad rose of all my days. We can express our feelings about bread and roses, guns and roses, the War of the Roses, or days of wine and roses. We can say that lovely is the rose, roses are red, or a rose is a rose is a rose. When we combine words, their arrangement is crucial: *Violets are red, roses are blue,* though containing all the ingredients of the familiar verse, means something very different. We all know the difference between *young women looking for husbands* and *husbands looking for young women,* and that *looking women husbands young for* doesn't mean anything at all.

Inside everyone's head there must be a code or protocol or set of rules that specifies how words may be arranged into meaningful combinations. Modern linguists call it a *grammar*, some-

times a *generative grammar* to distinguish it from the grammars used to teach foreign languages or to teach the dos and don'ts of formal prose.

A grammar assembles words into phrases according to the words' part-of-speech categories, such as noun and verb. To highlight a word's category and reduce visual clutter often it is convenient to omit the sound and meaning and put the category label on top:

N
|
rose

Similarly, the word *a*, an article or *determiner*, would look like this:

det
|
a

They can then be joined into the phrase *a rose* by a rule that joins a determiner to a noun to yield a noun phrase (NP). The rule can be shown as a set of connected branches; this one says 'a noun phrase may be composed of a determiner followed by a noun':

The symbols at the bottom of the branches are like slots into which words may be plugged, as long as the words have the same labels growing out of their tops. Here is the result, the phrase *a rose*:

With just two more rules we can build a complete toy grammar. One rule defines a predicate or verb phrase (VP); the rule says that a verb phrase may consist of a verb followed by its direct object, a noun phrase:

The other rule defines the sentence itself (S). This rule says that a sentence may be composed from a noun phrase (the subject) followed by a verb phrase (the predicate):

When words are plugged into phrases according to these rules, and the phrases are plugged into bigger phrases, we get a complete sentence, such as *A rose is a rose*:

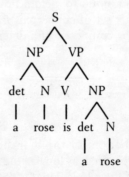

Other parts of the rules, not shown here, specify the meaning of the new combination. For example, the complete NP rule says that the meaning of *the yellow rose of Texas* is based on the meaning of *rose*, which is called the *head* of the phrase, and that the other words modify the head in various ways: *yellow* specifies a distinctive trait, *Texas* its location.

These rules, though crude, illustrate the fantastic expressive power made available by grammar. First, the rules are *productive*. By specifying a string of *kinds* of words rather than a string of *actual* words, the rules allow us to assemble new sentences on the fly and not regurgitate preassembled clichés – and that allows us to convey unprecedented combinations of ideas. Though we often speak of roses being red, we could talk about violets being red if the desire came over us (perhaps to announce a new hybrid), because the rule allows us to insert *violets* into the N slot just as easily as *roses*.

Second, the symbols contained by the rules are *symbolic* and hence *abstract*. The rule doesn't say, 'A sentence may begin with a bunch of words referring to a kind of flower'; rather, it says, 'A sentence may begin with an NP,' where NP is a symbol or variable that can be replaced by any noun, just as x or y in a mathematical formula can be replaced by any number. We can use the rules to talk about flowers and their colors and smells, but we can just as easily use them to talk about karma or quarks or floob-boober-bab-boober-bubs (who, according to Dr Seuss, bounce in the water like blubbery tubs).

Third, the rules are *combinatorial*. They don't just have a single slot, like a fill-in-the-blank exam question; *every* position in the sentence offers a choice of words from a lengthy menu. Say everyday English has four determiners (*a, any, one*, and *the*) and ten thousand nouns. Then the rule for a noun phrase allows four choices for the determiner, followed by ten thousand choices for the head noun, yielding $4 \times 10,000 = 40,000$ ways to utter a noun phrase. The rule for a sentence allows these forty thousand subjects to be followed by any of four thousand verbs, providing $40,000 \times 4,000 = 160,000,000$ ways to utter the first three words of a sentence. Then there are four choices for the determiner of the object (640 million four-word beginnings) followed by ten thousand choices for the head noun of the object, or $640,000,000 \times 10,000 = 6,400,000,000,000$ (6.4 trillion) five-word sentences. Suppose it takes five seconds to produce one of these sentences. To crank them all out, from *The abandonment abased the abbey* and *The abandonment abased the abbot*, through *The abandonment abased the zoologist*, all the way to *The zoologist zoned the zoo*, would take a million years.

Many such combinations are ungrammatical of course, owing

to various complications I haven't mentioned – for example, you can't say *The Aaron, a abandonment,* or *The abbot abase the abbey.* And most of the combinations are nonsensical: Abandonments can't abbreviate, and abbeys can't abet. Yet even with these restrictions the expressive range of a grammar is astonishing. The psychologist George Miller once conservatively estimated that if speakers keep a sentence perfectly grammatical and sensible as they choose their words, their menu at each point offers an average of about ten choices (at some points there are many more than ten choices; at others, only one or two).[7] That works out to one hundred thousand five-word sentences, one million six-word sentences, ten million seven-word sentences, and so on. A sentence of twenty words is not at all uncommon (the preceding sentence has twenty words before *and so on*), and there are about one hundred million trillion of them in English. For comparison, that is about a hundred times the number of seconds since the birth of the universe.

Grammar is an example of a combinatorial system, in which a small inventory of elements can be assembled by rules into an immense set of distinct objects. Combinatorial systems obey what Miller calls the Exponential Principle: The number of possible combinations grows exponentially (geometrically) with the size of the combination.[8] Combinatorial systems can generate inconceivably vast numbers of products. Every kind of molecule in the universe is assembled from a hundred-odd chemical elements; every protein building block and catalyst in the living world is assembled from just twenty amino acids. Even when the number of products is smaller, a combinatorial system can capture them all and provide enormous savings in storage space. Eight bits define 2^8 = 256 distinct bytes, which is more than enough for all the numerals, punctuation marks, and upper- and lowercase letters in our writing system. This allows computers to be built out of identical specks of silicon that can be in just two states, instead of the dozens of pieces of type that once filled typesetters' cases. Billions of years ago life on Earth settled on a code in which a string of three bases in a DNA molecule became the instruction for selecting one amino acid when assembling a protein. There are four kinds of bases, so a three-base string allows for $4 \times 4 \times 4 = 64$ possibilities. That is enough to give each of the twenty amino acids its own string, with plenty left over for the start and stop instructions that begin and end the protein. Two bases would have been too few ($4 \times 4 = 16$), four more than needed ($4 \times 4 \times 4 \times 4 = 256$).

Perhaps the most vivid description of the staggering power of a combinatorial system is in Jorge Luis Borges's story 'The Library of Babel.'[9] The library is a vast network of galleries with books composed of all the combinations of twenty-two letters, the comma, the period, and the space. Somewhere in the library is a book that contains the true history of the future (including the story of your death), a book of prophecy that vindicates the acts of every man in the universe, and a book containing the clarification of the mysteries of humanity. People roamed the galleries in a futile search for those texts from among the untold number of books with false versions of each revelation, the millions of facsimiles of a given book differing by a character, and, of course, the miles and miles of gibberish. The narrator notes that even when the human species goes extinct, the library, that space of combinatorial possibilities, will endure: 'illuminated, solitary, infinite, perfectly motionless, equipped with precious volumes, useless, incorruptible, secret.'

Technically, Borges needn't have described the library as 'infinite.' At eighty characters a line, forty lines a page, and 410 pages a book, the number of books is around $10^{1,800,000}$, or 1 followed by 1.8 million zeroes. That is, to be sure, a very large number – there are only 10^{70} particles in the visible universe – but it is a finite number.

It is easy to make a toy grammar that is even more powerful than the scheme that generates The Library of Babel. Suppose our rule for the verb phrase is enriched to allow a sentence (S) to appear inside it, as in *I told Mary he was a fool*, in which *he was a fool* comes after the object NP *Mary*:

Now our grammar is *recursive*: The rules create an entity that can contain an example of itself. In this case, a Sentence contains a

Verb Phrase which in turn can contain a Sentence. An entity that contains an example of itself can just as easily contain an example of itself that contains an example of itself that contains an example of itself, and so on:

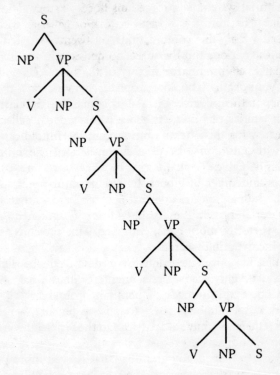

In this case a sentence can contain a verb phrase, which can contain a sentence, which can contain a verb phrase, which can contain a sentence, ad infinitum. For example, I think I'll tell you that I just read a news story that recounts that Stephen Brill reported that the press uncritically believed Kenneth Starr's announcement that Linda Tripp testified to him that Monica Lewinsky told Tripp that Bill Clinton told Vernon Jordan to advise Lewinsky not to testify to Starr that she had had a sexual relationship with Clinton. That statement is a Russian doll with thirteen sentences inside sentences inside sentences. A recursive grammar can generate sentences of any length, and thus can generate an infinite number of sentences. So a human being possessing a recursive grammar can express or understand an infinite number of distinct thoughts, limited in practice only by stamina and mortality.

∼

The idea that the creativity inherent in language can be explained by a grammar of combinatorial rules is usually associated with the linguist Noam Chomsky. Chomsky traced the idea to Wilhelm von Humboldt, a nineteenth-century pioneer of linguistics, who explained language as 'the infinite use of finite media.' According to Chomsky, the idea is even older than that; Humboldt was the last in a tradition of 'Cartesian linguists' dating back to the Enlightenment.[10]

Enlightenment philosophers were captivated by the dizzying range of thoughts made expressible by a combinatorial grammar. In his book *The Search for the Perfect Language* the semiotician Umberto Eco recounts the many Promethean schemes these philosophers came up with to perfect and harness their power.[11] Descartes noticed that the decimal system allows a person to learn in a day the names of all the quantities to infinity, and he suggested that a universal artificial language built on similar principles could organize all human thoughts. Leibniz, too, dreamed of a universal logical grammar that would generate only valid sequences of ideas, banishing irrationality and error forever.

Three hundred years later we still are fallible, and still take years to learn a Babel of local languages with their tens of thousands of arbitrary signs. Why has no modern language used the horse-power of combinatorial grammar to the fullest and abandoned the unprincipled, parochial, onerous-to-memorize laundry list called vocabulary? The answer becomes clear when we look at the most famous of the combinatorial schemes of the Enlightenment, the philosophical language of Bishop John Wilkins. The arbitrary name was an affront to Wilkins's sense of good design, and he strove for a way to eliminate it. He wrote, 'We should, by learning . . . the *Names* of things, be instructed likewise in their *Natures*.'

Wilkins's system, laid out in a lengthy 1668 opus, offered the user a *non-arbitrary* name for every thing by dividing the universe into categories and subcategories and sub-subcategories, and assigning a vowel or consonant to every branch in the tree. The first syllable identified one of the forty categories into which Wilkins had sorted all thinkable thoughts. For example, Z stood for 'sensitive species' (animals) and could be followed by *i* for 'beasts' (quadrupeds). The next consonant picked out a subdivision; *t*, for example, stood for rapacious terrestrial European canines. A final vowel pinpointed the species, yielding *Zita* as the name for dogs. By similar computations one could deduce another two thousand names for things. *Zana* is a scaly river fish with reddish flesh, in other words, salmon. *Sibα* is a type of public military relation, namely, defense. *Debα* is a portion of the first of the terrestrial elements (fire), to wit, flame. *Coba* is a consanguinous economic relation of direct ascendant, a.k.a. father.

Wilkins's philosophical language has been analysed insightfully by Borges and Eco, and we can see why no one today speaks Wilkish.[12] For one thing, it forces users to perform a chain of computations in their heads every time they want to refer to a dog. Every vowel and consonant is laden with meaning and acts as a premise in a lengthening deduction. Speakers of the language would have to play a game of Twenty Questions, inferring an entity from a description, for every word in a sentence. They could of course simply memorize the answers, such as that a portion of the first of the terrestrial elements is a flame, but that is not much easier than memorizing that the word for flame is *flame*.

A second problem is that there are more things in heaven and earth than were dreamt of in Wilkins's philosophy, which identified only two thousand concepts. Wilkins understood the exponential principle and tried to cope with the problem by lengthening the words. He provided suffixes and connectors that allowed *calf*, for example, to be expressed as *cow + young*, and *astronomer* to be expressed as *artist + star*. But eventually he gave up and resorted to using synonyms for concepts his language could not generate, such as *box* for *coffin*. Wilkins's dilemma was that he could either expand his system to embrace all concepts, which would require even longer and more unwieldy strings, or he could force his users to remember the nearest synonym, reintroducing the despised memorization process.

A third problem is that in a logical language words are assembled purely on information-theoretic principles, with no regard to the problems that incarnate creatures might have in pronouncing and understanding the strings. A perfect combinatorial language is always in danger of generating mouthfuls like *mxyzptlk* or *bftsplk*, so Wilkins and other language-designers of the Enlightenment all had to make concessions to pronounceability and euphony. Sometimes they defiled their systems with irregularities, for example, reversing a vowel and consonant to make a word more pronounceable. At other times they hobbled the system with restrictions, such as that consonants and vowels must alternate. Every even-numbered position in a word had to be filled by one of the nine vowels of English, and that restricted many categories, such as species in a genus, to nine apiece, regardless of how many species exist in the world.

Another problem is that Wilkins's words are packed tight with information and lack the safety factor provided by redundancy. The slightest slip of the tongue or pen guarantees misunderstanding. Eco catches Wilkins himself misusing *Gαde* (barley) for *Gαpe* (tulip).

Finally, all that power is not being put to any sensible use. The beauty of a combinatorial system is that it generates combinations that have never before been considered but that one *might* want to talk about some day. For example, the combinatorial system known as the periodic table of the elements inspired chemists to look for hitherto unknown chemical elements that should have occupied the empty slots in the table. Combinatorial grammar allows us to talk about a combinatorial world, a world in which violets could be red or a man could bite a dog. Yet familiar objects and actions around us often form a *non*combinatorial list of distinctive kinds. When we merely have to single out one of them, a combinatorial system is overkill. We never will have to refer to fish with an enmity to sheep or to military actions with scales and reddish flesh, and that's what a combinatorial system for words like Wilkins's allows us to do. To refer to everyday things it's easier to say *dog* or *fish* than to work through a complicated taxonomy that is just a fancy way of singling out dogs or fish anyway.

~

The languages of Wilkins and other Enlightenment thinkers show that combinatorial grammar has disadvantages as well as advantages, and that illuminates our understanding of the design of human language. No language works like Wilkins's contraption, with every word compiled out of meaningful vowels and consonants according to a master formula. All languages force their speakers to memorize thousands of arbitrary words, and now we can see why.[13] Many bodily organ systems are made from several kinds of tissue optimized for jobs with contradictory specifications. Our eyes have rods for night vision and cones for day vision; our muscles have slow-twitch fibers for sustained action and fast-twitch fibers for bursts of speed. The human language system also appears to be built out of two kinds of mental tissue. It has a lexicon of words, which refer to common things such as people, places, objects, and actions, and which are handled by a mechanism for storing and retrieving items in memory. And it has a grammar of rules, which refer to novel relationships among things, and which is handled by a mechanism for combining and analysing sequences of symbols.

To a parsimonious scientific mind, however, two mental mechanisms can be one too many. The poet William Empson wrote of the Latin philosopher,

Lucretius could not credit centaurs;
Such bicycle he deemed asynchronous.[14]

Today's skeptics also might wonder about a two-part design for language. Perhaps words and rules are two modes of operation of a single faculty. Simple, familiar thoughts need short noises, which we call words, and complicated, unfamiliar thoughts need long noises, which we call phrases and sentences. A single machine might make either short or long noises, depending on the kinds of thoughts it is asked to express. Or perhaps there is a gradual continuum between memory and combination rather than two distinct mechanisms, with words at the memory end of the continuum and sentences at the combination end.

To show that words and rules are handled by different machines we need to hold the input and output of the putative machines constant. We need side-by-side specimens in which the same kind of thought is packed into the same kind of verbiage, but one specimen shows the handiwork of a word regurgitator and the other shows the handiwork of a rule amalgamator. I believe that languages do provide us with such specimens. They are called regular and irregular words.

English verbs come in two flavors. Regular verbs have past tense forms that look like the verb with *-ed* on the end: Today I *jog*, yesterday I *jogged*. They are monotonously predictable: *jog–jogged, walk–walked, play–played, kiss–kissed*, and so on. (Regular nouns, whose plurals end in *-s*, such as *cats* and *dogs*, are similar.) The list of regular verbs is also open-ended. There are thousands, perhaps tens of thousands, of regular verbs in English (depending on how big a dictionary you consult), and new ones are being added to the language all the time. When *fax* came into common parlance a decade or so ago, no one had to inquire about its past-tense form; everyone knew it was *faxed*. Similarly, when other words enter the language such as *spam* (flood with E-mail), *snarf* (download a file), *mung* (damage something), *mosh* (dance in roughhouse fashion), and *Bork* (challenge a political nominee for partisan reasons), the past-tense forms do not need separate introductions: We all deduce that they are *spammed, snarfed, munged, moshed*, and *Borked*.

Even young children do it. In 1958 the psychologist Jean Berko Gleason tested four- to seven-year-old children with the following procedure, now known as the *wug*-test:

This is a wug.

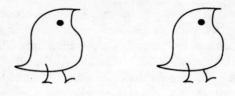

Now there is another one.
There are two of them.
These are two _____.

The children could have refused to answer on the grounds that they had never heard of a wug and had never been told how to talk about more than one of them. Instead, Berko Gleason wrote, 'Answers were willingly, and often insistently, given.' Three-quarters of the preschoolers and 99 percent of the first-graders filled in the blank with *wugs*. Similarly, when shown a picture of a man who knows how to *rick* or *bing* or *gling* and did the same thing yesterday, most children said that he *ricked* or *binged* or *glinged*.

The children could not have heard their parents say *wugs* or *binged* before entering the lab, because these words had been coined especially for the experiment. Children therefore are not parrots who just play back what they hear. And the children could not have been previously rewarded by parents for uttering those forms, because the children did not know the words before entering the lab. Children therefore are not like pigeons in a Skinner box, who increase or decrease the frequency of responses in reaction to the contingencies of reinforcement. Noam Chomsky and Eric Lenneberg, pioneers of the modern study of language and contemporaries of Berko Gleason in the Harvard-MIT community, pointed to children's ability to generalize constructions such as the regular past tense in support of their theory that language is actively acquired by a special rule-forming mechanism in the mind of the child.[15]

As it happens, all children are subjects in a version of Berko Gleason's experiment. Children often make up words or mangle them and are happy to put their new verbs in the past tense. Here are some examples:

spidered
lightninged
smunched
poonked
speeched
broomed
byed (went by)
eat lunched
cut-upped egg[16]

All children also make creative errors in their speech like these:

I buyed a fire dog for a grillion dollars.
Hey, Horton heared a Who.
My teacher holded the baby rabbits and we patted them.
Daddy, I stealed some of the people out of the boat.
Once upon a time a alligator was eating a dinosaur and the
dinosaur was eating the alligator and the dinosaur was eaten by
the alligator and the alligator goed kerplunk.[17]

Such errors bring us to the second flavor of a verb in English:
irregular. The past-tense form of an irregular verb is not simply
the verb decorated with an *-ed* ending. For example, the past tense
of *buy* is not *buyed*, but *bought*. Similarly, the past tense of *hear,
hold, steal*, and *go* are *heard, held, stole*, and *went*.

Irregular verbs contrast with regular verbs in almost every
way. Whereas regulars are orderly and predictable, irregulars are
chaotic and idiosyncratic. The past tense of *sink* is *sank*, and the
past tense of *ring* is *rang*. But the past tense of *cling* is not *clang*,
but *clung*. The past tense of *think* is neither *thank* nor *thunk*, but
thought. And the past tense of *blink* is neither *blank* nor *blunk*
nor *blought*, but a regular form, *blinked*. The language maven
Richard Lederer wrote a poem, 'Tense Times with Verbs,' that
begins:

The verbs in English are a fright.
How can we learn to read and write?
Today we speak, but first we spoke;
Some faucets leak, but never loke.

Today we write, but first we wrote;
We bite our tongues, but never bote.
Each day I teach, for years I taught,
And preachers preach, but never praught.
This tale I tell; this tale I told;
I smell the flowers, but never smold.
If knights still slay, as once they slew,
Then do we play, as once we plew?
If I still do as once I did,
Then do cows moo, as they once mid?[18]

Also in contrast to the regulars, irregular verbs form a closed list. There are only about 150 to 180 irregular verbs in modern English (depending on how you count), and there have been no recent additions.[19] The youngest irregular is probably *snuck*, which sneaked into the language over a century ago and is still not accepted by purists.[20] And the freewheeling children in Berko Gleason's study were downright stodgy when it came to irregular forms: Only one out of eighty-six turned *bing* into *bang*, and one other turned *gling* into *glang*.[21]

These differences suggest a simple theory. Regular past-tense forms are predictable in sound and generated freely because they are products of a rule that lives in the minds of children and adults: 'The past tense of a verb may be formed from the verb followed by the suffix *-ed*.' The rule would look just like the rules of syntax in the toy grammar we played with earlier,

and would generate a similar inverted-tree-like structure:

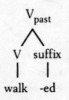

Irregular verbs, in contrast, are unpredictable in form and restricted to a list because they are memorized and retrieved as individual words. An irregular form would look just like the lexical entry we saw when considering the name of the rose. It would be linked with the entry for the plain form of the same verb and labeled as its past tense:

hold _____ held
    sound: *hōld*   sound: *hĕld*
    meaning: ✍   meaning: ✍
    part of speech: V   part of speech: V
                             tense: past

Two mechanisms trying to do the same job would get in each other's way unless something adjudicated between them, and there is indeed a simple principle: If a word can provide its own past tense from memory, the rule is blocked; elsewhere (by default), the rule applies.[22] The first part explains why we adults don't say *holded* and *stealed*; our knowledge of *held* and *stole* blocks the rule that would have added *-ed*. The second part explains why both children and adults say *Borked* and *moshed* and *ricked* and *broomed*; as long as a verb does *not* have a form in memory, the rule may be applied. The ability of a rule to apply *elsewhere* or by default – that is, to any word that does not already have a specified form in memory – is the source of its power. A speaker who needs to express a past tense or plural is never left speechless, even when a search in memory comes up empty-handed.

The theory that regular forms are generated by rule and irregular forms are retrieved by rote is pleasing not only because it explains the differences in productivity between the two patterns but also because it fits nicely into the larger picture of the design of language.

At first glance irregular verbs would seem to have no reason to live. Why should language have forms that are just cussed exceptions to a rule? What are they good for, besides giving children a way to make cute errors, providing material for humorous verse, and making life miserable for foreign language students? In Woody Allen's story 'The Kugelmass Episode' a humanities professor in a midlife crisis finds a magic cabinet that projects him into any book he takes in with him. After a tempestuous affair with Madame Bovary, Kugelmass tries again with another novel, but this time the cabinet malfunctions, and the professor 'was projected into an old textbook, *Remedial Spanish*, and was running for his life over a barren, rocky terrain as the word *tener* ("to have") – a large and hairy irregular verb – raced after him on its spindly legs.'[23]

But under the word-and-rule theory we need not suppose that evolution fitted us with a special gadget for irregularity. Irregular forms are just words. If our language faculty has a knack for memorizing words, it should have no inhibitions about memorizing past-tense forms at the same time. These are the verbs we call irregular, and they are a mere 180 additions to a mental lexicon that already numbers in the tens or hundreds of thousands. Irregular and regular forms therefore would be the inevitable outcome of two mental subsystems, words and rules, trying to do the same thing, namely, express an event or state that took place in the past.

Regular and irregular forms throw a spotlight on the advantages and disadvantages of words and rules, because everything else about them is the same: They both are one word long, and both convey the same meaning, past tense. The advantage of a rule is that a vast number of forms are generated by a compact mechanism. In English the savings are significant: The rules for *-ed*, *-s* and *-ing* (the three regular forms of the verb) cut our mental storage needs to a quarter of what they would be if each form had to be stored separately. In other languages, such as Turkish, Bantu, and many Native American languages, there can be hundreds, thousands, or even millions of conjugated forms for every verb (for different combinations of tense, person, number, gender, mood, case, and so on), and the savings are indispensable. The rule also allows new words like *mosh*, rare words like *abase*, and abstract words like *abet* to be supplied with a past tense (*moshed, abased, abetted*), even if there were no previous opportunities for the speaker and hearer to have committed the form to memory. On the other hand, a rule is more powerful than needed for words we hear so often that retrieval from memory is easy. As we shall see, it is the most common verbs, such as *be*, *have, do, go,* and *say,* that turn out to be irregular in language after language.

Rules have another shortcoming that invites the word system to memorize irregulars. Recall that one of the nuisances plaguing John Wilkins as he designed his perfect language was that flesh-and-blood humans had to pronounce and understand the products of the rules. A sequence of sounds that encodes a concept precisely and efficiently may be unresolvable by the ear or unpronounceable by the tongue. So it is with the rule for the past tense in English. The delicate tongue-tap that graces the end of a regular form may escape a listener and be omitted when he reproduces it, resulting in a solecism such as *suppose to, use to,* or *cut and dry,* or in signs and inscriptions like these:

Broil Cod
Use Books
Whip Cream
Blacken redfish
Can Vegetables
Box sets
Handicap Facilities Available

In certain older expressions *-ed* was omitted so often that the expression eventually lost the *-ed* altogether, even among careful speakers and listeners. That's how we ended up with *ice cream* (originally *iced cream*), *sour cream, mince meat,* and *Damn Yankees.*[24] Irregular verbs, in contrast, tend to use vowel changes such as *ring–rang, strike–struck,* and *blow–blew,* which are as clear as a bell.

Similarly, the very obliviousness to the details of the verb that makes a rule so powerful (it applies across the board to all verbs, whether they are familiar sounding or not) can let it blindly jam a suffix onto the end of an inhospitable sound. The result can be an uneuphonious tongue-twister such as *edited* or *sixths.* Monstrosities like these are never found among the irregulars, which all have standard Anglo-Saxon word sounds such as *grew* and *strode* and *clung,* which please the ear and roll off the tongue.[25]

Language works by words and rules, each with strengths and weaknesses. Irregular and regular verbs are contrasting specimens of words and rules in action. These are the themes of this book, but with many twists to come. It would be too good to be true if we reached a major conclusion about the most complicated object in the known universe, the human brain, simply by seeing how children name pictures of little birds. The word-and-rule theory for regular and irregular verbs is an opening statement in the latest round of a debate on how the mind works that has raged for centuries. It has inspired two alternative theories that are equally ingenious but diametrically opposed, and intensive research showing what is right and wrong about each of them – perhaps resolving the debate for good. The theory has solved many puzzles about the English language, and has illuminated the ways that children learn to talk, the forces that make languages diverge and the forces that make them alike, the way that language is processed in the brain, and even the nature of our concepts about things and people. But to reach those conclusions we first must put regular and irregular verbs under a more powerful magnifying glass, where we will find some unexpected fingerprints.

2

DISSECTION BY
LINGUISTICS

Regular and irregular words have long served as metaphors for the law-abiding and the quirky. Psychology textbooks point to children's errors like *breaked* and *goed* as evidence that we are a pattern-loving, exception-hating species, explaining everything from why children have trouble learning simple laws of physics to why adults make errors when using computers or diagnosing diseases. In *1984* George Orwell has the state banning irregular verbs as a sign of its determination to crush the human spirit; in 1989 the writer of a personal ad in the *New York Review of Books* asked, 'Are you an irregular verb?' as a sign of her determination to exalt it.

Science is not always kind to folklore from the natural world. Elephants do forget, lemmings don't commit mass suicide, two snowflakes can be alike, we use more than 5 percent of our brains, and Eskimos don't have a hundred words for snow. We had better give irregular and regular verbs a closer look before using them as evidence for a language faculty that works by words and rules, or more generally, a mind that works by lookup and computation.

Regular and irregular forms do not work in isolation; they are part of the integrated living system we call a language. This chapter will tease out regular inflection from the linguistic organs and tissues in which it is embedded. The next chapter, on irregular verbs, will have a different feel. Living creatures can be dissected, but creatures dead so long that only a trace of the living organs remain must be excavated. Our tour of the irregular verbs will uncover them from layers of historical sediment laid down over thousands of years.

Does language even *have* an anatomy? Many people think about language in the following way: We need to communicate, and language is the fulfillment of that need. For every idea there is a word and vice-versa, and we utter the words in an order that reflects the connections among ideas. If this commonsense view is true, there would be little need to speak of language being a complex system. The complexity would reside in the meanings, and language would reflect that complexity directly.

The point of this chapter is to show that this view is mistaken. I will put regular verbs under a microscope to reveal the delicate anatomy that makes them work. Language does express meaning as sound, of course, but not in a single step. Sentences are put together on an assembly line composed of mental modules, shown on the following page. One is a storehouse of memorized words, the mental lexicon. Another is a team of rules that combine words and parts of words into bigger words, a component called *morphology*. A third is a team of rules that combine words into phrases and sentences, a component called *syntax*. The three components pass messages about meaning back and forth with the rest of the mind so that the words correspond to what the speaker wants to say. This interface between language and mind is called *semantics*. Finally, the assembled words, phrases, and sentences are massaged by a set of rules into a sound pattern that we can pronounce when speaking or extract from the stream of noise when listening. This interface between language and the mouth and ear is called *phonology*.

Many people are suspicious of box-and-arrow diagrams of the mind. The walls of the boxes and the paths of the arrows often seem arbitrary, and could just as easily have been drawn differently. In the case of language, however, these components pop out as we tease apart the phenomena, and at least some of the divisions are now becoming visible in the living brain, as we will see in chapter 9.[1] This chapter will explore the kinds of discoveries that have led linguists to divide language into parts, using only the facts of regular and irregular words. First, we will see why the lexicon is different from the two boxes of rules to the right, then why morphology is in a different box from syntax, and finally, why phonology and semantics each gets a box.

~

The easiest boxes to keep separate ought to be the boxes containing words and rules. From the discussion in the preceding chapter, it should be clear that a simple word like *duck* belongs in the lexicon to the left in the diagram. Just as clearly, a sentence like *Daffy is a duck* is assembled by the rules of syntax in the box on the right. According to the words-and-rules theory, irregular forms such as *swam* are also words that come from the lexicon, because they are as arbitrary as *duck*. What do we do then with regular forms like *quacked*? They look like words and sound like words, but I have been insisting they don't have to be stored in the lexicon. They don't seem like words, but they don't seem like sentences either, which are the clearest products of rules.

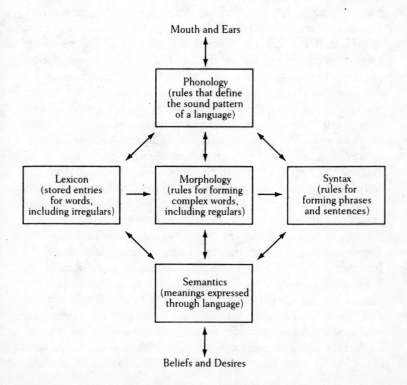

The problem is that the words *word* and *rule* come from everyday parlance and are as scientifically fuzzy as other vernacular terms, like *bug* and *rock*. On closer examination, the word *word* has two very different senses.[2] The first sense matches the everyday notion of a word: a stretch of sound that expresses a concept, that is printed as a string of letters between white spaces, and that may be combined with other words to form phrases and sentences. Some of these words are stored whole in the lexicon, like *duck* and *swam*; others are assembled out of smaller bits by rules of morphology such as *quacked* and *duck-billed platypus*. A technical term for a word in this sense is a *morphological object*, to be distinguished from phrases and sentences, which are syntactic objects.

The second sense of *word* is a stretch of sound that has to be memorized because it cannot be generated by rules. Some memorized chunks are smaller than a word in the first sense, such as prefixes like *un-* and *re-* and suffixes like *-able* and *-ed*. Others are larger than a word in the first sense, such as idioms, clichés, and collocations. Idioms are phrases whose meanings cannot be computed out of their parts, such as *eat your heart out* and *beat around the bush*. Collocations and clichés are strings of words that are remembered as wholes and often used together, such as *gone with the wind* or *like two peas in a pod*. People know tens of thousands of these expressions; the linguist Ray Jackendoff refers to them as 'the Wheel of Fortune lexicon,' after the game show in which contestants guess a familiar expression from a few fragments. A chunk of any size that has to be memorized – prefix, suffix, whole word, idiom, collocation – is the second sense of *word*. It is the sense of *word* that contrasts with *rule*, and the sense I had in mind when choosing the title of this book. A memorized chunk is sometimes called a *listeme*, that is, an item that has to be memorized as part of a list; one could argue that this book ought to have been called *Listemes and Rules*.

So *walked* is a word in the first sense (a morphological object) and not a word in the second sense (a listeme); its listemes are *walk* and *-ed*. These one-part listemes – prefixes, suffixes, and the stems they attach to, such as *walk* – are called morphemes, a term coined by the nineteenth-century linguist Baudouin de Courtenay to refer to 'that part of a word which is endowed with psychological autonomy and is for the very same reasons not further divisible.'[3]

~

What about the rules? Why divide the rules of morphology, which build complex words (including regular plurals and past-tense forms), from the rules of syntax, which build phrases and sentences? Both are productive, recursive, combinatorial systems, and some linguists see them as two parts of a larger system.[4] Yet all linguists recognize that they are not identical. This may seem of no interest to anyone but a student cramming for a Linguistics 101 final, but in fact it has been a source of countless barroom arguments, late-night dorm-room debates, and irreconcilable differences.

What is the correct word for people who pass by: *passerbys* or *passersby*? Do nervous fiancées dread the first meeting of the *mother-in-laws* or the *mothers-in-law*? Who did Richard Nixon force to resign: a series of *Attorney Generals*, or a series of *Attorneys General*? Here are a few real-life examples:

> Dear Ms Grammar,
> A member of the Friday Night Couples League . . . had a *hole in one* on the third hole and another on the fifth. Did he have two *holes in one* or two *hole in ones*? One of us believes that the pattern should be the same as in *attorneys general* and *passersby*. The other disagrees, believing that *holes in one* would indicate that the golfer gained multiple holes in one shot. A Diet Coke has been wagered on this, and we have agreed that Ms Grammar shall be the final authority.[5]

SPOONFULS

From a recipe: 'Now throw in two tablespoons full of chopped parsley and cook ten minutes more. The quail ought to be tender by then.' Never mind the quail; how are we ever going to get those tablespoons tender? The word, of course, is *tablespoonfuls*, no matter how illogical it seems. One dictionary contains the entry *spoonsful*, but this is not generally accepted.[6]

Gin and tonic season (no hyphens, please) is just about finished, but Joe Galeota of West Roxbury would still like to know how to order when he's having more than one. 'Friends advised me that the answer is "gins and tonic" because alcohol is the main ingredient,' he writes.[7]

Never has the U.S. faced a worse crisis than in 1887, after the invention of the Jack-in-the-Box. It had become a fad overnight, and everyone was having a whale of a time when someone asked, 'What is its plural?' 'Jack-in-the-Boxes!' claimed some. Others hotly insisted, 'Jacks-in-the-Box!' Civil war seemed inevitable, when Zeke Kelp's Crusade won a compromise on 'Jacks-in-the-Boxes.' Unthanked for forty-three years, Kelp will be honored next week when N.Y. City unveils a hydrant in his name.[8]

All right, the last example isn't from real life; it's from the *Early Cartoons and Writings by Dr Seuss*. The others are from well-known language columnists. *Hole-in-one* is from Ms Grammar, the nom de plume of Barbara Walraff when presiding over 'Word Court' in the *Atlantic Monthly*. *Spoonful* is from Theodore Bernstein, the late *New York Times* editor who wrote the syndicated column 'Bernstein on Words.' *Gin and tonic* is from Jan Freeman, who dispenses 'The Word' in the *Boston Globe*.

People disagree on how to pluralize nouns, and they care about who is correct. Purists insist that the *-s* belongs on the noun in the middle of the expression (*notaries public, runners-up*), and those with the common touch are content to leave it at the end (*notary publics, runner-ups*). 'Ms Grammar' advised her beseechers that *holes in one* is technically correct, but added, 'to say "two holes in one" is to ask to be misunderstood.' Her Solomonic suggestion was to say *a hole in one twice*, and to buy two Diet Cokes.

For my purpose – figuring out how the human mind deals with language – there is no correct answer. Most disputes about 'correct' usage are questions of custom and authority rather than grammatical logic (see 'The Language Mavens' in my book *The Language Instinct*), and in these disputes in particular, both parties have grammatical logic on their side. Their agony highlights the distinctions among lexicon, morphology, and syntax, and illustrates the theme of this book: that the mind analyses every stretch of language as some mixture of memorized chunks and rule-governed assemblies. How people pluralize an expression depends on how they tacitly analyse it: as a word or as a phrase.

With a simple word the plural suffix goes at the end: one *girl*, two *girls*. Now what happens in a compound word composed of two simple words, such as *cowgirl*? The plural still goes on the end: *two cowgirls*, not *two cowgirl* or *two cowsgirls*. That is because the word *girl* inside *cowgirl* is special. It is called the *head* of the word, and it stands for the word as a whole in determining its meaning (a cowgirl is a kind of girl) and in determining its plural: The -*s* goes on *girl*. A *phrase* also has a head, and it too determines the meaning and gets the plural. But now we discover the major difference between a word, the product of morphology, and a phrase, the product of syntax: In the phrase, the head is on the *left*, not the right. If you meet more than one girl from Ipanema (head = *girl*), they are *girls from Ipanema*, not *girl from Ipanemas*. With a word the plural is on the end (*cowgirls*); with a phrase the plural can be in the middle (*girls from Ipanema*).[9]

The seeds of the *mother-in-law* dispute were sown by a special option of English: Occasionally a phrase gets repackaged into a long word. For example, a hangover victim may complain of a *bottom-of-the-birdcage taste* in her mouth; the phrase *bottom of the birdcage* has been packaged as a word that modifies *taste*. When a word-made-from-a-phrase is new and fresh, speakers still can perceive the anatomy of the phrase inside the word. For example, we parse the modifier *bottom-of-the-birdcage* to understand that it means something as foul as the bottom of a birdcage.

But when the phrase is used as a word repeatedly, the original meaning can recede from collective memory. The phrase boundaries melt into a glob, and speakers no longer sense its parts. No one thinks of *Thursday* as *Thor's Day* anymore, or of *breakfast* as *breaking a fast*. Modern English has thousands of former phrases and complex words that have congealed into what people now perceive as simple words, such as *business* (busyness), *Christmas* (Christ's Mass), and spinster (one who spins). The meltdown, of course, does not happen overnight or in all speakers at once; there must have been a time when some English speakers still heard *Christmas* as *Christ's Mass* and others heard it as the arbitrary name of the holiday, just as today's older speakers hear the *awe* in *awesome* where younger speakers hear the whole word as a synonym for *good*.

Most of our disputed plurals originated as phrases and then became words. Long ago people might have thought, 'She is not my *mother in reality*; she is only my *mother in law*' (that is, according to canon or Church law). But the concept of a spouse's mother needs a word, and eventually the phrase got reanalysed as that word: 'She is my *mother-in-law*.' Similar meltdowns occurred in these phrases:

Jack is in the box → That is a *Jack-in-the-box*.
Phyllis completed that hole in one shot → She got a *hole-in-one*.
Barry passed by → He is a *passerby*.
I set aside a spoon full of parsley → I set aside a *spoonful*.

If some speakers still hear the phrase inside the word, they will be tempted to put the plural marker on the head of the phrase: two *mother* + *s in law*, *Jack* + *s in a box*, *hole* + *s in one*, *passer* + *s by*, *spoon* + *s full*. But if speakers glom the words together in their minds, they will be tempted to put the plural marker at the end: *motherinlaw* + *s*, *jackinthebox* + *es*, *passerby* + *s*, *holeinone* + *s*, *spoonful* + *s*.

It's not that phrase hearers interpret these expressions literally (for example, that a mother-in-law is a mother as recognized by the law), or that the phrase-deaf treat them as any old string of consonants and vowels; both surely recognize them as complex words built out of familiar words. It's just that they grow different kinds of connective tissue when piecing these expressions together. Those who would describe themselves as *sons-in-law* hear *mother* as the head of a phrase inside the word (shown in the left tree in the diagram); those who would describe themselves as *son-in-laws* hear a string of little words inside the big word (right tree):

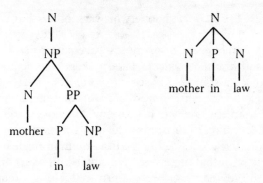

A proof that the *in-law* expressions have congealed into words may be found in the umbrella word *in-law*, which can stand alone and be pluralized in the usual way: *The in-laws are coming over*. It is a good bet that many of today's commonly used phrases will also become opaque some day and turn into words; the giveaway will be a plural at the end. Don't be surprised if one day you hear about *grant-in-aids*, *bill of ladings*, or *work of arts*.

This ambiguity – one stretch of sound, two ways of building a tree in the mind – also started the controversy raised by reports such as the following:

> While Mo Vaughn should finish well over .300 with close to 40 home runs and more than 100 RBIs, Mike Piazza has not been producing anywhere close to what he did last season, when he hit .362 with 40 homers and 124 RBIs.[10]

Baseball purists who deplore artificial turf and the designated hitter get equally incensed by the plural form *RBIs. RBI* is an acronym for *run batted in*, a run scored by a teammate as a consequence of one's batting the ball. An RBI and then another RBI are two runs batted in, and the acronym for *runs batted in* is just *RBI* – so it should be *124 RBI*, not *124 RBIs*. (The purists are not mollified by the sportscasters' common alternative, *ribbies*.) But the purists fail to recognize that acronyms, like phrases, can turn into bona fide words as a language evolves, as in *TV, VCR, UFO, SOB*, and *PC*. Once an acronym has become a word there is no reason not to treat it as a word, including adding a plural suffix to it. Would anyone really talk about *three JP* (justices of the peace), *five POW* (prisoners of war), or *nine SOB* (sons of bitches)?

An additional puzzle surrounds *governors-general, solicitors-general,* and *attorneys-general.* The speakers who bequeathed the plurals to us must have analysed the words as phrases, which have their heads on the left. Indeed, a governor-general is a general governor, namely, one who has several governors under him. The puzzle is, why didn't they simply call him a *general governor*? After all, the adjective comes before the head noun in English, not after it. The answer is that these words, together with many other terms related to government, were borrowed from French when England was ruled by the Normans in the centuries after the invasion of William the Conqueror in 1066. In French, the adjective can come *after* the head noun, as in *États-Unis* (United States) and *chaise longue* (long chair, garbled into the English *chaise lounge*). The earliest citation in the *Oxford English Dictionary* is from 1292: 'Tous attorneyz general purrount lever fins et cirrographer' (All general attorneys may levy fines and make legal documents). Anyone who insists that we eternally analyse (hence pluralize) these words as they were analysed in the minds of the original speakers of Norman French also should insist that we refer to more than one major general as *majors general,* because a *major-general* was once a general major (from the French *major-général*). Long ago our linguistic foreparents forgot the French connection and reanalysed *general* from a modifying adjective to a modified noun.

So if you are ever challenged for saying *attorney-generals, mother-in-laws, passerbys, RBIs,* or *hole-in-ones,* you can reply, 'They are the very model of the modern *major general.*' They come from reanalysing a phrase into a word, a common development in the history of English, and a nice demonstration that we treat stretches of language not as sounds linked directly to meanings but as structured trees. People who put different trees on the same sound will use the sound in different ways, even if the meaning is the same.

∼

Let's now peer into the morphology box. Morphology may be divided into *derivation* – rules that form a new word out of old words, like *duckfeathers* and *unkissable* – and *inflection* – rules that modify a word to fit its role in a sentence, what language teachers call conjugation and declension. The past tense and plural forms are examples of inflection.

English inflection is famous among linguists for being so boring. Other languages exploit the combinatorial power of grammar to generate impressive numbers of forms for each noun and verb. The verb in Spanish or Italian comes in about fifty forms: first, second, and third persons, each singular and plural, each in present, past, and future tenses, each in indicative, subjunctive and conditional moods, plus some imperative, participle, and infinitive forms. Languages outside the Indo-European family, such as those spoken in Africa or the Americas, can be even more prolific. In the Bantu language Kivunjo, for example, a verb is encrusted with prefixes and suffixes that multiply out to half a million combinations per verb.[11] But English speakers subsist on only four:

open
opens
opened
opening

Strangely enough, English grammar does not have only four roles for verbs to play. It has at least thirteen different roles, but it shares the four forms among them, as if suffixes were expensive and the designers of the language wanted to economize.

The first suffix is a silent bit of nothing, -∅, which when added to the stem *open* turns it into the inflected form *open∅*. You may wonder: Why say that speakers hallucinate an imaginary suffix at the end of a word? The reason is that it distinguishes the root or stem – the irreducible nugget found in the mental dictionary that captures the essence of a verb and upon which suffixes are hung – from a particular incarnation of that verb with a particular person, number, and tense. In English they can sound the same – *to open* and *I open* – which disguises the fact that they are different versions of the verb. In other languages the form of the verb that you look up in a dictionary cannot be pronounced. For example, in Spanish you can say *canto, cantéis, canten,* and so on, leaving *cant-* as the stem, but you can never say *cant-* by itself. Stems are therefore not the same things as pronounceable verb forms, and that distinction is useful to preserve in English – *to open* versus *open∅* – even though the two forms sometimes sound the same.

The suffix *-∅* is used in four variations of the verb in English:

Present tense, all but third-person singular: I, you, we, they *open* it.
Infinitive: They may *open* it, They tried to *open* it.
Imperative: *Open!*
Subjunctive: They insisted that it *open.*

The suffix *-s* is used for only one purpose:

Present tense, third-person singular: He, she, it *opens* the door.

The suffix *-ing* is used in at least four ways:

Progressive participle: He is *opening* it.
Present participle: He tried *opening* the door.
Verbal noun (gerund): His incessant *opening* of the boxes.
Verbal adjective: A quietly-*opening* door.

Finally we come to our friend *-ed,* which has four jobs:

Past tense: It *opened.*
Perfect participle: It has *opened.*
Passive participle: It was being *opened.*
Verbal adjective: A recently-*opened* box.[12]

Why make all these distinctions among verb forms that sound the same? One reason is that the list of phrases calling for a form such as *opened* have nothing in common: To capture the behavior of *-ed,* we have no choice but to list four phrase types separately. Another reason is that some distinctions that are inaudible for regular verbs are audible for irregular ones, and this shows that English speakers register these distinctions as they speak. About a third of the irregular verbs have different forms for the stem, the past tense, and the perfect participle: *I sing, I sang, I have sung; I eat, I ate, I have eaten.* A few make a further distinction and have a special form for the verbal adjective – *a newly wedded couple; a drunken sailor; a shrunken head; rotten eggs* – which is not used for the participle: People say *They have wed*, not *wedded*; *He has drunk*, not *drunken*; *It has shrunk*, not *shrunken*; *The eggs have rotted*, not *rotten*. And one verb comes in *eight* different forms:

Infinitive; subjunctive; imperative; To *be* or not to *be*; Let it *be*; *Be* prepared.

Present tense, first-person singular: I *am* the walrus.

Present tense, second-person singular, all persons plural: You/ we/they *are* family.

Present tense, third-person singular: He/she/it *is* the rock.

Past tense, first- and third-person singular: I/he/she/it *was* born by the river.

Past tense, second-person singular, all persons plural; subjunc- tive: The way we/you/they *were*; If I *were* a rich man.

Progressive and present participle; gerund: You're *being* silly; It's not easy *being* green; *Being* and Nothingness.

Perfect participle: I've *been* a puppet, a pauper, a pirate, a poet, a pawn and a king.

With nouns, too, different grammatical forms have to dip into the same small pool of suffixes. The naked stem *dog* must be distinguished from the singular *dog* + Ø because a *dogcatcher* doesn't catch just one dog and a *dog lover* doesn't just love one. The *dog* inside these compounds refers to dogs in general and thus differs in meaning from the singular form in *a dog*. The plural *dogs* uses *-s*, which we have already met in the verb system in *She opens the door*. The possessive form *dog's* (singular) and *dogs'* (plural) use it too; the three noun forms *dogs, dog's*, and *dogs'* differ only in punctuation.

All this redundancy suggests that regular inflection in English is remarkably simple. All the inflections are suffixes; none of the grammatical roles call for a prefix or some other way of decorating or tinkering with a word. And every word has at most one inflectional suffix. We never get *opensed* or *opensing*, nor do the plural *-s* and possessive *'s* stack up when several owners own something: *the dog's blanket*, not *the dogs's* (dogzez) *blanket*. Finally, each niblet of sound making up a suffix has a life of its own and combines with several verb forms, noun forms, or both, rather than being a slave to only one role. This suggests that instead of crediting English speakers with seventeen verbose rules like 'To form the past tense, add *-ed* to the end of the verb,' we can credit them with just *one* rule:[13] 'A word may be composed of a stem followed by a suffix,' like the simple rule shown on page 18. All the other details can be handled by assuming that suffixes are stored in the mental lexicon with entries like those for words, perhaps something like this:

-ed
  sound: *d*
  part of speech: suffix
  use 1: past tense of a verb
  use 2: perfect participle of a verb
  use 3: passive participle of a verb
  use 4: adjective formed from a verb

By factoring seventeen verbose rules into one austere rule and four lexical entries, one per suffix, we not only save ink but get some insight into the mental organization of language. English *could* have used seventeen different forms for its seventeen slots in the noun declension and verb conjugation: prefixes such as *ib-*, *tra-*, and *ka-*, suffixes such as *-og*, *-ig*, and *-ab*, and so on. Instead the slots share a few sounds (*-∅*, *-ed*, *-s*, *-ing*) and one position (immediately following the verb). This miserliness, called syncretism, is found in language after language. Syncretism suggests that the mind keeps separate accounts for the templates that build words (for example, 'word = stem + suffix'), for the scraps of sound that may be added to words (*-s*, *-ed*, and *-ing*), and for the roles these additions can play (for example, plural, participle, imperative).[14] A particular construction like the English past tense is a mix-and-match affair, assembled by hooking together parts also used in other constructions. No one knows why languages like to recycle their suffixes and other ways of modifying words. It's certainly not to save memory space, because the savings are trivial. Perhaps the reason is to help listeners recognize when a word is composed of a stem and a suffix rather than being a simple stem. Whatever its purpose, syncretism shows that in the language system, combination is in the blood; even the tiniest suffixes are combinations of smaller parts.

~

Syncretism – one form, several roles – is one kind of violation of the simplest conceivable system in which every sound has one meaning and vice-versa. The other kind of violation – one role, several forms – is rampant in languages as well; linguists call it allomorphy.[15] Take the regular past-tense suffix – or is it suffix*es*? Though always spelled *-ed*, it is pronounced in three different ways. In *walked*, it is pronounced *t*. In *jogged*, it is pronounced *d*. And in *patted*, it is pronounced *id*, where *i* is a neutral vowel called 'schwa.' We also find allomorphy in the regular plural: The suffix *-s* has three different forms in *cats*, *dogs*, and *horses*.

Are there in fact three past-tense suffixes and three plural suffixes? In some languages, we are forced to this messy conclusion. Dutch speakers, for example, select either *-en* or *-s* as the regular plural, depending on the sound of the end of the noun. But in English the three-way variation has a simpler explanation, worked out by the linguists Arnold Zwicky and Alan Prince. *One* past-tense suffix is stored in the lexicon, not three, and a separate module fiddles with its pronunciation: the rules of phonology, which define the sound pattern or accent of a language.[16]

Why do we pronounce the past-tense suffix as *t* in *walked*, *d* in *jogged*, and *id* in *patted*? The choice is completely predictable, and can be stated as a list of rules:

1. Use *id* if the verb ends in *t* or *d* (for example, in *patted* and *padded*).
2. If it doesn't, use *t* if the verb ends in an unvoiced consonant – that is, a consonant in which the vocal cords don't buzz,

namely *p, k, f, s, sh, ch,* and *th* (for example, *tapped, walked, sniffed, passed, bashed, touched* and *frothed*).

3. Use *d* for all other verbs: those ending in vowels, such as *played* and *glowed,* and those ending in the voiced consonants, *l, r, m, n, b, g, v, z, j, zh,* and *th* (for example, *pulled, marred, slammed, planned, scrubbed, pegged, saved, buzzed, urged, camouflaged,* and *bathed*).

This sounds like something out of the tax code. Let's see if we can do better.

The first thing to notice is that nothing in these rules is specific to the past tense. Other constructions that use *-ed* work the same way:

	t	*d*	*id*
Past tense:	kicked	flogged	patted
Perfect participle:	has kicked	has flogged	has patted
Passive participle:	was kicked	was flogged	was patted
Verbal adjective:	a kicked dog	a flogged horse	a patted cat

Outside the verb system entirely is yet another *-ed* construction that comes in the three variations; it turns a noun that means 'X' into an adjective that means 'having X':

	t	*d*	*id*
Nominal adjective:	hooked	long-nosed	one-handed
	saber-toothed	horned	talented
	pimple-faced	winged	kindhearted
	foulmouthed	moneyed	warm-blooded
	thick-necked	bad-tempered	bareheaded

The regular plural *-s* also comes in three forms, which you can hear in *hawks, dogs,* and *horses.* The variation mirrors the past tense uncannily. Use *iz* when the noun ends in a sibilant sound: *s, z, sh, zh, j,* or *ch.* If it doesn't, use *s* if the noun ends in an unvoiced consonant. Use *z* for all other nouns. In fact, not only does this pattern appear with the plural, it appears with the other *-s* suffixes as well:

	s	z	iz
Plural:	hawks	dogs	horses
3rd person singular:	hits	sheds	chooses
Possessive:	Pat's	Fred's	George's

The variation even appears in versions of *-s* that aren't genuine suffixes. English speakers commonly contract the verbs *has, is,* and *does* to their final consonant and glue it onto the end of the preceding word, as in *Mom's left* or *Dad's home*. Sure enough, the contraction is pronounced in three ways, depending on how the word ends:

	s	z	iz
has:	Pat's eaten.	Fred's eaten.	George's eaten.
is:	Pat's eating.	Fred's eating.	George's eating.
does:	What's he want?	Where's he live?	

That's not all. English has an *affective -s* that can be used to form nicknames in some dialects and argots, as in *Pops, Moms, Fats, Pats,* and *Wills* (the prince second in line to the British throne). That *-s* can also show up in emotionally colored slang such as *bonkers* and *nuts,* similar to the *-y* and *-o* that give us *batty* and *wacko.* (Sometimes the two suffixes are even used together, as in *Patsy, Bugsy, Mugsy, footsie, fatso,* and *Ratso.*) Still another version of *-s* appears in adverbial forms such as *unawares, nowadays, besides, backwards, thereabouts,* and *amidships.* A final use for *s* is as a meaningless link joining the words in compounds such as *huntsman, statesman, kinsman, bondsman, Scotsman,* and *grantsmanship.* And yes, all of these *-s*'s can be pronounced either as *s* or as *z,* depending on the preceding consonant (it's hard to come up with examples for the third column):

	s	z	iz
Affective:	Pops, Patsy	Wills, bonkers	
Adverbial:	thereabouts	towards, nowadays	
Link in compound:	huntsman	landsman	

So we have *fifteen* suffixes that show the same three-way or two-way variation. Forty-one suffixes that happen to fall into fifteen parallel sets of alternatives is too much of a coincidence to stomach. More likely, *one* set of rules creates the three-way variation, and the set applies in at least fifteen situations.

There is a second, equally striking set of coincidences that runs across the suffixes. If the variation came from any old set of *if . . . then* rules, we would expect to find all kinds of pairings between stems and suffixes: for example, 'Use *s* after the vowels *a* and *e* or after the consonants *th* and *g*,' 'Use *d* after a *k*,' and so on. But the rules are far more lawful than that. The *t* sound comes after unvoiced consonants, and the *t* itself is unvoiced. The *d* sound comes after voiced sounds, and the *d* itself is voiced. The *-s* suffixes show the same chameleonlike behavior: We find unvoiced *s* after unvoiced consonants, and voiced *z* after voiced consonants. It looks as if something is trying to keep the consonants at the end of a word consistent: All of them are voiced, or all of them are unvoiced.

Indeed, something is – the sound pattern of the English language. English never forces speakers to turn their vocal cords on for one consonant then off for the next, or vice-versa. We see the restriction in force in one-piece words that end in a cluster of consonants. These words never received a suffix; they just happen to be built that way, so any sound pattern they display cannot have come from a suffix rule, but rather from the way English speakers like to pronounce words in general. In all but one of these words, the vocal cord switch can be left in the 'off' position:

After *k* (unvoiced):	*s* can occur	*z* cannot occur
	ax, fix, box	—
	t can occur	*d* cannot occur
	act, fact, product	—
After *p* (unvoiced):	*s* can occur	*z* cannot occur
	traipse, lapse, corpse	—
	t can occur	*d* cannot occur
	apt, opt, abrupt	—
After *t* (unvoiced):	*s* can occur	*z* cannot occur
	blitz, kibitz, Potts	—
After *s* (unvoiced):	*t* can occur	*d* cannot occur
	post, ghost, list	—

In one English word, *adze*, the vocal cord switch is left in the 'on' position:

After *d* (voiced):	*s* cannot occur	*z* can occur
	—	*adze*

In *no* English word is the voicing switch toggled on and off, in an ending like *zt*, *gs*, *kz*, or *sd*.

These difficult-to-pronounce clusters *can*, however, be created by a dumb rule of morphology that pins a suffix onto the end of a word without regard for how the resulting train of consonants is to be pronounced. That is what happens when a rule adds a *d* sound to *walk* or an *s* sound to *dog*. English cleans up these awkward mismatches with a different kind of rule. The rule says, 'When there is a cluster of consonants at the end of a syllable, adjust the voicing setting of the last consonant to make it consistent with its neighbor on the left.' (In other words, change *kz* to *ks*, *pd* to *pt*, and so on.) The rule does not care whether the syllable was formed by a past-tense suffix, a plural suffix, a contracted *has*, a nickname with -*s*, or anything else. It kicks in *after* the syllable has been assembled, in the cleanup module we call phonology.

Can we now tell whether the suffix stored in the lexicon is -*d*, and is converted to a *t* when it finds itself at the end of *walk*, or whether it is -*t* and is converted to *d* when it finds itself at the end of *jog*? A little detective work can settle the question. Not every sound cares about the consonant that follows it. Those that do are consonants in which the airstream is obstructed, namely *p*, *b*, *t*, *d*, *k*, *g*, *s*, *sh*, *ch*, *z*, *zh*, and *th*. But the vowels, and the vowel-like consonants *r*, *l*, *n*, and *m*, are indifferent to what comes after them; they tolerate either *s* or *z*, either *t* or *d*, as we see in these one-piece words:

After *n*:	*s* can occur	*z* can also occur
	fence	*lens*
	t can occur	*d* can also occur
	lent	*lend*
After *r*:	*s* can occur	*z* can also occur
	force	*furze*

After *r*: cont.	*t* can occur	*d* can also occur
	fort	*ford*
After *l*:	*s* can occur	*z* can also occur
	pulse	*Wales*
	t can occur	*d* can also occur
	guilt	*guild*
After a vowel:	*s* can occur	*z* can also occur
	niece	*sneeze*
	t can occur	*d* can also occur
	goat	*goad*

Here we have laissez-faire environments in which the suffixes can show their true colors, untouched by rules of phonology. What do we find? That the virgin suffixes are pronounced -*d* and -*z*, not -*t* and -*s*:

After *n*:	we don't say *s*	we say *z*
	—	*grins* (grĭnz), *pins* (pĭnz)
	we don't say *t*	we say *d*
	—	*grinned*
After *r*:	we don't say *s*	we say *z*
	—	*wears* (wĕrz), *cores* (kŏrz)
	we don't say *t*	we say *d*
	—	*feared*
After *l*:	we don't say *s*	we say *z*
	—	*calls* (kŏlz), *balls* (bŏlz)
	we don't say *t*	we say *d*
	—	*smiled, well-heeled*
After a vowel:	we don't say *s*	we say *z*
	—	*flees* (flēz), *fleas* (flēz)
	we don't say *t*	we say *d*
	—	*flowed*

The -*t* and -*s* we hear in words with choosy sounds such as *walked* and *cats* must be the aftermath of the rule.

Finally, what about the funny extra vowel in *patted* and *horses*? Here again the change in sound is not some random act of vandalism. The vowel appears when *d* follows *t* or *d*, and when *z* follows *s* or *z*. The word endings that trigger the extra vowel are similar in pronunciation to the suffixes themselves, and that can't be a coincidence. Apparently a rule is trying to separate too-similar adjacent consonants by pushing a vowel between them: between *t* and *d, d* and *d, s* and *z, z* and *z, sh* and *z,* and so on. In many languages the rules of phonology do *something* when a rule of morphology leaves two identical or near-identical consonants in a row, presumably because there's no natural way to pronounce them. Some languages drop the second consonant, others merge the two into one long consonant, and still others, like English, wedge a vowel between them. As with the rule that fiddles with voicing, the rule that inserts a vowel must live in a phonology module separate from rules that stick on the various suffixes, because the rule is oblivious to what kind of suffix it manipulates.

We even can deduce which of the two rules applies first, the one that changes the voicing setting or the one that inserts the vowel. The devoicing rule is triggered by adjacent consonants; the vowel rule breaks up adjacent consonants. If the voicing rule came first, it would convert *pat + d* to *pat + t,* and only then would the vowel be inserted, yielding *pătĭt*:

Morphology:	*păt + d*
↓	
Devoicing:	*păt + t*
↓	
Vowel insertion:	*păt + ĭ + t*

But that is not how we pronounce it; we say *pătĭd.* This means that the vowel rule must have come first, creating *patted*; now the voicing rule is no longer compelled to do anything, because the *td* sequence that would trigger it has been broken up:

Morphology:	*păt + d*
↓	
Vowel insertion:	*păt + ĭ + d*
↓	
Devoicing:	not triggered

The ordering makes sense when you think about how the phonology module should be organized. It has some rules that edit the string of vowels and consonants composing a word (phonology proper), and other rules that convert the string into actual sounds or muscle movements (phonetics). The vowel-insertion rule makes a major change in the stuff that makes up a word, and belongs in the first subcomponent; the voicing rule does a last-minute adjustment of pronunciation for the benefit of the muscles, and belongs in the second.[17]

This completes the analysis of the three versions of the past-tense suffix. When we started, we needed forty-odd rules, each stipulating that some suffix be placed next to some word ending. We have ended up with just two rules. Best of all, what the rules do, why they do it, and in what order they do it all make sense in the light of the sound pattern of English. Indeed, this kind of layering may be found in languages all over the world.

Incidentally, there is corroborating evidence of a completely different kind that shows that the three forms of *-ed* and *-s* are created on the fly by a phonological rule. Some psycholinguists keep a pad and pencil in their pockets and write down every slip of the tongue they hear. People make one or two such errors for every thousand words they say, and many of the errors consist in deleting, repeating, or switching around vowels or consonants.[18] The last kind of error is called a Spoonerism, in honor of the Reverend William Spooner (1844–1930), Warden of New College at Oxford, who came out with surprises such as *Our queer old dean*, *You have hissed all my mystery lessons and tasted the whole worm*, and *It is now kistomary to cuss the bride*. They sound too good to be true, but I have heard similar errors myself. After I spoke at a scientific symposium the chair wrapped up the session by saying *I would like to spank the speakers*, and when I asked a friend how he liked his new condominium, he said *It seats my nudes*.

Speech errors provide clues on how the speech system is organized. For example, when a person intends to say *grapefruits* but accidentally leaves out the *t*, how does he pronounce the plural? If there were a distinct plural suffix pronounced *-ss*, he would say *grapefrooss*, since this is what the *t* in the *grapefruit* entry would have demanded. In fact he says *grapefrooz* – pronouncing the plural as *z*, which is appropriate to words ending in a vowel.[19] Similarly, a person may say *The infant tucks* –

touches the nipple, not *tuckɨz*, or may say *Did you buy enough breakfasɨz?*, not *breakfass*. The errors show that the form of the suffix must be computed after the vowels and consonants of the noun or verb were placed on the chute to the vocal tract.

English did not always have single-consonant suffixes and a rule that separates them from a too-similar word ending. Our current system is the result of a reorganization that began around the time of the origin of Modern English in the seventeenth century. Before that, *-ed* and *-s* suffixes were pronounced (and spelled) with vowels all the time, not just with words ending in *t* or *d* or in *s* or *z*. For centuries, English speakers had been concentrating stress on the first syllables of words, which shriveled the later syllables, and speakers began to leave out the vowels in the suffixes of many words. Writers called attention to the new, clipped pronunciations by spelling them phonetically with an apostrophe in place of the deleted vowel, as in Shakespeare's play about 'a pair of star-cross'd lovers':

> Death, that has suck'd the honey of thy breath,
> Hath no power yet upon thy beauty:
> Thou art not conquer'd; beauty's ensign yet
> Is crimson in thy lips and in thy cheeks.

The guardians of the English language deplored the change, as they do all changes. In 'A Proposal for Correcting, Improving, and Ascertaining the English Tongue,' Jonathan Swift wrote:

> What does your lordship think of the words 'drudg'd,' 'dis-turb'd,' 'rebuk'd,' 'fledg'd,' and a thousand others everywhere to be met with in prose as well as verse? Where, by leaving out a vowel to save a syllable, we form so jarring a sound, and so difficult to utter, that I have often wondered how it could ever obtain.

His contemporary, Samuel Johnson, who was standardizing the spellings of English words in a way that reflected the morphemes that composed them, recognized that *'d* and *-ed* were the same morpheme, and obliterated the distinction in their spelling, making *ed* the spelling for both.[20] It is unclear why he chose to leave the *e* in *-ed* across the board (*mapped* and *matted*), but opted to spell *-s* either with or without an *e*, depending on how it is pronounced (*maps* and *masses*).

Today the old syllabic suffix survives in a handful of adjectives: *accursed, aged, beloved, bended* (in the expression *on bended knees*), *blessed, crooked, cussed, dogged, jagged, learned, naked, ragged, wicked,* and *wretched.* (A few more survive in rural dialects, such as *forkèd, peakèd, streakèd,* and *stripèd.*)[21] Many of them are archaic or poetic and are used mainly in self-conscious speech. The psychologist Melissa Bowerman, a researcher of child language, had this exchange with her four-year-old daughter about a class trip to a natural history museum.[22]

MOTHER (*playfully*): Maybe you'll see something wingèd.
DAUGHTER: Maybe we'll see something snakèd!

~

We've seen why the syntax box, which builds phrases and sentences, has to be separated from the morphology box, which builds words. We also have seen why the phonology box, which massages words into a pronounceable stream of sound, has to be separated from syntax, morphology, and the lexicon. But why do we need separate boxes for semantics (the thoughts expressed in language) and the lexicon? Could we reduce the difference between regular and irregular verbs to a difference in meaning between the two kinds of verbs, rather than putting one kind in the morphology box and the other in the lexicon? Do we even need to talk about an 'entry in the mental lexicon,' the address in memory that holds a link to a sound and a link to a meaning? Or could we connect thoughts to sounds directly, eliminating the middleman? Here are some facts that suggest that we do need to credit the human mind with something like dictionary entries.

First, the English irregular verbs could not have arisen simply from a communal effort to optimize clarity. While irregular forms on average are harder to mistake for their base forms than regular forms are (*bring* doesn't sound like *brought*, nor *take* like *took*), many irregulars are *identical* to their base forms: *Today I hit, yesterday I hit; Today I put, yesterday I put.* A sentence such as *On*

Wednesday I cut the grass could mean last Wednesday, next Wednesday, or every Wednesday. If *cut* were regular, the ambiguity would never arise: *On Wednesday I cutted the grass* would single out the preceding Wednesday. Despite the potential ambiguity, however, twenty-eight English verbs insist on remaining unchanged in the past tense.

Also, irregular forms do not correlate with any kind of meaning. Many verbs are similar in meaning but have completely different past-tense forms. For example, *hit, strike,* and *slap* all refer to hitting. *Hit* is an irregular verb that does not change in the past tense: *Today we hit golf balls; Yesterday we hit golf balls. Strike* is an irregular verb that changes its vowel, yielding *struck.* And *slap* is a regular verb, with past tense *slapped.*

Not only are there verbs with similar meanings and different past-tense forms, there are verbs with different meanings and the same past-tense forms. English has a class of verbs linguists call *light verbs,* such as *come, go, do, take, have, set, get, put,* and *stand.* Compared to ordinary verbs they are less filling; a light verb doesn't have a meaning that stays with it, but takes on dozens of meanings, especially in combination with particles such as *in, out, up, off, over,* and *around*:

> *come* (move to here), *come around* (agree), *come into* (inherit), *come* (reach orgasm), *come off as* (appear), *come out* (divulge homosexuality), *come to* (awaken)
>
> *go* (move to there), *go out with* (date), *go nuts* (dement), *go in for* (choose), *go off* (explode), *go off* (spoil)
>
> *do* (act), *do in* (kill), *do up* (decorate), *do a number on* (overwhelm), *do lunch* (eat together)
>
> *take* (cause to go with), *take in* (swindle), *take off* (launch), *take in* (welcome), *take over* (usurp), *take up* (commence), *take a leak* (urinate), *take a bath* (lose money), *take a bath* (bathe), *take a walk* (walk), *take a look* (look)
>
> *have* (possess), *have* (eat), *have* (seduce), *have a heart* (sympathize), *have over* (entertain), *have a cow* (be angry)
>
> *get* (retrieve), *get* (become), *get over* (survive), *get out* (divulge), *get off on* (enjoy), *get a life* (self-improve)
>
> *set* (place), *set off* (ignite), *set up* (arrange), *set up* (trick), *set up* (introduce), *set right* (rectify), *set the stage* (prepare)

put (cause to be at), *put off* (procrastinate), *put off* (offend), *put
one over on* (fool), *put down* (insult), *put down* (euthanize),
put in for (request), *put out* (extinguish), *put out*
(inconvenience), *put out* (consent to sex)

stand (rise), *stand out* (impress), *stand up for* (defend), *stand in*
(replace), *stand off* (repel)

But in every instance they retain their irregular past-tense forms
in the extended meanings: *Barney came around, Barney came
out, Barney came off as* (never *comed*); *Joan took him in, Joan
took a bath, Joan took over* (never *taked*); and so on. All the
meanings march in lockstep with the same irregular past-tense
forms, no matter how tenuous the semantic thread that links
them. The mind links an irregular sound such as *took* not with
the meaning of a word directly but with the word's *root* – a
unique address in the mental lexicon, like the boldfaced entry
for a word in a dictionary, which can have several meanings
listed under it.[23]

An even more curious demonstration comes from families of
words with the same stem and different prefixes. Words with
prefixes keep the past-tense form of the stem: *eat–ate* becomes
overeat–overate; *make–made* becomes *remake–remade*. That is not
surprising, because we all hear the *eat* inside *overeat* – overeating
is, after all, a kind of eating, namely, eating too much. What *is*
surprising is that the same thing happens when the meaning of
the combination is opaque. Few people sense the meaning of the
stand inside *understand*, the *get* inside *forget*, or the *come* inside
become. Nonetheless no one is tempted to say *understanded,
forgetted*, or *becomed*; the irregular forms persist, giving us
understood, forgot, and *became*. Here are some examples:

come–came, become–became, overcome–overcame
go–went, undergo–underwent
get–got, forget–forgot
take–took, mistake–mistook, overtake–overtook, partake–partook,
    undertake–undertook
set–set, beset–beset, upset–upset
stand–stood, understand–understood, withstand–withstood
draw–drew, withdraw–withdrew

hold–held, behold–beheld, uphold–upheld, withhold–withheld
give–gave, forgive–forgave

Irregular forms stick like glue to their verb roots, even when reduced to meaningless little tokens inside a bigger verb. Speakers of English seem to analyse *become* as *be-* + *come* and *understand* as *under-* + *stand*, even though the meaning of *become* is not computable from the meaning of *be-* and the meaning of *come*, and *understand* has nothing to do with standing. This is not something we have to learn in school. When we acquire language, our minds analyse sets of words, looking for their parts as if they were clues in a combinatorial puzzle. We mentally arrange them in a matrix according to overlap:

	be-	over-	under-	up-	with-
come	become	overcome			
draw					withdraw
hold	behold			uphold	withhold
set	beset			upset	
stand			understand		withstand
take		overtake	undertake		

and use the common denominators in the rows and columns to make incisions in the words, thinking of them thereafter as amalgams of parts: *become* = *be-* + *come*, *withdraw* = *with-* + *draw*, and so on.[24]

Of course, it was English speakers of centuries past, our linguistic ancestors, who first analysed *become* as *be-* + *come* and extended the *come–came* pattern to it, and it is possible that to them the words were as transparently built out of parts as *overeat* or *remake* are to us today. Even so, it is unlikely that we have been stupidly memorizing *became, overcame, withdrew,* and so on as structureless strings of vowels and consonants. If we were to come across a new complex word, such as *undercome, bestand, overhold,* or *withset,* and were unaware of its meaning, we would almost certainly use the irregular forms of the words inside them: *undercame* (not *undercomed*), *bestood, overheld, withset.* Moreover, it is not pure sound that carries the irregular form: The past of *succumb* and *encumber* are *succumbed* and *encumbered,* not *succame* and *encameber*, because people don't perceive them as containing a prefix followed by the word *come*, only the sound *kŭm*.

Clearly the perception of an embedded word comes from its spelling: *become* contains *c-o-m-e*; *succumb* doesn't. But spelling does not directly inform speakers how to form the past tense; it merely assigns a distinct visual signature to every root, and speakers choose the past-tense form that goes with the root. Samuel Johnson, who standardized the spellings of thousands of modern words, used people's perception of the anatomy of words as a rationale in his decisions, and that is one of the reasons that the spellings of English words notoriously do not always reflect their sounds; often they reflect morphological structure instead. We see this in the many words that sound alike but are not perceived as being the same word (that is, as having the same root), and are not given the same past-tense form:

meet–met	versus	*mete–meted*
ring–rang	versus	*wring–wrung*
bear–bore	versus	*bare–bared*
steal–stole	versus	*steel–steeled*
break–broke	versus	*brake–braked*

In the last three cases the spellings divulge the presence of words that are recognizable in other guises – the adjective *bare*, the noun *steel*, the noun *brake* – and we will see in chapter 6 that this makes an especially big difference in how we compute their past-tense forms.[25]

~

The English system of inflection, we have seen, dissects cleanly into a few simple components. The past-tense rule belongs to a component, morphology, that builds things out of parts using rules. The rule itself is a masterpiece of minimalism – 'a word can be composed of a stem and a suffix' – with all other details distilled out and collected in the lexical entry for the suffix. The suffix itself is shared among several inflections (past tense, participle, and so on), and its variant pronunciations (*t, d, ɨd*) do not wastefully multiply listings but are computed automatically by two ubiquitous rules of phonology. The distinction between the lexicon (including irregular inflection) and grammar (including regular inflection) is a distinction between a list of entries and an algorithm for combining them, rather than a side effect of a general yearning to distinguish meanings.

That leaves the irregular words. Every irregular tells a story, and they are the topic of the next chapter.

3

BROKEN TELEPHONE

In the game known as Broken Telephone (or Chinese Whispers) a child whispers a phrase into the ear of a second child, who whispers it into the ear of a third child, and so on. Distortions accumulate, and when the last child announces the phrase, it is comically different from the original. The game works because each child does not merely degrade the phrase, which would culminate in a mumble, but *reanalyses* it, making a best guess about the words the preceding child had in mind.

All languages change through the centuries.* We do not speak like Shakespeare (1564–1616), who did not speak like Chaucer (1343–1400), who did not speak like the author of *Beowulf* (around 750–800). As the changes take place, people feel the ground eroding under their feet and in every era have predicted the imminent demise of the language. Yet the twelve hundred years of changes since *Beowulf* have not left us grunting like Tarzan, and that is because language change is a game of Broken Telephone.

A generation of speakers uses their lexicon and grammar to produce sentences. The younger generation listens to the sentences and tries to infer the lexicon and grammar, the remarkable feat we call language acquisition. The transmission of a lexicon and grammar in language acquisition is fairly high in fidelity – you probably can communicate well with your parents and your children – but it is never perfect. Words rise and fall in popularity as the needs of daily life change, and also as the hip try to sound different from the dweebs and graybeards. Speakers

*For a chart that summarizes the history, dates, and family affinities of the English language, see page 236.

swallow or warp some sounds to save effort, and enunciate or shift others to make themselves understood. Immigrants or conquerors with regional or foreign accents may swamp the locals and change the pool of speech available to children.

Children, for their part, do not mimic sentences like parrots but try to make sense of them in terms of underlying words and rules. They may hear a mumbled consonant as no consonant at all, or a drawn-out or mispronounced vowel as a different vowel. They may fail to discern the rationale for a rule and simply memorize its outputs as a list. Or they may latch on to some habitual way of ordering words and hypothesize a new rule to make sense of it. The language of their generation will have changed, though it need not have deteriorated. Then the process is repeated with their children. Each change may be small, but as changes accumulate over centuries they reshape the language, just as erosion and sedimentation imperceptibly sculpt the earth.

That is how irregular forms, in particular, come down to us. Most of the forms were originally created by rules, but a later generation never grasped the rules and instead memorized the forms as words. They were words for every generation thereafter, and each irregular was free to accumulate its own quirks from subsequent distortions and reanalyses. Because irregulars originated from rules they are not a random grab-bag but rather display patterns, fossils of the long-dead rules. A. L. Kroeber, a founder of modern anthropology, reminisced that his 'first remembered purely intellectual pleasure' was seeing patterns in English irregular verbs, a foretaste of his search for systematicity in culture more generally.[1]

This chapter is a guided tour of the irregular nouns and verbs of English, with commentary on where they came from and where they are going. These words all will have their turns on stage throughout the book, so it's helpful to get to know them individually. This is also a lively way to come to understand how language changes, including how it is changing today.

People often ask me how linguists know the way people pronounced things in centuries past. After all, Chaucer, unlike Nixon, did not secretly tape his conversations for the benefit of future historians. Old pronunciations can be painstakingly inferred from a diverse set of clues. One of them is spelling. Before Samuel Johnson standardized English orthography, people spelled more or less as they pleased, trying to capture the sounds of language as they heard them. Spellings were more phonetic, and changes in spelling give clues to changes in pronunciation. For example, when writers started to spell Old English *bi-healfe* (behalf) as *behaf*, one can guess that people had stopped pronouncing the *l*. Other clues come from wordplay. For example, Shakespeare rhymed or punned *case* and *ease, hate* and *eate, say* and *sea*, and *shape* and *sheep*, suggesting that speakers of Early Modern English pronounced the vowels in each pair in the same way (clues from spelling suggest it was $\bar{a}$). A third kind of clue is found in the writings of language snobs who criticize or lampoon the speech of their contemporaries, inadvertently immortalizing it to the good fortune of modern linguists. Other clues exist as well, and together they can triangulate on the most common and most probable pronunciations.

We can never say for sure what *the* pronunciation of a given word at a given time actually was. Just as there are regional accents today (London, Boston, Texas, and so on), there were regional varieties of English centuries ago; indeed, many more of them, because people did not move around as much as we do, did not send their children to melting-pot schools, and had no dictionaries to consult. Also, the written record is haphazard. Most words and pronunciations were in use long before the first literate person chanced to write them down, and many others went to the grave along with their speakers. When word histories can be reconstructed, invariably they are convoluted, eye-glazing yarns. This is to warn you that the word histories presented here have been simplified to highlight the kinds of psychological processes that cause words to *have* histories.[2]

~

Words aren't regular or irregular across the board. Words are regular or irregular only with respect to certain inflections, some more tolerant of irregularity than others.

The present progressive suffix-*ing*, as in *The joint is jumping*, is 100 percent regular. There isn't a single exception to the rule, not even the rebellious *be*, which meekly submits and shows up as *being*. Why, when it comes to *-ing*, does no verb hear a different drummer? One reason is that the progressive construction came into English relatively recently, late in the Middle English period of 1100 to 1450. It borrowed the *-ing* suffix from the gerund (a construction that turns a verb into a noun, as in *the changing of the guard*), and the newly cloned *-ing* suffix had the progressive all to itself and did not have to compete with alternative forms hanging around from earlier periods. Another reason is that *-ing* is found in a separate syllable, which makes it easy for listeners to hear a word such as *breaking* as *break + ing*. That is an advantage over *-s* and *-ed*, which sound as if they are part of a stem, like *act, box,* or *maze*. As we shall see, the camouflage of *-s* and *-ed* can invite listeners to misanalyse a regularly inflected combination as a one-piece irregular word.

One other suffix is completely regular: the possessive *'s*. Any noun can take it, even the irregular nouns that cannot appear with an *s* sound when it is a plural suffix, such as *mouse* and *man*. We have no trouble saying *the man's hat, the mouse's mother,* or *the goose's egg*, even though we never say *the mans, the mouses,* or *the gooses*. Why no irregulars? The possessive is unusual because it attaches to a *phrase* rather than to a *word*. One can talk not just about *the cat's pajamas* but about *the cat in the hat's pajamas*, where the pajamas belong to the cat, not to the hat:

The plural *-s* attaches to a word:

The possessive *'s* attaches to a phrase:

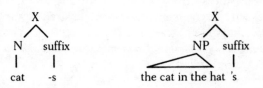

A former student, Annie Senghas, once said to someone at a conference, 'The woman sitting next to Steven Pinker's pants are like mine.' I was fully clothed; the woman sitting next to me had pants like Annie's. Dave Barry's column-within-a-column 'Ask Mr Language Person' once had the following exchange:

Q: Recently, did your research assistant Judi Smith make a grammatically interesting statement regarding where her friend, Vickie, parks at the Miami Herald?

A: Yes. She said, quote, 'She comes and parks in whoever's not here's space that day.'

The word *here* is not even a noun! Since *'s* is perceived not to be attached to an adjacent noun, it cannot unite with that noun in people's minds, and therefore never evolves into an irregular word. The exceptions that prove the rule are the possessive pronouns *my, your, his, her, our,* and *their,* which are, in a sense, irregular replacements for *me's, you's, him's, her's, us's,* and *them's.* Pronouns are one-word phrases; in any sentence position where you can say *the man in the gray suit* you can also say *he* or *him.* A pronoun, being a phrase, is the only kind of word that *could* form a cohesive amalgam with *'s,* which in effect is what possessive pronouns are.

The third-person singular *-s,* as in *Dog bites man,* steps aside for irregular forms in only four verbs: *be–is* (not *be's*), *have–has, do–does* (pronounced *dŭz*), and *say–says* (pronounced *sĕz*). These, by the way, are the four most frequent verbs in the English language.[3] In chapter 5 we will see that this is not a coincidence.

Nouns embrace several kinds of irregular plurals.[4] Many nouns ordinarily don't take any plurals: mass nouns such as *mud, celery, furniture,* and *evidence* are treated as seamless stuff rather than countable things. (A former graduate student who is a Russian emigré was teased by fellow students for saying, 'I hev three evidences for thees theory.') Of the count nouns, which do take plurals, exactly seven change their vowel instead of adding *-s:*

man–men, woman–women (pronounced *wĭmĭn*), *foot–feet, goose–geese, tooth–teeth, mouse–mice, louse–lice*

Why do we flip the vowels in these nouns? Originally they took plural suffixes, just like regular nouns, though the suffixes were different from today's *-s.* For example, *foot,* originally *fot,* had the plural *foti.* But as we saw in chapter 2, you can't just force a consonant or vowel onto the end of a word and hope that nothing else happens. People adjust their pronunciation of a sound in anticipation of the sounds to come. In many modern English dialects, for instance, speakers pronounce the *i* differently in *write* and *ride* and the *ou* differently in *shroud* and *about*. In *keep cool* the first *k* sound is pronounced toward the front of the mouth, the second one toward the back. In words like *find* and *sound* the *n* vanishes and the vowel reminds us of the vanished consonant by being sounded through the nose. Most of us are unaware that we make these adjustments and are puzzled when children spell *find* as *fid*, though it is an accurate transcription of the *n*-less word they hear. Some of these adjustments come from the way we control our muscles, but others get standardized into phonological rules, which define what we hear as an accent.

In the Germanic languages that were ancestral to English there was a phonological rule that changed the pronunciation of a vowel from the back of the mouth to the front of the mouth if the next syllable contained a vowel pronounced high and in front. The rule spared people from having to jerk their tongue backward and then forward while pronouncing the words. So in *foti*, the plural of *fot*, the back *o* was altered to the front *e*, harmonizing with the front *i* in the suffix: roughly, *feti*. The process is called umlaut and it is still visible in our linguistic cousin, German, as the two little dots over some vowels: *die Kuh* (the cow), *die Kühe* (the cows).

In the Middle English period, speakers began to mumble the unstressed syllables at the ends of words and then began to drop them outright. At that point people must have been hearing the altered vowel in *feti* as a different vowel altogether, not as a tweaked *o*, because when the suffix was dropped, the speakers kept the altered vowel in the stem, even though nothing was there to tweak it anymore. The eventual result was *feet*. It reminds me of the explanation of why there is a basketball team in arid Los Angeles called the Lakers and a team in pious Utah called the Jazz. Originally the teams were based in Minneapolis, The Land of Lakes, and in New Orleans, The Birthplace of Jazz. When the teams moved, they kept their names, even though the names no longer made sense.

Another three irregular plurals take the old Anglo-Saxon suffix -*en* rather than -*s*:

  child–children, ox–oxen, brother–brethren

Of the three, only *children* is part of the standard American vernacular (though the others are preserved in some nonstandard dialects, together with archaic plurals such as *eyen, shoon,* and *hosen*). Most Americans meet *oxen* mainly in writing, and commonly say *oxes* instead.[5] Similarly they perceive *brethren* as an inkhorn term for monks and parishioners. As a result, the *-en* sounds archaic and lends itself to silly wordplay. Shortly after the appointment of Ruth Bader Ginsburg to the U.S. Supreme Court, where she joined fellow person of gender Sandra Day O'Connor, *Newsweek* reported, 'The brethren – and now, two sistren – had to know that the swiftness and scope of their ruling would be viewed as a landmark victory for working women.'[6] In the argot of computer hackers, who try to outdo each other with logical extensions of irregular patterns, the plural of the computer called the *VAX* is *VAXen,* and there also have been sightings of *faxen, boxen, soxen,* and *Macintoshen.*[7]

Several names for gregarious animals that are hunted, gathered, or farmed are identical in the singular and plural:

> *fish, cod, flounder, herring, salmon, shrimp*
> *deer, sheep, swine, antelope, bison, elk, moose*
> *grouse, quail*

These forms may have come from constructions in which the singular is used to refer to potential quarry in the aggregate, as in *We went hunting for duck.*

A fourth class of nouns takes the regular *-s* ending but changes its final consonant, usually *f* but sometimes *th* or *s,* from unvoiced to voiced:

> *calf–calves;* also *elf, dwarf, half, hoof, knife, leaf, life, loaf, self,*
> *scarf, sheaf, shelf, thief, wife, wharf, wolf*
> *mouth–mouths;* also *truth, sheath, wreath, youth*
> *house–houses*

Something familiar is going on here: A voiced consonant *z* is being shoved against an unvoiced consonant, and one of them bends to make the cluster consistent. We saw this happening in the regular nouns, where *-s* is pronounced differently in *dogs* and *cats.* But strangely, in these nouns the suffix *z* keeps its voicing, and the noun surrenders it – a right-to-left smearing that violates the usual left-to-right smearing of English phonology. Some linguists have posited a special rule, *regressive voicing*, to generate these examples. The rule, though, would have to be handcuffed to these two-dozen-odd words, because most nouns ending in *f* or *th* are regular and would have to be left untouched. The plural of *reef* is regular (*reefs*, not *reeves*), and the same is true for nouns such as these:

> *birth, booth, earth, faith, growth, hearth, length, month, tenth*
> *belief, brief, chief, proof, safe, spoof, turf*

Even many of the so-called irregular nouns are questionable; many speakers simply pronounce *hoofs, wharfs, oaths,* and *truths* in the ordinary way. I prefer a different theory: that some nouns have two stems, one for the singular, one for the plural, and that the plural stem is tagged as incomplete without a suffix: *knive-, loave-, wolve-,* and so on. After all, if *-ed* and *-ing* are tagged as suffixes that cannot be pronounced unless they are attached to a stem, why can't there be stems that cannot be pronounced unless they have a suffix attached to them? The regular suffix *-s* then applies, generating the plural form without further ado.[8]

Finally, there are nouns that take Latin or Greek plurals. As the singer Alan Sherman has pointed out, 'One hippopotami/Cannot get on a bus. Because one hippopotami/Is two hippopotamus.' Here are five families with Latin plurals:

> *alumnus – alumni;* also *bacillus, cactus, focus, fungus, locus,*
> *nucleus, radius, stimulus*
> *genus–genera, corpus–corpora*
> *alga–algae;* also *alumna, antenna, formula, larva, nebula, vertebra*
> *addendum–addenda;* also *bacterium, curriculum, datum,*
> *desideratum, erratum, maximum, medium, memorandum,*
> *millennium, moratorium, ovum, referendum, spectrum,*
> *stratum, symposium*
> *appendix–appendices;* also *index, matrix, vortex*

And here are two families with Greek plurals:

> *analysis–analyses*; also *axis, diagnosis, ellipsis, hypothesis,*
>   *parenthesis, synopsis, synthesis, thesis*
> *criterion–criteria*; also *automaton, ganglion, phenomenon*

These nouns come from science and academia, and the plurals were borrowed directly from Latin or Greek together with the singulars. They must be irregular forms that are memorized as a list, not the products of a rule attaching *-i* or *-ae*, because most nouns shun these plurals except in the speech of people with an attitude:

> *apparatus–apparatuses*; also *bonus, campus, caucus, census,*
>   *chorus, circus, impetus, prospectus, sinus, status, virus*
> *area–areas*; also *arena, dilemma, diploma, drama, era,* etc.
> *album–albums*; also *aquarium, chrysanthemum, forum, museum,*
>   *premium, stadium, ultimatum*

Latin- and Greek-inspired plurals in a sense are still not part of the English language. They are not acquired as part of the mother tongue in childhood, and are uncommon in everyday speech among nonacademic adults. Instead they are learned in school together with the Pythagorean theorem and the dates of the Peloponnesian War. Since they follow no living rule, and people couldn't have memorized them unless they went to the right schools and read the right books, they are shibboleths of membership in the educated elite and *gotcha!* material for pedants and know-it-alls (the kind of people who insist that the millennium begins January 1, 2001).

Admittedly, I cringe when I hear *this phenomena, those criterias*, and *the media is*, and I could barely contain myself during the speech from the president of the alumni association who kept thanking *the alumnis*. I also get a perverse pleasure from correcting students who refer to *an important piece of data* or write that *this data is important*. (*Data* is the plural of *datum*, I tell them, so one ought to say, *The datum is important*; *The data are important*.) Yet by the same logic I ought to correct myself when I refer to *an agenda, two candelabras, this insignia,* or *that propaganda*, which are the plurals of *agendum*, *candelabrum*, *insignium*, and *propagandum*. And I refuse to hear a word about *genii*, *termini*, *aquaria*, *podia*, *lexica*, *fora*, *stadia*, or *apices*. In any case, whenever pedants correct, ordinary speakers hypercorrect, so the attempt to foist 'proper' Greek and Latin plurals has bred pseudo-erudite horrors such as *axia* (more than one *axiom*), *peni*, *rhinoceri*, and this one:

THE FAR SIDE   By GARY LARSON

"Fellow octopi, or octopuses ... octopi? ... Dang, it's hard to start a speech with this crowd."

It should be 'Fellow octopuses.' The *-us* in *octopus* is not the Latin noun ending that switches to *-i* in the plural, but the Greek *pous* (foot). The etymologically defensible *octopodes* is not an improvement.

The flip side of plural pomposity is playful punning that deflates it, and for decades wags have seen the opening. In a *Peanuts* cartoon, Linus had to bring eggshells to Miss Othmar's class so he could make *igli*. The comedian Shelley Berman has talked of *stewardi* wearing *blice*. Wayne and Shuster performed a

skit in which Julius Caesar nibbled on a *spaghettus*. In Richard
Lederer's 'Foxen in the Henhice,' Farmer Pluribus reached for
some *Kleenices* while being serenaded by *tubae, harmonicae,
accordia, fives*, and *dra*.[9] Henry Beard and Roy McKie's *A
Gardener's Dictionary* contains the following entry:[10]

> **Narcissus:** wonderful, early-blooming flower with an unsatisfac-
> tory plural form. Botanists have been searching for a suitable
> ending for years, but their attempts – *narcissi* (1947), *narcissusses*
> (1954), *narcissus* for both singular and plural (1958), and *multi-
> narcissus* and *polynarcissus* (1962, 1963) – haven't enjoyed any real
> acceptance, and thus, gardeners still prefer to plant the easily
> pluralized daffodil or jonquil.

This may seem silly and inconsequential, but the following
story appeared in *The New Republic* on December 12, 1994: 'In Las
Vegas, The Flying Elvi sued The Flying Elvises for trademark theft.
Both organizations leap from airplanes in Elvis Presley (late
period) costumes and dance and pretend to sing upon landing.'

The masterpiece in the underappreciated genre of irregular
plural humor comes from the National Puzzlers' League, the
association of twisted geniuses who devise impossibly clever word
problems. One kind of puzzle, the falsie, begins by finding a
pair of words that look as if they are related by a morphological
rule:

> False iteration: *bus–rebus, bozo–rebozo, ally–really*
> False predecessor: *lope–antelope*
> False feminine: *butter–buttress, car–caress, under–undress*
> False comparative: *ling–linger*
> False plural (from Hebrew): *inter–interim*

The puzzle itself takes the form of a poem (called a *flat*) that uses a
pair of falsely related words. The words are deleted from the poem
and their locations are marked with placeholders. The object of
the puzzle is to guess the pair of words from the context of the
poem. The following flat by the puzzler known as Trazom (in real
life Joshua Kosman, the senior music critic of the *San Francisco
Chronicle*) contains a seven-letter singular noun in the place

marked ONE and its six-letter false plural in the place marked
MANY. Try it (the answer is in the notes).

False Plural (7, 6)
Turn over on your side, my dear,
And tuck your foot behind your ear;
And I, meanwhile, will crouch like this
And give your neck a tender kiss.
Let's see now – let your arms go slack
And clasp your hands behind my back;
I'll reach around and drape my knee
Across your shoulder – goodness me!
I must confess, this is a stretch,
But honeybunch, you mustn't kvetch.
I know it hurts, I know it smarts—
But these arcane erotic arts
Don't yield their secrets right at first;
And now, I think, we're past the worst.
So please don't throw a ONE, sweet miss—
The MANY says we'll soon reach bliss.[11]

~

Now we come to the irregular verbs. A menagerie of nearly two
hundred words coming in many shapes and sizes, they are a vivid
demonstration of how the human mind, reacting to the events of
history, reshapes a language over centuries and millennia.[12]

The verbs *be, have, do,* and *go* are irregular in many of the
world's languages. They are the most commonly used verbs in
most languages and often pitch in as auxiliaries: 'helper' verbs that
are drained of their own meanings so that they may combine with
other verbs to express tense and other grammatical information,
as in *He is jogging, He has jogged, He didn't jog, He is going to jog.*
Many language scientists believe that the meanings of these verbs
– existence, possession, action, motion – are at the core of the
meanings of *all* verbs, if only metaphorically. For example, the
mind treats *telling him a story* as causing the story to go to him
resulting in him having it, and it treats *dying* as going out of
existence.[13]

In English we saw that *be* stands out from all other verbs with its eight-way conjugation. Its irregular past-tense form stands out too. Together with *go*, it is the only verb whose past tense is a completely unrelated word, a relation that linguists call suppletion:

> *be–was/were–been*
> *go–went–gone*; also *undergo, forgo*

Suppletion arises from a merger of two verbs. Old English, spoken from about 400 to 1100, had three verbs for *be*: *beon, esan,* and *wesan.* They probably differed in meaning, with *beon* referring to permanent states and the other *be*s to temporary ones. (The distinction is similar to the one in modern Spanish between *ser* and *estar*: *Yo soy Americano* [I am American], a long-term trait, contrasts with *Yo estoy contento* [I am happy], a temporary state.) Adding to the surfeit, different sets of *be*s were used in different parts of England. In the Middle English period (1100–1450) they merged into one verb. As in a corporate merger, in a linguistic merger the workers scramble to fill a smaller number of positions, because a verb generally permits only one form in every slot in its conjugation. *Beon* supplied the base form *be*; *esan* supplied *am, is,* and *are*; *wesan* supplied *was* and *were*.

For mysterious reasons, in the Middle English period the verb *go* usurped the past-tense form of another verb, *wend* (as in *to wend one's way*), namely, *went*. Today the verb *wend*, bereft of its old past-tense form, has the regular past *wended*, but its original form followed a pattern that can be seen today in other irregular verbs, such as *bend–bent, send–sent,* and *spend–spent.*

Have, also irregular in many languages, is one of two English verbs that drops its final consonant and replaces it with a *d*:

> *have–had, make–made*

Originally these were *haved* and *maked*, but enough lazy speakers swallowed the consonants that at some point in the Middle English period speakers didn't hear them and assumed that they were not there at all.

The verb *do* does something slightly different – it takes on a *-d*, and changes its vowel: *do–did–done*. Its participle form *done* (as in *You've done it again*) is a contraction of the verb with an old suffix, *-en*. (The same thing happens in *be–was–been* and *go–went–gone*.) The *-en* suffix is found in about fifteen English participles (such as *spoken, sworn, chosen, blown*, and *written*), but the suffix is not attached by a rule. New verbs, such as the neologisms *fax, Bork, spam*, and *mosh*, never get *-en* participles; no one says:

> I've already faxen it.
> That's the third nominee the Republicans have Borken this session.
> The company has spammen its customers with ads once too often.
> Not tonight, dear; I'm sore from having moshen all night.

Putting aside weird *be*, what do all these verbs – *had, made, did*, and the *bent–sent–spent* family – have in common? They all end in *t* or *d*. These, of course, are the same consonants that make up the pronunciation of the *regular* suffix *-ed*. About half the irregulars end in *t* or *d*, because they originally took some version of the regular *-ed* suffix but then fell off the regular bandwagon for one reason or another. These lapsed regulars, together with the regulars themselves, were dubbed *weak* in 1819 by Jacob Grimm of Grimms' Fairy Tales fame; Grimm was also one of the first historians of the Germanic languages. Grimm called the verbs 'weak' because they were too wimpy to hold on to their own unique past-tense forms. We will meet the more macho *strong* verbs later in the chapter.

Some version of the weak past-tense suffix *-ed* can be found in all the Germanic languages, including English, German, Dutch, and the Scandinavian languages. The suffix originated in an ur-language, Proto-Germanic, spoken by a tribe that occupied most of northern Europe in the first millennium B.C. Linguists call it the *dental suffix* because it was pronounced with the tongue against the gum ridge behind the teeth.

Why didn't the weak verbs make life simple and just stay regular? It is because combinatorial rules of grammar have a cost, as we saw in chapter 1: They blindly join things together without looking at what they are made of, and thus can create ungainly chimeras. Two strange things can happen when a verb finds itself with a suffix grafted onto its rear end. One of them is illustrated by the largest class of irregular verbs in English, the no-change verbs:

hit–hit; also *slit, split, quit, knit, fit, spit, shit*
rid, bid, forbid
shed, spread, wed
let, bet, set, beset, upset, wet
cut, shut, put
burst, cast, cost, thrust
hurt

(Some of these verbs have alternative past-tense forms: irregular *bid–bade, forbid–forbade/forbad, spit–spat*, the mainly British *shit–shat*, and regular *slitted, knitted, fitted, wetted* and *thrusted*.)

Note that all twenty-eight verbs end in *t* or *d*. Most of them arose in Middle English and Early Modern English (1450–1700) when the regular ending was often *-de* or *-te*. Throughout the language, *es* at the ends of words, formerly pronounced, were dropping like flies; the 'silent *e*' in the modern spelling of words such as *bake* is a souvenir of the earlier period. Thus a form such as *hitte* got reduced to *hit*. But why did speakers stand by as these past-tense forms shrank into confusing copies of their stems, rather than making the verbs regular, which would have given them the more distinctive *hitted*?

If I may be permitted to psychoanalyse speakers who have been dead for centuries, it probably came from a widespread human habit: We don't like to put or keep a suffix on a word that looks like it already has the suffix.[14] In this case, people don't like to put a version of *-ed* on a verb that already ends in *t* or *d*. Psycholinguists have offered several explanations. Perhaps speakers develop a stereotype for 'past-tense form,' namely, 'ends with *t* or *d*,' and unconsciously think that a stem that fits the stereotype has already been inflected and stop themselves from adding the suffix again. Perhaps when the mind assembles past-tense forms, it gets confused between the *it* or *ed* or *ut* that is already at the end of the stem and the *-t* or *-d* it is trying to add and merges them into a single sound, like the girl who said, 'I know how to spell *banana*, but I don't know when to stop.' Perhaps the suffix *-d* is applied, and the unpronounceable result, *hitd*, is cleaned up, not by the ordinary phonological rule that inserts a neutral vowel between the *t* and *d*, but by a special rule that deletes the *d*. Perhaps several of these explanations are correct.

In any case the no-extra-suffix habit is alive and well in modern speakers. The psycholinguists who jot down speech errors have found that people are prone to leaving out *-ed* on *regular* verbs that end in *t* or *d*. For example, they say, *So we test 'em on it*, intending to say *tested*, or *That's what I need to do*, intending to say *needed*.[15] The same thing happens when people are brought into the lab, given a list of verbs, and asked to say them aloud in the past tense as quickly as they can.[16] Children, too, don't like to add *-ed* to verbs ending in *t* or *d* – they make their signature error, *breaked*, less often with verbs that end in *t* or *d*, such as *hitted*, *putted*, *builded*, and *meeted*, than with verbs with other endings, such as *bringed* and *buyed*.[17] These habits are leaving their mark on English as it continues to evolve: Even in careful speech and writing, many people use no-change past and participle forms for verbs like *bust*, *pet*, *shred*, and *tread*, as in *She got the fleas when she pet the dog* and *This is an area where few psychologists have tread*.[18]

The phobia of adding a surplus suffix extends beyond the past tense. Gardening scriveners often cannot bring themselves to write *crocuses*, *gladioluses*, and *narcissuses* (as we learned in the *Gardeners' Dictionary* entry for *Narcissus*), and write headlines such as 'Hardy Gladiolus Have Long Been a Favorite,' and 'Dutch Crocus Herald the Arrival of Spring.' (No doubt these are symptoms of a Latin-conscious *-us/-i* anxiety as well.) I have seen an ad for a sprayer that *fits all hose* and another one for *the pantyhose that last*, and still another announcing *All fax on sale*. People treat the *sh* sound as similar to *s*, leading the *Boston Globe* handyman to write about adjusting *window sash*,[19] and leaving every professor baffled as to how to refer to more than one *prefrosh* (pre-freshmen). Many people have trouble keeping up with the *Joneses* and instead merely try to keep up with the *Jones*. When it comes to the possessive *'s*, hardly anyone follows the advice in Strunk and White's famous style manual to refer to *Charles's hat* (charlzis) or *the Jones's car*; it's usually *Charles' hat* and *the Jones' car*, both in writing and speech. And what do you say to someone who has a *daddy-long-legs* climbing up each shoulder?

When a word has a verbatim replica of a suffix inside it, rather than just a reminder of one, the attempt to add the real suffix often results in clumsiness or unintelligibility. When there is rain or snow or hail or thunder coming down from the skies, it is said to be raining, snowing, hailing, and thundering. What about when there is lightning? Is it *lightninging*? Not very likely, and some speakers snip out an *-ing* and say *It is thundering and lightning*. Many adjectives can be turned into adverbs by adding *-ly*, such as *softly, surely,* and *happily*. What about those adjectives that already end in *-ly*, such as *ugly, friendly, heavenly,* or *leisurely*? *Uglily*? *Friendlily*? *Heavenlily*? *Leisurelily*? Pthack. (The *Atlantic Monthly*, perhaps hoping no one would notice, once ran a story entitled 'Friendily Yours.') Sometimes brand names can be turned into colloquial verbs for traveling or sending:

We Chevy'd up and down Main Street.
I FedExed the package last night.
Down to their last thirty dollars, they Greyhounded home.
Because of his fear of flying he Amtrak'd to New York.

But even if your frequent flyer plan is with United Airlines, it is unlikely that you have ever *United* to San Francisco.

Sometimes people can get into trouble by speaking as if a word that appears to contain an affix really does contain it. An interstate trucking company must have lost the business of the literate when it proudly painted its trucks with the slogan 'Faster than rail, regular than mail.'[20] Former President George Bush used to tell reporters that he spent his vacation *bonefishing*, leading them to wonder what the best bait is for catching bones, and presumably his heart was in the right place when he explained, 'I hope I stand for anti-bigotry, anti-Semitism, anti-racism.'

Back to the verbs. Repeated-suffix phobia is also the explanation for the class that originally contained *wend–went*:

bend–bent; also *send, spend, lend, rend, build*

These verbs devoice their final consonant, *d*, into *t*. They began as *bend + de*, and the double *d* was fixed by trimming the final consonant of the stem, yielding past tense *ben + de*. The extra twist is that the phonological rule that today turns *-d* into *-t* in words like *walked* and *passed* used to be triggered by words ending in *l, m, n,* and *v* as well. *Bende* became *bente*, which then lost its *e* to give us *bent*. The overeager *-d → -t* rule can also be blamed for these verbs:

> *burn–burnt*; also *learn, dwell, spell, smell, spill, spoil*

The irregular forms ending in *-t* show the English language changing before our eyes: Most of them are on their way out. American speakers mainly use *burnt* as an adjective, not a past-tense form – *The toast is burnt because Bernie burned it* – and would not be caught dead saying *learnt, dwelt, spelt, smelt, spilt,* or *spoilt*. *Rent* is used only for emotional resonance, as in *The Vietnam war rent the fabric of American society*, and *lent* is giving way in American English to *loaned*, the past of *to loan*. In 'Childe Harold's Pilgrimage' (1812) Byron describes a battlefield using three verbs in the class that range from the moribund to the dead:

> The thunder clouds close o'er it, which when rent,
> The earth is covered thick with other clay
> Which her own clay shall cover, heaped and pent,
> Rider and horse – friend, foe, in one red burial blent!

Blend–blent, of course, has become completely regular, as have most of the other verbs with *-t* in their past-tense forms, such as *wend–went, pen–pent, gird–girt, geld–gelt,* and *gild–gilt*. Like many obsolete irregulars, *gilt* and *pent* have left relics among the adjectives: *a gilt-edged book, pent-up energy*.

〜

Another reason that regular forms can go to seed is the Los Angeles Lakers effect that gave us irregular plurals such as *feet* and *mice*. Grafting a suffix onto a stem can trigger changes in the pronunciation of the stem, and sometimes the change can stay in the word long after the trigger has vanished.

Many languages distinguish a vowel sound pronounced quickly from the same sound drawn out; they are called short and long vowels. The vowels traditionally called 'short' and 'long' in English, such as the ones in *bet* and *beet*, used to differ in this way, as we see in their spellings: The long vowel was symbolized by writing two short vowels in a row, as if it took twice as long to pronounce.

Starting around the year 1000, English speakers shortened their pronunciation of a vowel when extra phonetic stuff (such as a consonant or syllable) was added, pushing new consonants into the syllable.[21] Here are some examples that have survived in modern English:

bone–bonfire
break–breakfast
child–children
Christ–Christmas
deep–depth
five–fifth
know–knowledge
sheep–shepherd
wide–width
wise–wisdom

Shortening a vowel is a natural reaction when material is added to the end of the syllable. A syllable is a unit of timing, taking up a constant tick of the speech clock. If material is added to the end of a syllable, the vowel is often shortened to maintain the rhythm.[22] This habit of pronunciation could easily have turned into a full-fledged rule. In his April 5, 1997 column, the language maven William Safire ventured that the pronunciation of *seminal* as 'SEM-uh-null' in place of 'SEE-muh-null' was an instance of academic bowdlerization – prissy professors covering up the fact that the word *seminal* comes from the word *semen*. Safire's theory, however, would have to go to lengths worthy of Oliver Stone to explain why those professors, presumably hatching plots in their SEEminars, have also changed the pronunciations of *vanity*, *sanity, cleanliness, brevity,* and *criminal* to hide the fact that they come from *vain, sane, clean, brief,* and *crime*. All, of course, are products of a phonological rule in English that shortens vowels at the beginning of many three-syllable words.

Take a verb with a long vowel like *keep*. Add the regular suffix and spell it phonetically: *keept*. Shorten the vowel in response to the extra stuff at the end. We end up with something pronounced *kept* – one of a number of modern irregular past-tense forms that would be regular but for their shortened vowels:

> *keep–kept*; also *creep, leap, sleep, sweep, weep*

Add some other habits of Middle English speakers that we have come across – using *-t* more widely, dropping suffixes – and you understand many other irregular verbs in modern English:

> *feel–felt*; also *deal, kneel, dream, leave*
> *bleed–bled*; also *breed, feed, lead, mislead, plead, read, speed, meet*
> *hide–hid*; also *slide, bite, light, alight*
> *flee–fled, say–said, hear–heard, lose–lost, shoot–shot*
> *sell–sold*; also *tell, foretell*
> *do–did*

(As before, some of these verbs allow regular past-tense forms, such as *kneeled, dreamed, speeded, lighted,* and especially, *pleaded*. For some of the verbs – particularly *sell, tell,* and *do* – the reasons for the vowel changes are a bit more complicated.)

Kept, of course, isn't simply *keeped* pronounced with a clipped vowel; neither is *hid* just a short version of *hide* nor *shot* a short version of *shoot*. The pairs of vowels traditionally called 'long' and 'short,' and spelled as if they are double and single scoops of the same sound, are in fact very different vowels. How did that happen?

The perpetrator is a process of language change that is the opposite of the various slurrings and swallowings and cutting of corners that we have seen so far. All of those changes make it easier for the speaker to speak but do nothing for the listener, who would rather have the speaker enunciate clearly. Sometimes listeners do get their way; speakers *enhance* the difference between a pair of vowels by adding, exaggerating, or embroidering each in a different manner.[23]

For many centuries speakers of Old and Middle English enhanced the difference between short and long vowels by making the long vowels tense: that is, the muscle at the root of the tongue is tensed up, changing its shape and making the vowel in *great* sound different, as well as longer, than the vowel in *get*. Enhancement went wild, however, during the dawn of Early Modern English in the fifteenth century, when the pronunciation of the long vowels was scrambled in a linguistic revolution called the Great Vowel Shift. Before the shift, *keep* had been pronounced something like *cape*, *hide* like *heed*, *boot* like *boat*. After the shift, the English spelling of the long vowels no longer made much sense, nor did the pairings of 'short' and 'long' vowels in siblings like *keep* and *kept*. Since the children of Early Modern English could not have heard a relationship between the vowels, the past-tense forms struck them as a ragbag that just had to be memorized outright, and so they remained for subsequent generations. Thus verbs that entered the popular language after the Great Vowel Shift, such as *peep* (1460) and *seep* (1790), and verbs whose pronunciations eventually drifted into rhyming with the *keep* verbs, such as *reap* and *heap*, did not undergo a vowel change; they remained intact when they first submitted to *-ed*, giving us *peeped*, *seeped*, *reaped*, and *heaped*, not *pept*, *sept*, *reapt*, and *heapt*.

Here is a small mystery: What is the verb that goes with the past-tense form *wrought*, as in *The Watergate scandal wrought great changes in American politics*, and the participle form in Judges 23:23, *What hath God wrought!*, quoted by Samuel Morse in the first intercity telegram? According to the theory that irregulars are pairs of memorized words, an irregular past-tense form could, in principle, survive in memory without a corresponding stem. *Wrought* appears to be an example: Most people have no idea what the verb is. Many guess *wreak* (based on an analogy with *seek–sought*) or *wring* (based on an analogy with *bring–brought*), but both guesses are wrong. The answer is *work*: *Wrought iron* is worked iron, and a person who is *all wrought up* is a person who is all worked up. (Old theater saying: 'Plays are wrought, not written.') *Wrought* belongs to a family of verbs that replace their rhyming parts with *ought* or *aught*:

buy–bought; also *beseech, bring, catch, fight, seek, teach, think*

How do you get *wrought* from *work* or *sought* from *seek*? The connection is less mysterious when we realize that the now silent *gh* used to be pronounced, somewhat like the *ch* of *Bach, loch,* and *Chanukah.* Start with *work* (actually *wyrcan,* but I will use modern spellings to make the changes clearer). Add the suffix *-t* to get *workt.* Soften the *k* sound to *gh,* yielding *worght* – an old phonological trick to avoid the strenuous *-kt.* A vowel and an adjacent *r* often switched places in the history of English, because *r* sounds a lot like a vowel, which makes its order with respect to a vowel hard to hear. Thus *brid* became *bird, thrid* became *third, hross* became *horse,* and *worght* became *wroght.* We no longer pronounce the *gh*; and recall that many English vowels were shuffled during the Great Vowel Shift (the vowel spelled *ou* was once pronounced *ō*), and that vowels often get shortened when a suffix is added (so *ō* becomes *ŏ*). The result is *wrought* and the mystery is solved.

~

In the 1980s the irascible *New York Times* book reviewer Anatole Broyard wrote that he doubted that English had 'any life left in it, any flavor or idiosyncrasy.' His colleague Maggie Sullivan followed up in a column of her own:

> Anatole Broyard is right to sound the alarm. We are losing this idiosyncrasy; as a language changes, strong verbs tend to become weak. For example: although once shepherds shore their sheep, sheep are no longer shorn, they are sheared.
>
> This issue should arouse lovers of the English language. Weakening the verbs can only weaken the language itself. To keep English from becoming a feeble tongue, we must reinforce our verbs. Fortunately, I have come up with a two-part plan. First, we must not allow new verbs to enter the language in a weak state. We must ensure, for example, that *to clone* is established as *clone, clewn, clown,* as in: *Future generations of booksellers may reproach us for not having clown Joyce Carol Oates and Isaac Asimov . . .* And *to gentrify* as *gentrify, gentrifo,*

gentrifum, as in: *The newcomers gentrifo one block and now the whole old neighbourhood is gentrifum.*

Since new verbs are few and far between, I offer the second part of my plan – creating new strong verbs. English has some strong verbs with unique patterns for their principal parts, such as *go, went, gone.* Individuality makes them particularly vulnerable. Their patterns would hold up better if each pattern had more representatives. If we create allies for our unique strong verbs, we can buttress them and increase their number. Here are suggestions for new strong verbs:

Conceal, console, consolen: After the murder, Jake console the weapon.

Subdue, subdid, subdone: Nothing could have subdone him the way her violet eyes subdid him.

Fit, fat, fat: The vest fat Joe, whereas the jacket would have fat a thinner man.

Displease, displose, displosen: By the look on her face, I could tell she was displosen.

Sullivan's plan to 'strengthen' the language captures two hallmarks of the second kind of irregular verb in English, the so-called strong verbs. They belong to alliances with similar sounds, and despite this solidarity, they have been dwindling for millennia.

The families of strong verbs have a history stretching back more than 5500 years. Most of the languages of Europe, Iran, and the northern half of India, and many current and extinct languages of Turkey, western Asia, and China, show similarities in vocabulary and grammar that suggest they are descendants of a single language spoken by an expansive and mysterious prehistoric tribe. The most popular theory is that they were a late-neolithic farming people with domesticated horses, wheeled vehicles, and a military leadership, who expanded from a homeland in southern Russia around 3500 B.C.[24] An alternative is that they were the people that first brought farming to Europe, beginning in 7000 B.C. from a homeland in eastern Turkey.[25] Though we don't know who they were or where they came from, we know a lot about how they spoke. Their language, Proto-Indo-European, has largely been reconstructed by historical linguists working backward from the commonalities in the daughter languages.[26]

Many Indo-European languages have echoes of the strong-verb patterns seen in English, such as *bear–bore, tear–tore,* and *sink–sank, drink–drank.* Some of these verbs and their past-tense forms actually existed in the ancestral language, such as *bher--bhor-* and *senk^w-–sonk^w-*. Proto-Indo-European apparently had a set of rules for forming the past tense, not by adding a suffix as in modern English, but by changing the vowels, as in modern Hebrew – a kind of rule called gradation, apophony, or ablaut. There were probably seven ablaut rules, more or less as follows: If the verb has *ei* followed by a consonant, change the *ei* to *a*. If the verb has *e* followed by a vowel-like consonant, change the *e* to *ă* – and so on for the other five classes.

When the Indo-Europeans started to spill out over Eurasia, the daughter tribelets lost touch, and games of Broken Telephone began in each one. Eventually the language radiated into the ancestors of our familiar languages and language families such as Germanic, Romance, Slavic, Celtic, Greek, Iranian, and Sanskrit. For example, the verb *werg-* (to do) ended up in Germanic as *werkam* (work), and in Greek as *erg-* (action) and *org-* (tool), which eventually crossed over into English as *energy, organ,* and *orgy.* When a word meaning 'do' turns into a word meaning 'orgy,' the change wrought by the chain of whisperers must have been considerable. It is remarkable that the seven classes of Indo-European strong verbs came through, tattered but recognizable, in Proto-Germanic, then in the West Germanic language spoken by the Angles and Saxons, and then in Old English, Middle English, and Modern English. That is why the strong verbs fall into clusters of similar-sounding forms today.

The rules themselves, however, did not survive. Imagine a rule that replaced *ĭ* with *a*, and suppose that people started pronouncing *ĭ* as *ī* in some verbs, *e* in others, and *ĭ* in still others, depending on the consonants following the vowel and many other factors. Children would have a hard time making sense of the rule, and at some point they would stop trying and simply memorize the past-tense forms as a list. By the time of Old English, the Indo-European vowel-change rules were extinct and their products had been mangled in different ways by the slings and arrows of outrageous fortune. At least a fifth of the verbs no longer obeyed the rules of their original class, and in the following centuries so many verbs joined, left, or switched classes that today the classes no longer correspond very well to the organization of the verbs in the minds of modern speakers.

Here is one Old English class, Class I, that has come through in recognizable shape:

> *rise–rose–risen*; also *arise, write, smite, ride, stride, dive, drive, shine, strive, thrive*

The list highlights a key feature of the strong verbs. While dictionaries happily list irregular forms such as *smite–smote–smitten*, *stride–strode–stridden*, *strive–strove–striven*, and heaven help us, *thrive–throve–thriven*, in the minds of real English speakers these forms are muzzy: People vaguely recognize them from books but are uncomfortable using them in their own speech, and are tempted to default to regular forms like *smited*, *strided*, *strived*, and *thrived*. Sometimes strong and weak forms live side by side in a person's mind, forming doublets like *strove* and *strived* or *dove* and *dived*.[27]

Doublets usually arise when an irregular form (such as *strove*) hovers in a twilight zone in memory and people are not sure whether they have heard the form or are confusing it with a similar form, like *drove*. Other doublets arise for the same reason that you say *tomayto* and I say *tomahto*: Britain and America are divided by a common language. The British prefer *dived*, the Americans prefer *dove*, and people who encounter both dialects, such as Canadians, are unsure. Often the members of a doublet will diverge in meaning, grammar, or formality, like twins who strive not to be confused. *Shone*, for example, is intransitive (without a direct object), as in *The stars shone in the sky*, and a touch poetic, whereas *shined* is an everyday form that may be used in transitive sentences such as *Melvin shined his shoes*. (It would sound silly to say *Melvin shone his shoes*.) For many people regular *hanged* means 'suspended by the neck until dead,' irregular *hung* merely 'suspended.' Sometimes a muzzy participle will enjoy full vigor as an adjective, often with its own meaning. For example, *smitten* is doing fine as an adjective that means 'infatuated,' not literally 'walloped' (though the original meta-phor is clear enough, and visible in related metaphors such as *stunning* and *lovestruck*).

Some of the past-tense forms originally in this class became muzzier and muzzier until they faded out entirely and their verbs became regular. *Abode* used to be the past tense of *abide* and today survives only as a noun meaning 'residence.' No speaker of modern standard English uses *chide–chode, glide–glode, gripe–grope,* or *writhe–wrothe,* though some examples, such as *climb–clomb,* cling to life in rural areas of Britain and America. Many of the wayward verbs did not fall into the arms of regularity but were attracted to *other* irregular patterns. For example, the short vowel *ĭ* is common in participles like *driven, risen,* and *written,* and in many weak verbs, and it inspired *bit* and *hid* in the standard dialect of English. In nonstandard dialect we find *clim, writ, strid, smit, div, driv,* and the forms immortalized in the Negro spiritual 'Joshua fit the battle of Jericho' and in the doggerel 'Spring has sprung/The grass is ris/I wonder where the boidies is.'[28]

The pairing of *ī* and *ō* in *rise–rose, drive–drove,* and other descendants of Class I can be seen, with variations, throughout the strong verbs, where *ī*-like vowels are frequently replaced by *ō*-like vowels:

> *find–found;* also *bind, grind, wind*
> *freeze–froze;* also *speak, bespeak, steal, heave, weave*
> *wear–wore;* also *bear, forbear, swear, forswear, tear*
> *take–took;* also *mistake, partake, forsake, shake*
> *wake–woke;* also *awake, break*

Forsook and *hove* are pretty recherché these days, with *hove* appearing mainly in nautical contexts such as *The ship hove to;* other uses, such as *Irving hove his lunch,* could only be said in jest. Like the other strong classes, the *swear–swore* class used to embrace more verbs, but many defected to the regular side:

> But unburied whiten the bones of the crew;
> Ah! would that the widow and orphan but knew
> The place where their dirge by deep billows is sighed,
> The place where unheeded, unholpen, they died.[29]

Some of the old irregular forms survive in rural dialects, such as *help–holp, tell–tole, melt–molt,* and *swell–swole,* and others survive in adjectives in the standard dialect such as *molten* and *swollen.*

If you shorten both vowels of the *e–o* pattern you get:

> *get–got*; also *forget, beget, tread*

which also beget some muzziness. The participle *has got* is British, *has gotten* American. As with many differences between the dialects, it was the Mother Country that corrupted the mother tongue; *gotten* was the form used in England when the first colonists left in the seventeenth century, and the Americans preserved it while it vanished in the British Isles. *Trod* and *trodden* sound vaguely Winnie-the-Poohish to American ears, because Americans seldom use the verb to *tread*: Where the British say *tread on,* Americans say *step on* (notwithstanding one of the slogans of the American Revolutionary War, 'Don't Tread on Me'). When *tread* is used, it is regular: *He treaded water;* not *He trod water. Begot* suffers because of the familiarity of *begat* in the King James Bible and the countless satires based on it.

Strangely enough, three common verbs undergo these vowel changes in reverse:

> *come–came*; also *become, overcome* (compare *wake–woke, take–*
>   *took*)
> *fall–fell*; also *befall* (compare *get–got*)
> *hold–held*; also *behold* (compare *swear–swore*)

Came came from a very old irregular whose origins are obscure, but *hold* (and maybe *fall*) really did get reversed. Originally *to hold* was *to held* (actually, *healdan*) with past tense *hold* (*heold*). Similarly, *fall* used to have the forms *feallan–feoll.* Some ancient, influential, and confused group of speakers managed to mix up these verbs with their past-tense forms. This is not as addled as it may seem; today people occasionally confuse the parts of a verb when the past tense or participle is more commonly used than the stem:

Even as environmentalists speak of a seamless web of life, and the artery advocates speak of a seamless city, the designs on the drawing board still rent the land from the sea and undermine its urbanity [from *rend–rent*].[30]

The videophone is the same size as a regular phone but includes a 3.3 inch color screen with a tiny camera and lens. The . . . company hopes to smitten prospective buyers by renting the phones for less than $30 a day [from *smite–smote–smitten*].[31]

REEBOK KICKS ITSELF OVER NAME WITH BAD FIT . . . For a company that made its reputation by helping to shod the women's aerobics movement, the Incubus name would definitely seem out [from *shoe–shod*; Reebok had named a women's running shoe *Incubus*, not realizing that the word refers to an evil spirit that has sex with women while they are asleep].[32]

Producer Harvey Weinstein hoves his boorish bulk up to the mike for his moment in the sun for the callow 'Shakespeare in Love' – but is miraculously sent packing by the deus ex machina of the orchestra [from *heave–hove*].[33]

Similarly, *hoist* was originally the past tense and participle of *hoise* (as in *For 'tis the sport to have the enginer Hoist with his own petar*, from *Hamlet*), but it has since been reanalysed as the stem in *hoist–hoisted*.

The following family is a freeze-frame of the process by which neat classes can get messier over the centuries:

> *blow–blew*; also *grow, know, throw, draw, withdraw, fly, slay*

What do they have in common? All end with a vowel, and all begin with a cluster of consonants except *know*. In fact even *know* begins with a consonant cluster in its spelling, and that tells a story. Spellings usually reflect old pronunciations, and the *k* in *know* was originally spoken aloud; the word was pronounced *k'nawa*. So these verbs used to be *completely* consistent. Owing to the disappearance of *kn* and *gn* at the beginning of spoken English words, one member no longer fit the membership requirements and had to be kept in the class by sheer stipulation. In the history of languages many law-abiding classes become more and more ragged as general pronunciation shifts mangle their members, until eventually the criteria become indiscernible to children and the words are memorized individually.

With only one nonconformist member thus far, the *blow* class has not yet disappeared, though it has suffered losses. *Slay–slew* has a biblical feel and may be on the way out, if we are to judge by recent usages such as *Burr slayed Alexander Hamilton in a duel*.[34] *Crow–crew* survives in the bookworm expression *The cock crew*; even small changes in the expression, such as *The rooster crew*, sound peculiar, and *Harvey crew over his victory* is unintelligible. Regional dialects have added or preserved a few more, such as *show–shew, saw–sew, sow–sew*, and *snow–snew*; in 1942 the *Chicago Sun* wrote of the weather, *It blew and snew and then it thew*. These forms are rarely heard today, however, and the trend is in the opposite direction: attrition into the regular class. Children make errors such as *blowed* and *knowed* more often than for any other kind of irregular verb.[35] The journalist H. L. Mencken was an assiduous student of the vernacular speech of the United States and documented many common nonstandard past-tense forms in his magisterial volumes *The American Language*. Among them are *blowed, knowed, throwed, drawed*, and one made famous by a character in Harriet Beecher Stowe's *Uncle Tom's Cabin*. Theodore Bernstein, in *The Careful Writer*, comments on her oft-quoted words:

TOPSY
'In the absence of such reorganization, the city's court structure as a whole has just "growed," like Topsy'; 'Like Topsy, that Government-held surplus of farm commodities "just keeps growin'."' Once and for all, Topsy's exact words, punctuated variously in different editions and in different books of quotations, were: 'I 'spect I grow'd.' No 'just,' no 'jes',' no 'growin',' no nuffin'. Anyway, Topsy, Queen of the Clichés, should drop dead. *See* Clichés.[36]

A few verbs besides *came* take an *ā* in the past tense:

eat–ate; also *give, forgive, bid, forbid, lie*

Bade is a somewhat stilted past-tense form of *bid* in the sense of 'ask' or 'command to,' though not in the sense of poker, bridge, or defense contracts – no one says *He bade three clubs*. *Lie–lay* is a trap seemingly designed to lure speakers into errors and to provide material for the lamentations of language lovers (including me, in private moments). A recent article by Cullen Murphy in the *Atlantic Monthly*, 'The Lay of the Language,' was devoted to the verb,[37] and even the Muppets have been dragged into the controversy. In 1999 the talking doll Sing and Snore Ernie had to be reprogrammed after purists objected to his statement, 'It feels good to lay down'[38] (the biggest hooha over a talking doll since Barbie set back the cause of gender equality by whining, 'Math is hard'). What's wrong with *lay*? Officially, it belongs to two verbs. One is an intransitive irregular verb, *lie–lay–lain*, meaning 'recline':

Stem: *Please lie down and tell me about your childhood.*
Past tense: *He lay down on the couch.*
Participle: *He has lain down on the couch.*

The other is a transitive regular verb, *lay–laid–laid*, meaning 'set down':

Stem: *Lay your cards on the table.*
Past tense: *He laid his cards on the table.*
Participle: *He has laid his cards on the table.*

Like Ernie, many casual speakers use *lay* for both – as in *I'm going to lay down* – and who can blame them? As if the sharing of *lay* in the two conjugations weren't confusing enough, the two verbs *ought* to be one, according to the grammatical logic of English. *Lay* means 'cause to lie,' and is one of a handful of verbs meaning 'cause to X' that differ by a vowel from a related verb meaning 'to X.' The others are *sit–set*, *rise–raise*, *fall–fell* (as in *to fell a tree*), and believe it or not, *drink–drench*. In most other cases, the verb that means 'to X' and the verb that means 'cause to X' sound the same:

The stick leaned against the house.
I leaned the stick against the house.

The planter stood on the deck.
I stood the planter on the deck.

The baby sat on the bed.
I sat the baby on the bed.

The 'ungrammatical' intransitive *lay* follows the pattern of *lean, stand,* and *sit* perfectly. Many purists believe that intransitive *lay* is a recent corruption, disseminated by rock lyrics such as Bob Dylan's *Lay Lady Lay* and Eric Clapton's *Lay Down, Sally.* But a rule of thumb in language is that any so-called corruption that occurs frequently enough for the guardians to notice it will turn out to have been common in the language for a century or more. Intransitive *lay* was unexceptionable in the seventeenth, eighteenth, and nineteenth centuries; for example, in 1812 Byron wrote, 'There let him lay' in 'Childe Harold's Pilgrimage.'[39] The historical linguists Thomas Pyles and John Algeo report:

> The brothers H. W. and F. G. Fowler (1931, p. 49) cite with apparently delighted disapproval 'I suspected him of having laid in wait for the purpose' from the writing of Richard Grant White, the eminent nineteenth-century American purist – for purists love above all to catch other purists in some supposed sin against English grammar.[40]

Another long-term trend reshaping the English language is most apparent in our final class of irregular verbs, illustrated in the greeting card by Suzy Becker on the opposite page.

The *ing–ang–ung* pattern often is generalized in dialects and affectations of dialects, as in the jocular *Who would have thunk?* In 1998 the Texan columnist Molly Ivins entitled a book *You've Got to Dance with Them What Brung You,* allegedly a backwoods aphorism though more likely an urbanite's attempt at hick-chic. The baseball pitcher and sportscaster Dizzy Dean was said to have narrated a play as follows:

> The pitcher wound up and flang the ball at the batter. The batter swang and missed. The pitcher flang the ball again and this time the batter connected. He hit a high fly right to the center fielder.

The center fielder was all set to catch the ball, but at the last minute his eyes were blound by the sun and he dropped it![41]

Dave Barry, defending himself against enraged Neil Diamond fans after making a joke at the singer's expense in a prior column, describes the results of a reader survey:

> Unfortunately, a lot of survey voters are not so crazy about Neil's work, especially the part of 'Play Me' where he sings '. . . song she sang to me, song she brang to me . . .' Of course I think those lyrics are brilliant; however, they brang out a lot of hostility in the readers.

The *ing–ang–ung* pattern came down to us from another class of strong verbs in Old English, Class III, which included *singan–sang–sungen.* Many modern verbs follow it to varying degrees:

ring–rang–rung; also *sing, spring, drink, shrink, sink, stink, swim, begin*
cling–clung; also *fling, sling, sting, string, swing, wring, slink, stick, dig, spin, win*

run–ran–run
hang–hung
strike–struck, sneak–snuck
sit–sat, spit–spat

Most of the *ing–ang–ung* verbs end in *-ing* or *-ink*. Two of the others deserve comment.

Begin has the distinction of being the only common irregular verb that is neither monosyllabic nor built around a monosyllabic root. (The common, Anglo-Saxon words we use every day tend to be monosyllables, and the irregular verbs are no exception.) *Begin* is formed with the prefix *be-*, as in the similar irregulars *become, befall, beget, behold, beset,* and *bespeak.* In *begin*'s case, however, the residue, *-gin,* is not an English word; it came from a now-defunct Proto-Germanic verb meaning 'open.' (There are two other irregular past-tense forms, both somewhat unusual, whose stems cannot stand alone as verbs: *forsake–forsook* and *beseech–besought.*)

Snuck has the distinction of being the most recent irregular to enter the standard language, with a first citation in the *Oxford English Dictionary* from 1887. According to a recent survey, most younger Americans have no problem with *snuck,* though most older Americans frown on it.[42] William Safire quotes a letter from Doris Asmundsson, a professor emerita of English: 'Words like *creak, critique, eke, freak, leak,* and *tweak,* do not, in the past tense, become *cruck, crituck, uck, fruck, luck,* and *twuck.* Why then *snuck*? Eventually a *sneaker* might turn into a *snucker.*'[43] According to one theory, *snuck* sneaked into English via sound symbolism. Its connotation of quickness, furtiveness, and mild disreputability brought to mind the sound pattern of *slunk* and *suck,* especially since all three end in a suitably crisp *k.*[44] A less far-fetched explanation is that *sneak* is close in pronunciation to *sting, strike, dig,* and especially *stick –* an *ĭ* is just a lax, short *ē,* and *n* is basically *t* or *d* pronounced through the nose, as any cold-sufferer can tell you. The failure to rhyme with *creak* and *tweak* was no impediment, because similarity in the gestures of articulation matter more than similarity in sound, and that makes it tempting to analogize *stick–stuck* to *sneak–snuck.*

Many dialectal past-tense forms that don't rhyme exactly with *cling* or *slink* still take the *ŭ* vowel in the past tense. Mencken and others report *climb–clumb, shake–shuck, take–tuck, dive–duv*, and *drive–druv*, also heard in the English proverb 'Sussex won't be druv.' One speaker described what they used to do to endangered species in the olden days as follows: *They killed 'em and skun 'em out.* Dizzy Dean was famous for saying *He slud into second*, and some baseball fans say, 'If Dykstra hadn't dropped the ball, the runner wouldn't have tug' (tagged).[45] At the foot of this page is another common example.

The *ing–ang–ung* verbs are a bellwether of a millennium-old and still ongoing trend in the English language. In the fourteenth century the egalitarian preacher John Ball roused the rabble with the slogan 'When Adam delved and Eve span/Who was then the gentleman?' *Span* was the past tense of *spin*, following the *i–a–u* pattern of verbs such as *sing, swim*, and *begin*. But eventually the participle *spun* usurped the past-tense slot, relegating *span* to the dustbin of history. These takeovers are still going on. When the Walt Disney corporation released a film called *Honey, I Shrunk the Kids*, English teachers were up in arms: It should be *Honey, I Shrank the Kids*, they said. Nonetheless, most people say *shrunk, sprung, sunk*, and *stunk*, not *shrank, sprang, sank*, and *stank*. (Some go the other way: At an infamous moment in the O. J. Simpson murder trial, the prosecutor Christopher Darden said hopefully, 'The gloves appear to have shrank somewhat.')

The teachers are fighting a losing battle because even the language mavens are losing their grip on the distinction. William Safire got an earful from the Gotcha! Gang and the Uofallpeople Club when he wrote, '*Trivialize* had its moment in the vogue-verb sun, until the usage of this older verb shrunk to the very

ARLO & JANIS® by Jimmy Johnson

ARLO & JANIS reprinted by permission of Newspaper Enterprise Association, Inc.

occasional.'[46] The *Boston Globe*'s language maven, Jan Freeman, wrote that she once did a double-take upon hearing *They sort of sprang it on me*, momentarily thinking it was incorrect.[47]

Shrank, together with the other *ank* and *ang* words, is under assault from two directions – from its own past participle *shrunk*, and from the many *ing* verbs that have already lost their *ang*s and really do take the *ung* form in the past tense as well as in the participle: *He slung* (not *slang*) *the hash*, *They strung* (not *strang*) *him up with a rope*, *He flung* (not *flang*) *the ball at the batter*. Surely and steadily, *ing–ung–ung* is displacing *ing–ang–ung*, part of a larger erosion of the distinction between participles and past-tense forms throughout the verb system.

Regular verbs fail to distinguish pasts and participles at all – *I walk, I walked, I have walked* – and fewer than half of our irregular verbs continue to distinguish them; most are like *mean–meant–has meant* or *find–found–has found*. In nonstandard dialects the distinction is even feebler. *I seen it* and *A man come into the bar* are absolutely standard outside the upper and middle classes, even in urban areas, and *He begun to cry*, *She done it*, and *They gone home* are also common. In the early decades of the twentieth century, Mencken reported the past-tense forms *div, driv, riz, swole, taken, thrown*, and *writ*, and the participle forms (*has*) *ate, blew, broke, did, drank, drive, froze, gave, rode, rose, ran, stole, swam, took, tore, woke, wore*, and *wrote*. Nonetheless, many people assume that the erosion is a recent development:

Dear Ann Landers:
Have Americans forgotten there is such a thing as verb tense? I am shocked when I hear people say 'woulda came,' 'coulda went,' 'shoulda did,' 'woulda took,' 'had went,' 'hadn't came,' and so on.

Don't they realize 'woulda' and 'coulda' are slang versions of 'would've' and 'could've' – which are contractions for 'would have' and 'could have'?

I heard a narrator say, 'I seen' in a political commercial, and a TV reporter say, 'We haven't spoke.' . . . A TV anchorwoman said, 'had threw it' and 'between you and I.'

I am a secretary for almost 50 years and am thankful that, with only a high-school education, my English is impeccable. You will

do a lot of folks a big favor if you print this letter and bring it to
their attention.
　E. E.
　Wood Ridge, N. J.

Dear E. E.:
Thanks for taking the time and trouble to write. I shoulda thunk
to tell them off myself.[48]

Confusions of past and participle forms are easy to explain.
Some may originate in mishearings. As E. E. pointed out, the
auxiliaries *has* and *have* that signal the perfect construction are
often contracted to *he's, we've, could've, should've,* and *would've,*
or even *coulda, shoulda,* and *woulda.* (Anyone who has graded
student papers or dipped into internet discussion groups is also
familiar with *could of, should of,* and *would of.*) That makes the
haves easy to miss in rapid speech; *He's seen it,* in particular, is
easily reanalysed as *He seen it.*

Yet the main reason for the decline of the *ang–ung* distinction is
that *all* distinctions in English inflection have been declining for
the past thousand years; syntax has been shouldering the load
formerly borne by morphology. Old English and Middle English
had separate verb forms not only for present, past, and participle,
but also for different persons (*I, you, he/she*) and numbers
(singular and plural) within the past tense. The past forms for
sing, for example, would have been:

I sang	We sungon
Thou sunge	You sungon
He/she sang	They sungon

When the person and number distinctions collapsed, every verb
had to end up with a single past-tense form, and a game of
musical chairs broke out, with the different stems competing for
the remaining chair. With some verbs the singular won, such as
sing–sang–sung; with others the plural won, such as *sling–slung–*
slung (the past plural was usually similar to the participle – the
phenomenon of syncretism we met in chapter 2). Another free-
for-all took place among the participles of the verbs that kept their

-*en* and had to grab a stem from the collapsing conjugation. Some took the stem of the base form, such as *take–took–taken*, others took the stem of the past form, such as *break–broke–broken*, and still others kept their own stem, such as *swell–swelled–swollen*. Some participles can jump ship to another pattern: If you apply the *break–broke–broken* pattern to *shake* and *take*, you get the somewhat cutesy *shooken* and *tooken*. I have been advised that *tooken* has become standard in Generation X circles, but if it is, do not blame it on their ethos of ironic detachment; it was used as early as 1946 in 'Put That Kiss Back Where You Found It,' a song recorded by Benny Goodman: 'Took it when I wasn't lookin'/And my heart you've also tooken.'[49] The steady erosion of distinctions in English inflection helps us understand why we continue to be confused by verbs such as *shrink* and *spring* in the second millennium after the end of Old English.

For the sake of completeness, here are the remaining irregular forms. *Shorn* and *swollen* belong to a small group of verbs that are regular except for their participles:

  swell–swelled–swollen, shear–sheared–shorn
  show–showed–shown; also *sow, sew, prove, strew*

(A few other irregular participles were orphaned from their verbs and survive only as adjectives, most of them somewhat unusual: *bereft, unbidden, clad, cleft, cloven, drunken, forlorn, girt, gilt, misbegotten, hewn, beholden, laden, molten, mown, pent, misshapen, clean-shaven, shod, sodden.*) I couldn't figure out where to put these:

  beat–beat–beaten
  choose–chose–chosen
  see–saw–seen
  stand–stood; also *understand, withstand*

The *stand–stood* pattern is heard in the oft-cited plaint of the fight promoter Joe Jacobs, 'I should have stood in bed,' and in the modal auxiliary verb

  can–could

which retains a present-past contrast in usages like *I can't polka now, but I could before I broke my leg.* Other pairs of modal auxiliaries – *may–might, will–would,* and *shall–should* – began life as different tenses of the same verb, but the couples divorced long ago and *might, would,* and *should* are no longer past-tense forms.

Exactly how many irregular verbs are there in the English language today? If we don't double-count prefixed families such as *get* and *forget,* don't count dialectal forms such as *drug* and *brung,* do count verbs that are irregular either in standard American or standard British English, and do count the muzzy but widely recognizable forms, we end up with 164 modern irregular verbs: 81 weak (ending in *t* or *d*), 83 strong. Compare this to Old English, with 325 strong verbs alone, and it is clear that English is indeed becoming 'weaker.' In later chapters we will see whether the surviving but endangered irregular verbs are sustainable.

~

We have seen how the weak past-tense forms can be traced to Proto-Germanic about 2000 years ago, and the strong forms can be traced back to Proto-Indo-European at least 5500 years ago. But where did they come from? They certainly were not designed by a committee, and presumably did not arise from divine revelation. No one knows the answer, but a few brave linguists have speculated.

The dental suffix in Proto-Germanic, the ancestor of our *-ed,* may have come from a reduced form of the verb *to do.*[50] Many languages use an empty verb like *do* as an auxiliary verb that carries information about the statement as a whole, such as tense, degree of completion, and negation. Indeed, Modern English uses *do* for that purpose in *yes-no* questions (*Do you want to dance?*) and in negations (*Alice doesn't live here anymore*). In the history of a language, prefixes and suffixes often arise from the erosion of verbs such as *do, take, be,* and *have,* a process called grammaticalization.[51]

If the dental suffix came from *do,* it would explain why it has the sound *d* or *t.* In Proto-Germanic, *do* could come after a noun or another verb, very roughly like *He hammer-did* or *She walk-did.*

The *do* could have eroded to the stub *d* and attached itself to the verb, giving us the ancestor of *-ed*.

The theory also explains why *-ed* has become the *regular* suffix, applying freely to any new or strange verb. The phrase containing *do* and a verb would have been created by the rules of syntax, the combinatorial system par excellence, which allows almost anything to combine with anything else. A promiscuous auxiliary verb would have been a natural ancestor of a promiscuous suffix: Just as a verb like *do* can combine with any verb at all (*He did abandon, He did abate, He did abbreviate*, and so on), so its descendant *-ed* might have retained this habit, allowing *it* to combine with any verb at all (*abandoned, abated, abbreviated*, and so on).

The Indo-European ablaut or vowel-change patterns, the ancestors of our strong verb forms, change an *e* (a sound between *Ed* and *aid*) or a neutral vowel to *a* (as in *father*), or to *ō* (as in *hoe* or *horse*). The *e* is pronounced with the tongue hump toward the front of the mouth, the *a* and *ō* with the tongue low and toward the back. This contrast, between a higher front vowel and a lower back vowel, survives in the majority of modern English irregular verbs. The base forms have sounds like *ĕ* and *ē* and *ĭ* and *ī* and *ā*, and the past-tense forms have sounds like *ă* and *ŏ* and *ŭ* and *o͝o*.

That may not be a coincidence. Three of the great linguists of the middle decades of the twentieth century, Roman Jakobson, Jerzy Kuryłowicz, and Morris Swadesh, noticed that in many languages the vowels pronounced with the tongue high and at the front of the mouth tend to be used for the basic forms of nouns and verbs (such as the singular form of a noun and the infinitive of a verb), whereas the vowels pronounced with the tongue lower and farther back tend to be used for the specially marked forms (such as plural nouns and tensed verbs).[52] Moreover, the higher and farther front vowels have different connotations from the lower and farther back vowels in pairs of contrasting words. The high front vowels come first in expressions such as *pitter-patter* and *dribs and drabs*; we don't say *patter-pitter* or *drabs and dribs*. And in pairs such as *this* and *that*, *here* and *there*, and *me* and *you*, the higher and farther-to-the-front vowels are found in the word that means 'self' or 'near the self,' the lower and farther-to-the-back word means 'other' or 'far from the self.' That is true not only in English but in many families of languages.[53]

Perhaps this ubiquitous vowel contrast is a case of sound symbolism. The linguist Roger Wescott has pointed out that high front vowels are pronounced with a constricted mouth cavity and the tongue close to the visible part of the vocal tract, whereas low back and central vowels are pronounced with a large mouth cavity and the tongue buried from view. That may call to mind the conceptual distinction between presentness and pastness. Pastness may remind people of a cavity or space, because a past event is separated by an interval of time from the present moment, and metaphorically speaking time equals space. It may also remind people of remoteness or distance, because metaphorically speaking long ago equals far away. Perhaps as Indo-European was developing, speakers vaguely felt that lower and farther back vowels fit better with the concept of an event separated in time from the present, and that higher and farther front vowels fit better with an event in the here and now.[54] Of course, the Indo-Europeans had to pick *some* vowel contrast if they were to mark tense with a vowel, and for all we know they could just have easily gone the other way. But the fact that the vowel contrast appears in many unrelated languages with similar roles, and was preserved and embellished in our own 5500-year game of Broken Telephone, hints that it might have some semantic resonance for human minds.

4

IN SINGLE COMBAT

Now that we know all about regular and irregular verbs, how well does the words-and-rules theory hold up? In some ways quite well; in other ways not so well. According to the theory, the ingredients of language are a list of memorized words, each an arbitrary pairing between a sound and a meaning, and a set of productive rules that assemble words into combinations. Regular and irregular forms exemplify the two ingredients: Regular forms are generated by rule, irregular forms are memorized by rote.

The dissection of language in chapter 2 showed that the organs that secrete regular forms are elegant combinatorial systems, just as the theory had led us to expect. But the excavation in chapter 3 showed that the depository for irregular verbs is not disorganized and inert, which is what the theory had led us to expect. The irregular verbs are shot through with patterns:

blow–blew, grow–grew, know–knew, throw–threw
bind–bound, find–found, grind–ground, wind–wound
drink–drank, shrink–shrank, sink–sank, stink–stank
bear–bore, swear–swore, tear–tore, wear–wore

This is not what we would expect if they were a laundry list of arbitrary items. Suppletive pairs such as *go–went* and *be–was* are the exception, whereas if the irregulars were truly acquired one by one they could just as easily have been the rule.

We have seen that many irregular patterns are fossils of extinct rules that lived in the heads of speakers long ago, but history can explain only part of the patterning. If the irregulars got all of their patterning from old rules and have degenerated steadily since the rules became extinct, today's patterns should be tattered versions of the ancient Indo-European strong classes, with a core of surviving verbs that fit the old patterns and a halo of distorted verbs that have drifted off in various directions. But many verbs have *joined* the strong classes, or have jumped from class to class, and today's irregular families cross-classify the original ones. *Ring–rang* originally was weak (with a past tense like *ringed*) and was attracted to the *ing–ang–ung* class by analogy to verbs like *sing*. The same happened to *dig–dug, stick–stuck, wear–wore, show–shown,* and many others. Some verbs that entered the language after the Old English period also were seduced into strong classes, such as *fling–flung* and *sling–slung*. Others switched from one strong class to another, such as *slay–slew* and *draw–drew* (originally *drough*). Still others found the *weak* irregular patterns appealing, such as *light–lit* and *creep–crept*.[1]

The words-and-rules theory would be off the hook if these attractions and conversions were a thing of the distant past. Perhaps the vowel-change rules did not die out completely with the Indo-European or Germanic tribes, but lingered in weakened form in parts of England for a few centuries before giving up the ghost. Unfortunately, some of the conversions are fairly recent. *Kneel–knelt, dive–dove, catch–caught,* and *quit–quit* became popular only in the nineteenth century; George Washington, for example, used *catched*, and Jane Austen used *quitted*. And as we saw, *snuck* came into English a century ago and is only now becoming standard.

The funny irregulars in the nonstandard dialects of English add to the worries. While some rural irregulars are quaint holdovers of old standard forms, such as *help–holp, climb–clim,* and *creep–crope*, many, if not most, are homegrown products of the creativity of local speakers: *bring–brang–brung, dive–div, chide–chode, snow–snew, climb–clomb, drag–drug, slide–slud, fling–flang,* and literally hundreds of others.[2]

As an experimental psychologist I have been trained not to believe anything unless it can be demonstrated in the laboratory on rats or sophomores. To my knowledge no one has yet studied irregular verbs in rats, but the linguists Joan Bybee and Carol Moder have studied them in sophomores and have shown that they are all too happy to generalize irregular patterns to novel verbs. They asked the students in a university linguistics course (not only sophomores, of course) to write down the past-tense forms of existing and made-up verbs, completing sentences such as *Sam likes to spling. Yesterday he* _____. Almost 80 percent of the subjects offered *splang* or *splung*. Even when given real words, some of the subjects were tempted by irregular patterns and came up with creative forms such as *dig–dag, sting–stang, slink–slank, streak–strack, skid–skud,* and *clip–clap*. Bybee and Moder may have duplicated in the laboratory the process that gave us forms like *fling–flung* and *drag–drug* in the history of English.[3]

But maybe not. University students treat anything that looks like a test as a test, and perhaps they saw the task as a challenge to their ingenuity in coming up with creative forms. We know that outside the lab, people ruminate over irregular patterns and sometimes deduce ways in which they ought to be generalized. Here is Leo Rosten and Leonard Ross's character Hyman Kaplan, a Jewish immigrant learning English at night school:

> 'I tought de pest time "bite" should be – "bote."'
> Miss Mitnick gave a little gasp.
> '"Bote"?' Mr Parkhill asked in amazement. '"Bote"?'
> '"Bote"!' said Mr Kaplan.
> Mr Parkhill shook his head. 'I don't see your point.'
> 'Vell,' sighed Mr Kaplan, with a modest shrug, 'if is "write, wrote, written" so vy isn't "bite, bote, bitten"?'
> Psychic cymbals crashed in Mr Parkhill's ears.
> 'There is not such a word "bote,"' protested Miss Mitnick, who took this all as a personal affront. Her voice was small, but desperate.
> 'Not-soch-a-void!' Mr Kaplan repeated ironically. 'Mine dear Mitnick, don' *I* know is not soch a void? Did I say *is* soch a void? All I'm eskink is, isn't logical *should be* such a void?'[4]

The students in the *spling* experiment also may have thought to themselves, 'Isn't logical *should be* soch a void?' and may have concocted forms they never would have used when speaking naturally.

A related worry is that people make note of irregular forms, especially in other people's dialects, and consciously use them in wit and wordplay, as we discovered in the previous chapter and can see in these old jokes:

> A woman gets into a taxi in Boston's Logan airport and asks the driver, 'Can you take me someplace where I can get scrod?' He says, 'Gee, that's the first time I've heard it in the pluperfect subjunctive.'

> A friend of mine came across some cut flowers that were so spectacularly red she thought they must be fake. 'These are amazing,' she said. 'Are they dyed?' The florist shook his head. 'No, no, not at all,' he said. 'Just put 'em in water and they'll be fine.'[5]

> A man was on trial for pulling a woman down the street by her hair. The judge asked the arresting officer, 'Was she drugged?' The policeman replied, 'Yes, sir, a whole block.'[6]

Verbal humor depends on the audience's recognizing that an odd word *is* odd; if the words were unexceptional, there would be no joke. So just because some wiseguy finds a novel past-tense form logical or amusing, it does not mean that the form is a natural product of the system he uses in everyday speech. Instead the form may be a product of one's intellectual faculties reflecting back onto one's language, an ability called metalinguistic awareness.

But one kind of generalization cannot be a product of conscious cogitation, namely, the errors preschool children make in their spontaneous speech:

> It was neat – you should have sawn it!
> Doggie bat me [bit].
> The cheerios got aten by the Marky.
> I know how to do that. I truck myself [tricked].
> He could have brang his socks and shoes down quick.
> And they swang into a roller coaster and we went with their cars and they were sliding and they did a leap.
> Elsa could have been shotten by the hunter, right?
> So I took his coat and I shuck it [shook].
> This is the best place I ever sot [sat].

I bate Paul up [beat].
You mean just a little bitty bit is dranken?[7]

The psychologist Fei Xu and I combed through transcripts of the speech of nine children in an electronic archive and pulled out all the past-tense and participle forms, 20,000 in all.[8] We found numerous irregular forms that are not standard English:

beat–bate	*crush–crooshed*	*fling–flang*	*say–set*
bite–bet	*fight–fooed*	*jump–janged*	*sleep–slep*
bring–brid	*fit–feet*	*lift–left*	*swing–swang*

Children don't make these errors very often – only in two tenths of 1 percent of the opportunities – but eight of the nine children made at least one while the tape was running, and we know that children continue to make them well into the school-age years. I have a drawing from a seven-year-old girl with the caption *Win I Wit to Git a crismis chre and it wus Sowing so i braing my umbrella*[9] and I remember that when I was twelve I had a persistent urge to use *raught* as a past tense of *reach*, on the analogy of *teach–taught*.

~

The irregular patterns refuse to die. Irregular verbs are supposed to be a list of arbitrary words memorized by rote, just like *duck* and *walk*, with only a trace of patterning left behind by long-defunct rules. Instead, people extract the patterns and extend them to new words, just as they do with the *regular* pattern in errors like *breaked*, in neologisms like *moshed*, and in the *wug*-test. The distinction between regular and irregular inflection, and therefore between words and rules, is not so clear anymore. Either the irregular patterns are generated by rules, just like the regular pattern, or linguistic productivity does not depend on rules in the first place but can arise from words via some ability to associate the patterns in known words with the patterns in new ones.

Both alternatives have been developed into famous, full-blown theories of the English past-tense system. Their clash is one of the most vigorous controversies in the modern study of the mind, echoing through psychology, linguistics, philosophy, computer science, and neuroscience. According to the theory of *generative phonology* developed by Noam Chomsky and Morris Halle, rules rule.[10] Every drop of patterning in past-tense forms, regular or irregular, is squeezed out into rules, and only the compressed, desiccated residue is stored in the mental lexicon. According to the theory of *parallel distributed processing* or *connectionism* developed by David Rumelhart and James McClelland, there *are* no rules: People store associations between the sounds of stems and the sounds of past-tense forms, and generalize the associations to new words if they are similar to old words.[11] Both theories invoke a single kind of mental computation to explain how people generate regular and irregular forms, but for generative phonology it's rules all the way down, whereas for connectionism it's memory all the way up.

A clash over irregular verbs may call to mind the remark that academic debates are heated because so little is at stake. But in this case something is at stake. The past-tense debate is the latest battle in the centuries-old disagreement over two very different ways of understanding the mind:

> When a man reasons, he does nothing else but conceive a sum total from addition of parcels, or conceive a remainder from subtraction of one sum from another; which, if it be done by words, is conceiving of the consequence of the names of all the parts to the name of the whole, or from the names of the whole and one part to the name of the other part . . . For REASON is nothing but *reckoning*.[12]

In this passage from *Leviathan*, written in 1651, Hobbes uses *reckoning* in the original sense of counting, calculating, or computing. For example, suppose the definition of 'man' is 'rational animal.' Then if we are told that something is 'rational' and an 'animal' (names of parts) we can deduce it is a 'man' (name of whole), and if we are told that something is a 'man' (name of whole) and that it is 'rational' (name of one part) we can deduce that it is a rational 'animal' (name of the other part). These steps could be laid out as mechanical instructions to recognize and copy words, a kind of symbol, and therefore could be 'reckoned' or computed by someone who has no idea what the concepts 'rational' and 'animal' even mean. If the symbols are patterns in the brain rather than words on a page, and the patterns trigger other patterns because of the way the brain is wired, then we have a theory of thinking.

Among the people influenced by Hobbes was Leibniz, who was inspired as well by John Wilkins and other designers of artificial languages discussed in chapter 1. Leibniz took Hobbes literally when he said that reason is nothing but reckoning. He devoted much of his life to inventing a scheme that would perfect the computations underlying thought, turning arguments into calculations and making fallacies as obvious as errors in arithmetic. 'Once this has been done,' he wrote, 'if ever further controversies should arise, there should be no more reason for disputes between two philosophers than between two calculators. All that will be necessary is that, pen in hand, they sit down together at a table and say to each other . . . "let us calculate." '[13] In one version of Leibniz's scheme, 'man' is assigned the number 6, 'animal' is assigned 2, and 'rational' is assigned 3. Since $2 \times 3 = 6$, a rational animal must be a man; since $6 \div 3 = 2$, a man is not just any old rational being but specifically a rational animal. If the number for 'monkey' is 10, one may calculate that monkeys are not men or vice versa, and that monkeys, while animals, are not rational.[14]

The idea that intelligence arises from the manipulation of symbols by rules is a major doctrine of the school of thought called rationalism, generally associated with Leibniz and Descartes. When the symbols stand for words and the rules arrange them into phrases and sentences, we have grammar, the subject of Cartesian linguistics, which later inspired Humboldt and then Chomsky. When the symbols stand for concepts and the rules string them into chains of inference, we have logic, which became the basis for digital computers, the artificial intelligence systems that run on them, and many models of human cognition.[15]

But symbol manipulation is not the only way the mind might work:

> There appear to be only three principles of connection among ideas, namely, *resemblance, contiguity* in time or place, and *cause or effect*. Experience shows us a number of uniform effects, resulting from certain objects. When a new object, endowed with

similar sensible qualities, is produced, we expect similar powers and forces, and look for a like effect. From a body of like color and consistence with bread we expect like nourishment and support.[16]

In this passage from his 1748 *Enquiry Concerning Human Understanding*, David Hume summarizes the theory of *associationism*, a major tenet of the school of thought called empiricism. The mind connects things that are experienced together or that look alike (Hume later eliminated cause and effect as a separate principle) and generalizes to new objects according to their resemblance to known ones. Just as the rationalists were obsessed by combinatorial grammar, the associationists were obsessed by memorized words. In his 1689 *Essay Concerning Human Understanding*, John Locke pointed to the arbitrary connection between words and things as the quintessential example of how the mind forms associations by contiguity in time: We learn *dog* when Mother says 'dog' in the presence of a dog. Replace Locke and Hume's 'ideas' or 'sensible qualities' with 'stimuli' and 'responses,' and you get the behaviorism of Ivan Pavlov, John B. Watson, and B. F. Skinner. Replace the ideas with 'neurons' and the associations with 'connections,' and you get the connectionism of David Rumelhart and James McClelland.

The great debate between rationalism and empiricism is familiar to everyone who has taken a course in philosophy, psychology, or the history of ideas. It embraces such issues as whether the mind is packed with innate structure or is a blank slate on which the environment writes, and whether knowledge comes from making deductions using theories or gathering data from observation. But the issue that concerns us here is the nature of our mental machinery, in particular, whether intelligence arises from the manipulation of symbols or from associations between sensory qualities. And the competing theories of the English past tense provide us with an unusual opportunity.

It's been three hundred and fifty years since *Leviathan*, and scholars are still debating rationalism and empiricism. Both sides appeal to theoretical coherence, intuitive plausibility, political ramifications, and harmony with the morals of modern science, but these are pretty squishy criteria, and the debate rages on. We need a concrete, richly studied instance of human psychology in which the two grand theories can go head-to-head in explaining the same facts.

In ancient warfare, an army sometimes would send its mightiest warrior to face his counterpart from the opposing army in single combat. The outcome would embolden one side in the actual battle that followed, or might pre-empt it altogether, sparing unnecessary bloodshed. The most familiar example is the biblical story of David and Goliath, and in *The Right Stuff* Tom Wolfe suggested that a contemporary example may be found in the space race of the 1960s, in which the Mercury astronauts were treated like single combat warriors against the cosmonauts of the Soviet Union. Scientific debates sometimes work like single combat, not because the combat metaphor is particularly apt or appealing, but because it is easier to compare two great big theories when each is vested in a highly specific hypothesis and the hypotheses compete on the same ground.

The English past tense is the perfect site. The phenomenon is circumscribed and therefore tractable to study. Its history, acquisition, and patterns of use have been abundantly documented. It has obvious rulelike and memory-like features that serve as hurdles that any theory must clear. And it has brought out the best in the ingenuity of contemporary theoreticians, with each side devising a clever, elegant, detailed, and surprising model. The past tense is the only case I know in which two great systems of Western thought may be tested and compared on a single rich set of data, just like ordinary scientific hypotheses.

～

What are the facts to be explained? Irregular verbs defy the suggestion that they are memorized by rote because they show three kinds of patterning.

First, irregular past-tense forms are similar in sound to their base forms. For example, *drink* and *drank* share *d, r,* a vowel, *n,* and *k*; the only difference is that *drank* has the vowel *a* where *drink* has the vowel *i*. Similarly similar are *swear* and *swore, sleep* and *slept, freeze* and *froze*. In fact, all the irregulars except *go–went* and *be–was* share material with their stems. It didn't have to be that way. One can imagine a hypothetical language in which most of the verbs are like *go–went*, with nothing in common between stem and past, each stem and each past stuffed into its own memory slot. We need an explanation of why English does not look like that. Let's call this pattern *stem-past similarity*.

Second, a few kinds of change from a stem to its past are seen over and over among the 164 irregular verbs. The *ĭ-ă-ŭ* pattern in *drink–drank–drunk*, for example, is found, with variations, in *sing–sang–sung, sit–sat–sat, begin–began–begun, shrink–shrank–shrunk*, and twenty other verbs. Similarly, we have *freeze–froze, speak–spoke*, and *steal–stole*; *bleed–bled, breed–bred*, and *feed–fed*; *teach–taught, fight–fought*, and *bring–brought*. One can imagine a language in which every verb picked its own substitution of vowels and consonants from among the thousands that are logically possible. But generations of learners have passed down an English language that is very different from that possibility. Let's call this pattern, in which the change from stem to past in one verb is similar to the change from stem to past in another verb, *change-change similarity*.

Third, the verbs undergoing a given irregular change are far more similar than they have to be. If you are a verb and want to undergo the *ĭ-ă-ŭ* pattern, all you really need is an *ĭ*. But the verbs that do follow the pattern (*drink, spring, shrink*, and so on) have much more in common; most begin with a consonant cluster like *st-, str-, dr-, sl-*, or *cl-*, and most end in *-ng* or *-nk*. Similarly, the verbs whose pasts end in *-ew* (*blow, grow, throw, slay, draw*, and *fly*) tend to begin with a consonant cluster and end with a vowel. Verbs of a feather change together, and not just in sightings by word watchers. People extend old patterns to new verbs, as in *bring–brang, fight–fit*, and *spling–splung*, only when the new verb is highly similar to old ones in memory. We need an explanation of why the human mind is so impressed by similarity in sound; let's call this pattern *stem-stem similarity*.

A theory with rules for *irregular* verbs, as well as regular ones, could explain all three kinds of patterning. Imagine a rule that said, 'If a verb has the sound *consonant-consonant-ĭ-ng*, change *ĭ* to *ŭ*.' Notice that the rule doesn't spell out the past-tense form letter by letter; it just says, 'Change the vowel.' The rest of the input – the consonants before and after the vowel – come through in the output untouched. We have an explanation for stem-past similarity.

Now suppose that the mind prefers simple grammars, with a few rules, to complex grammars, with many rules. If there are fewer rules than verbs, many verbs will have to share a rule, such as 'Change ĭ to ŭ.' We have an explanation for change-change similarity.

Finally, notice that the rule has a condition on it: apply only to verbs that have two consonants before the vowel and an *ng* sound after it. The condition is a gatekeeper that allows in verbs that are similar to the *cling–clung* family and filters out those that merely contain ĭ. This could explain stem-stem similarity.

Now, any drudge can go over the list of irregular verbs in the preceding chapter and write down a set of tedious rules for them: 'If a verb begins in *s* and ends in *ee*, change the *ee* to *aw*,' and so on. But that would be no improvement over the original list of verbs. A theory invoking rules must be more than a summary of the patterns among the verbs. It should be a *psychological* theory: a hypothesis of the format in which children acquire words, and an explanation of why verbs have the kinds of patterns they do. The trick is to find a compact set of rules that captures the generalizations the mind likes to make.

By far the most ambitious theory of this kind comes from Noam Chomsky and Morris Halle's 1968 magnum opus *The Sound Pattern of English*, later refined by Halle and the linguist K. P. Mohanan.[17] Their rules for irregulars are part of a larger set of rules that capture the sound pattern of a language (its accent). Clearly the speakers of a language know more than the list of words that happen to be in the language at a given moment. For example, English speakers intuit that *blicket, dax,* and *fep* are not English words but could be, whereas *ftip, rtut,* and *nganga* are not English words and could not be (though speakers of other languages might recognize them as possible words in their languages). English speakers also know that when *divine* is joined to -*ity* to create *divinity*, the *i* vowel in -*in*- changes from ī to ĭ, and that when *Canada* is joined to -*ian* to become *Canadian*, the final -*a* vanishes, the stress shifts from the first syllable *Ca* to the second syllable *na*, and the vowel in that syllable changes to ā. Chomsky, Halle, and Mohanan accounted for the patterns of thousands of

English words with just a few dozen phonological rules, each assigned to one or more of the boxes in the diagram on page 25: lexicon, morphology, syntax. Their theory comes from a field called *generative phonology*, a division of generative linguistics, the approach to language founded by Chomsky.

By positing rules that replace consonants and vowels (phonemes) in the irregular verbs, Chomsky and Halle enjoy the advantage of rules in general: accounting for patterns among verbs and changes, and for speakers' ability to generalize them. Amazingly, Chomsky, Halle, and Mohanan handled most of the dizzying patterns among the 165-odd irregular verbs with only *three* rules. Virtually all their other rules are needed to explain the sound pattern of English in general.

Chomsky, Halle, and Mohanan roundly reject the words-and-rules dichotomy. Verbs sit on 'a continuum of productivity and generality that extends from affixation of the *-ed* suffix in *decide–decided* to total suppletion in *go–went*,' with families like *sing–sang*, *ring–rang*, and *bind–bound*, *wind–wound* in between.[18] At one end of the continuum are the regular verbs, which are handled by a general rule that says nothing about the words it can apply to. At the other end of the continuum are suppletive verbs such as *go* and *went*, which are simply listed as pairs. In between are the other irregulars, which are handled by a smaller set of rules, each tagged to apply to certain verbs. By stipulating which verb may be touched by which rule, the theorists circumvented the problem of crafting the rules to single out verbs by their sounds – no small matter, given that *shrink* has the past tense *shrank*, *sling* has the past tense *slung*, *bring* has the past tense *brought*, and *blink* has the regular past tense *blinked*.

Another rein on rules keeps each one in a *stratum* (a component or subcomponent) so that it does not run wild and apply where it shouldn't. To generate *keep–kept*, for example, Chomsky, Halle, and Mohanan invoked a rule that shortens a long vowel (changing $\bar{e}$ to $\breve{e}$) when it occurs before a consonant cluster, such as *pt*. But that rule cannot be allowed to apply across the board or it would turn *seeped* into *sept*, *wipe* into *wipped*, and so on. So Halle and Mohanan proposed that the *-t* and *-d* found in weak irregular verbs like *kept* is not the same as the *-ed* found in regular verbs, despite their similar pronunciations. Whereas the regular *-ed* is attached in the morphology box on page 25, the *-t* or *-d* in the irregular verbs is attached in the lexicon box, which is also where the shortening rule is confined. This may seem like cheating, but there are independent grounds for it. Other semiregular sort-of-rules that have nothing to do with the past tense, like the ones creating *serene–serenity* and *volcano–volcanic*, need the shortening rule too. That supports the idea that several rules are sequestered together in their own little community.

One move that allowed Chomsky, Halle, and Mohanan to get away with so few rules was factoring apart each complex change into several simple ones and allowing the simple ones to be assigned in different combinations to different verbs. For example, *tell–told* is produced by at least two rules, one that changes the vowel, the other that adds the *-d*. The rule that changes the vowel is also put to work in *swear–swore*, and the rule that adds *-d* is also put to work in *flee–fled*. Looking at the crisscrossing patterns in *tell–told, swear–swore, flee–fled, bend–bent, burn–burnt, deal–dealt, breed–bred*, and *hit–hit*, one can readily see that sharing minirules that add *t*s and *d*s, that delete extra ones, and that fiddle with vowels is far more economical than building a special rule for each family of rhyming verbs.

A monumental contribution of generative phonology was to slice rules even more finely so that they apply not to vowels and consonants but to the *components* of vowels and consonants, called *features*. The idea goes back to Roman Jakobson and is a striking, universal trait of human language. In chapter 2 we noticed that the three pronunciations of *-ed* in *walked, jogged*, and *patted* are determined by the final consonant of the stem:

id if the verb ends in *t* or *d*
t if the verb ends in *p, k, f, s, sh, ch*, or unvoiced *th*
d if the verb ends in a vowel or in *l, r, m, n, b, g, v, z, j, zh* or
  voiced *th*

Each list is not just any old connection of consonants. The two consonants in the first line, *t* and *d*, are pronounced with the same parts of the mouth (tongue tip against gum ridge) and in the same way (stopping the flow of air and then releasing it). These features – 'place = tongue-tip,' and 'manner = stopping' – are also found in the *-ed* suffix. A similar pattern of sharing also is found among *s, z,* and the plural suffix, all of which are sibilants. What we never see is a rule such as 'Add *-og* if the word ends in *z, r,* or *k*,' and other ragtag assemblies of phonemes.

All this can be captured if we have rules apply to *features* rather than to *phonemes.* One rule can state, 'At the end of a word, insert *i* to separate adjacent consonants that have similar features for place and manner of articulation' – no listing of *t, d, s, z, -ed,* or *-s* is necessary. Similarly, it's obtuse to have a rule that puts a *t* after consonants in the list *p, k, f, s, sh, ch,* and *th* – all these consonants obstruct the air stream (they are *obstruent*) and all are unvoiced. The rule can simply say, 'At the end of a syllable, copy the voicing feature from one obstruent consonant to the next.' All the unvoiced consonants like *p* and *k* will automatically get a *t*, while all the voiced consonants like *b* and *g* automatically get a *d*. Moreover, this rule also generates the *s* and *z* variants of the plural suffix.

Simple rules that inspect, flip, or excise features are ubiquitous in the world's languages. They not only are more economical than rules that hack away at arbitrary lists of consonants and vowels, but also allow speakers to generalize. The German sound *ch* is not normally found in English words, but any English speaker who labors to pronounce the celebrated composer's name as *Bach* knows that if there were a verb *to out-Bach*, as in *Handel out-Bached Bach*, the past tense would be pronounced *bacht*, not *bachd* or *bachid*. A rule that specifies *t* after 'unvoiced consonants' automatically embraces the unvoiced *ch* and tells a speaker what to do, even if the speaker had never learned that *ch* belongs on the list.

This economy and power also accrues to rules that fiddle with *vowels* if the vowels are dissolved into features such as these:

Is the tongue hump at the front or the back of the mouth?
Is the tongue hump high or low in the mouth?
Are the lips rounded or not?
Is the vowel long or short?
Is the vowel tense (tongue root scrunched forward) or lax?

Take *sing–sang* and *sit–sat.* The *i* is not replaced by some random vowel, such as the one in *say* or *boat* or *shoe*; it is replaced by ă, which is identical to ĭ except for tongue height: ĭ is front, unround, short, lax, and high; ă is front, unround, short, lax, and low. A rule that simply said, 'For the following irregular verbs, lower the vowel' would have to tinker with only one feature, not five; it would explain why ĭ was replaced by a similar vowel, not just any old vowel; and it would work right out of the box to yield *eat–ate* and *choose–chose*, which also lower a vowel. This simple rule, Lowering Ablaut, is one of the three irregular rules that Halle and Mohanan's theory gets away with. The other two are Backing Ablaut, which replaces mid front *e* in *bear* with mid back *o* in *bore*, and Shortening Ablaut, which replaces the long vowels in *flee* and *shoot* with their short counterparts in *fled* and *shot*.

If you have been skimming this last paragraph silently, every-thing should look shipshape – the vowels in *flee–fled* and *shoot–shot* are literally spelled as long and short versions of the same thing: *ee* versus *e*, *oo* versus *o*. But if you have been pantomiming the sounds with your mouth and listening to them with your mind's ear, or better yet, pronouncing them out loud, you should be thinking: *Now wait a minute!!* As we saw in the last chapter, 'long' and 'short' have been misnomers in English at least since the Great Vowel Shift in the fifteenth century, when people scrambled the pronunciations of vowels. The sound in *flee* is *not* a drawn-out version of the sound in *fled*, nor is *shoot* just a lengthy *shot*. In pronouncing *ee* (ē) the tongue is higher than it is when pronouncing ĕ and more tensed up, and the vowel glides up to a little *y* at the end, making it a diphthong (a succession of two vowels pronounced as if they were one). Likewise, in pronouncing *oo* (ū), the tongue is tenser and higher than when pronouncing ŏ, the lips are rounded, and there is a little *w* sound at the end. Those vowels are not particularly similar, and a rule capable of replacing one with another is capable of doing almost anything. To say that a Shorten-the-Vowel rule simplifies the hairy irregular verbs sounds like a hoax.

Chomsky and Halle realized this of course, and their solution is the most radical claim of the theory. When it comes to syntax, Chomsky is famous for proposing that beneath every sentence in the mind of a speaker is an invisible, inaudible *deep structure*, the interface to the mental lexicon. The deep structure is converted by transformational rules into a 'surface structure' that corresponds more closely to what is pronounced and heard. The rationale is that certain constructions, if they were listed in the mind as surface structures, would have to be multiplied out in thousands of redundant variations that would have to have been learned one by one, whereas if the constructions were listed as deep structures, they would be simple, few in number, and economically learned (see *The Language Instinct*, chapter 4, 120–124). Less well known is that Chomsky and Halle made a similar proposal for the sounds of words. Each word has a deep structure – in jargon, an *underlying form* – that may not sound like the way it is pronounced; indeed, it may be unpronounceable. Phonological rules then convert it to the surface form that is articulated and heard.

In the case of the so-called long vowels in English, Chomsky and Halle proposed that the underlying forms really *are* long versions of the short vowels. That is, in the mental lexicon the vowels in the following pairs are *identical* in every respect except how long it would take to pronounce them, with the long vowels taking about twice as long:

Short Vowels	Their Long Counterparts
d*i*n	div*ii*n (*divine*)
d*e*n	ser*ee*n (*serene*)
p*a*t	s*aa*n (*sane*)
f*u*nd	prof*uu*nd (*profound*)
sh*o*t	sh*oo*t (*shoot*)
b*o*mb	c*oo*n (*cone*)

In their actual pronunciations these pairs do differ in how long it takes to say the vowel, but they differ in many other ways besides, so why assume that the mind lists only the difference in length? The reason, according to Chomsky and Halle, is that the other differences are redundant and predictable, hence unnecessary to list. No pair of English words differs *only* by vowel length. Long vowels are also tense and diphthongs (they glide to a different vowel at the end); short vowels are lax and not diphthongs. A theory that gave the mind the ability to store every nuance of an English vowel – length, tenseness, tongue position, lip rounding, diphthongs, and so on – would falsely predict that English could contain long lax vowels, lax diphthongs, short tense vowels, and so on; and it would allow the language to contain short and long versions of otherwise identical vowels – all counter to fact. In a better theory, the vowel inventory would contain only the number of vowels necessary to distinguish words (for example, to distinguish *bit* from *bet*), and other rules would fill the vowel out into a full set of stage directions for the mouth and throat. The best way to do this is to list certain pairs of vowels as differing only by being long or short, and to have the long versions trigger obligatory rules that flesh out the rest of their pronunciation.

Chomsky and Halle therefore proposed that English has a rule of Long Vowel Tensing, which tenses all long vowels, and a rule of Diphthongization, which adds the little *y*s and *w*s that give us the two-part vowels in *lake* (leh-eek), *glide* (gla-eed), *need* (nee-y'd), *loud* (la-ood), and *road* (ro-ood). All this is more or less unexceptionable – *something* in the mind of an English speaker enforces a correlation among length, tenseness, and being a diphthong, and the predictable details of pronunciation need not be stored in individual lexical entries.

The theory that a word is stored as an abstract, not-directly-pronounceable deep structure has another advantage, pointed out by the linguist Aditi Lahiri and the psychologist William Marslen-Wilson.[19] Consider the seemingly simpler idea that memory holds the actual pronunciation of a word. The problem is: *which* actual pronunciation of the word? The word *hand*, for example, might be pronounced *h-a-n-d* when we enunciate it carefully and distinctly, but in natural conversation it comes out quite differently. The *nd* is pronounced as *nj* in *hand you*, as *m* in *hand me*, and as the *ng* sound in *hand care*. (That is one of the reasons why computer speech recognition systems, though pretty good at recognizing words in isolation, are still poor at recognizing words in connected speech.) But if, as Chomsky and Halle proposed, the dictionary entries for words are schematic – so that the last segments of *hand* are listed, say, as 'nasal' and 'dental' rather than as *n* and *d*, and they are fleshed out into full consonants by rules that work in different ways in different contexts – then a single representation could embrace the *hand* that appears in *hand, hand me, hand you,* and *hand care.*

But Chomsky and Halle went much further than merely claiming that words have a form in memory that is not identical to their pronunciations. They proposed that the underlying form of a word can be *wildly different* from its pronunciation. In particular, they proposed a complicated rule of Vowel Shift that raises or lowers the long vowels, reenacting in the minds of modern English speakers the Great Vowel Shift of the fifteenth century. In other words, they claim that ontogeny recapitulates phylogeny, and that the deep structures of words in our mental dictionaries correspond to the way Chaucer would have pronounced them (even though Chaucer, if he traveled through time to our century, would sound like a German to our ears). According to the Chomsky-Halle theory, the mental representations of words in different centuries over the past millennium, and in all the modern dialects of English, are the same; English has changed primarily by adding phonological rules. And English spelling, which did not track the Great Vowel Shift or other changes in pronunciation as the dialects evolved, captures our underlying mental representations of words.

Chomsky and Halle pursue the implications. Everyone agrees that a good spelling system ought to be stable across time and space. We should be able to read the writings of our great-great-grandparents, and of people on the other side of the Atlantic, even if they pronounce words differently from the way we do. Also, a spelling system ought to encode only the information necessary to identify the content of a word, not the trajectories of lip pursing and tongue flicking that can be predicted from the content and that people automatically execute as they talk. By these criteria, Chomsky and Halle concluded, English spelling is not only exonerated of the charge that it is an illogical, sadistic mess, but 'comes remarkably close to being an optimal orthographic system.'[20] Optimal for us, optimal for other modern English dialects, and optimal for all the recorded dialects of the past several centuries![21] Hear that, all you orthographically challenged, spell-checker-dependent, solecism-prone students and writers? Forget *cough* and *rough* and *dough* and *plough,* and *ghoti* spelling *fish,* and George Bernard Shaw's campaign to reform English spelling, and all the other complaints about crazy English:

> A *moth* is not a *moth* in *mother*,
> Nor *both* in *bother*, *broth* in *brother*,
> And *here* is not a match for *there*,
> Nor *dear* and *fear* for *bear* and *pear*.
> And then there's *dose* and *rose* and *lose*
> Just look them up – and *goose* and *choose*,
> And *cork* and *work* and *card* and *ward*,
> And *font* and *front* and *word* and *sword*,
> And *do* and *go* and *thwart* and *cart*—
> Come, come, I've hardly made a start!
>
> A dreadful language? Man alive!
> I'd mastered it when I was five.
> And yet to write it, the more I tried,
> I hadn't learned at fifty-five.[22]

What led Chomsky and Halle to this shocking conclusion? It was the drive to extirpate any trace of needless redundancy and complexity in their grammar for English sound patterns. Now we really do have a clean Shortening Ablaut rule for *breed–bred, flee–fled, shoot–shot,* and *lose–lost.* The underlying form of *breed* has a double-length version of the *e* in *bred,* so the shortening rule creates *bred* in a single step. (The formerly inconvenient fact that *breed* itself is not double-length *bred* is now handled by Vowel Shift, which makes it *briid,* followed by Tensing, which makes it *breed,* followed by Diphthongization.) Now if it was only a handful of irregular verbs that benefited from the Vowel Shift rule, the savings would be paltry compared to simply stipulating that '$\bar{e}$ changes to $\breve{e}$,' and so on. The savings begin to mount, however, when we look at *other* rules that can be simplified in exactly the same way, that is, by applying to the deep, pre-Shifted versions of vowels. With Vowel Shift available to handle the details of the long vowel, each of the following processes can be captured as a simple change-the-length rule:

Trisyllabic

Shortening:	*divine–divinity*	*serene–serenity*	*sane–sanity*
Cluster Shortening:	*crucify–crucifixion*	*intervene–intervention*	
-*ic* Shortening:	*satire–satiric*	*kinesis–kinetic*	*volcano–volcanic*
C*i*V Lengthening:	*study–studious*	*manager–managerial*	*Canada–Canadian*

Once you have the freedom to equip people with abstract underlying forms for their words, the irregulars get simpler and simpler. *Run–ran–run* can be handled by the rules for *sing–sang–sung, drink–drank–drunk,* and so on – if you suppose that the underlying form of *to run* is really *to rin,* and that Backing Ablaut and other rules apply to the *stem,* not just the participle, to make it surface as *run.* Likewise, the past-tense forms of *come, give, slay,* and *catch* are better behaved if their underlying stems are *kēm, gēv, slē,* and *kĕch.*

Most creatively of all, Chomsky, Halle, and Mohanan proposed that the *ch* of *Bach* is a covert English phoneme that lives underground in the lexical entries for *buy* and *fight* – namely, *bēch* and *fēcht* – and in the half-baked past-tense forms for *seek* and *teach.* Of course the *ch*s must be assassinated before they see the light of day, but not until they have triggered a rule that makes the past-tense form come out right. *Bēch* gets a -*t,* changes to *bōcht* by the Lowering and Backing Ablaut rules, at which point the *cht* triggers Cluster Shortening to yield *bŏcht* before *ch* makes the ultimate sacrifice, resulting in the form we spell *bought.*

~

What are we to make of this bold theory? As I mentioned at the start, a theory that posits rules for irregulars can account for the similarities between stems and their past-tense forms, such as why *swing* and *swung* are 80 percent the same: The rule targets a vowel for change, and leaves the rest of the verb alone. The Chomsky-Halle-Mohanan theory pushes the performance of rules to new heights, because their rules target only certain *features* of a vowel for change (such as tongue height or vowel length) and leave the rest of the vowel alone, too. Similarly, a theory positing irregular rules can account for the similarities in changes undergone by different verbs, for example, why the *ĭ-ă* pattern in *sing–sang* is also found in *drink–drank* and *sit–sat*. A few rules are shared by many verbs. Here the Chomsky-Halle-Mohanan theory succeeds with a vengeance, forcing almost 165 verbs to share only three rules.

Any theory that can tame the quintessentially unruly English irregular past-tense system with only three rules, each delicately adjusting a single feature, is undeniably brilliant. But is it true? Not necessarily. One problem comes from the assumption that every scintilla of patterning in the verb system needs an explanation in terms of the psychology of speakers, in particular that the patterns are distilled out into rules in the mind. Chomsky, Halle, and Mohanan's rule-by-rule derivations often recapitulate the history of a past-tense form in English over the centuries – deliberately – and that brings to mind an alternative explanation used throughout chapter 3: that the patterns are fossils of rules that died long ago. The surviving past-tense forms, semilawful though they are, could simply be memorized by today's generation without any help from the rules.

The defunct-rule explanation has an advantage over the Chomsky-Halle-Mohanan theory. Children don't hear underlying forms, and they are not provided with lessons about the rules that turn them into audible surface forms. They hear only the surface forms. If the rules and underlying forms are to play some role in mental life, children must infer the cascade of rules that generated the surface form, run it in reverse, and extract the underlying form. And the suggestion that English-speaking children hear *run* and infer *rin* or hear *fight* and infer the German-sounding *fěcht* is, frankly, beyond belief.

First, why would the child bother if the rules are there only to generate the surface form, and the child already *has* the surface form? (It's different with sentences, where the child needs rules to generate an infinite number of new ones; with word roots, there are only a finite number to learn.) And even if the child wanted to ferret out rules and underlying forms, how could they ever *find* the right ones if the crucial clues – the ones linguists themselves use to discover the rules – are found in pairs of words the children will learn only in adulthood if ever, such as *serene* and *serenity, manager* and *managerial, kinesis* and *kinetic*? At one point Chomsky and Halle concede the problem and say that their grammar is only what children *would* construct if, hypothetically, they could hear the entire vocabulary in one sitting before figuring out the rules, rather than learning the everyday words first. But then it's not clear what their theory is a theory *of* – it is not, by their concession, a theory of how real children acquire words or how real adults represent them. It may be interesting to indulge in a thought experiment of what an optimal child ought to do if he or she had the entire language to mull over at once. But that exercise would be useful only if the hypothetical child were a good idealization of a real child, and that is far from clear. Why would real children be equipped with an ability to extract intricate chains of rules and arcane word entries if they could never put that ability to use in the real world, and if the net result is the same language as the one they do acquire in the real world? It is more likely that children store words in the mental dictionary in a form that is not radically different in content from what they hear (though it may be more schematic).

Worse, it's not so clear that the thought experiment would come out the way Chomsky and Halle suppose it would. The word pairs that motivate the strange underlying entries, such as *kinesis–kinetic* and *intervene–intervention*, are inkhorn words encountered in writing or in the conversations of literate professionals. Anyone who needs to use these vowel patterns in a new word has the advantage of having seen similar words in print. People who are literate in English have been trained, usually with much weeping and gnashing of teeth, to associate the sounds ă and ā with the letter *a*, the sounds ĕ and ē with the letter *e*, and so on, when they learned the alphabet. That means that when speakers have to make a choice from among the short vowels in pronouncing new words such as in *contravene–contravention* or *elide–elision*, they may be guided by their knowledge of the alphabet, not by a naturally acquirable rule of English phonology.

But perhaps the biggest problem is that the Chomsky-Halle-Mohanan theory cannot explain the third kind of similarity running through the irregulars: the similarities among stems, as in *sting, string, sling, stink, sink, swing*, and *spring*. In their theory the rule of Lowering Ablaut for the participle is connected to these verbs by fiat – the entry for each verb says, 'Apply Lowering Ablaut to Me.' But then it is an unexplained coincidence that all the verbs are so alike. The list of verbs assigned to the rule could

just as easily have been *till–tull, wish–wush, fib–fub*, and *pith–puth*.
How can a theory that relentlessly soaks up every droplet of
redundancy between stems and pasts, and between the changes
applying to one stem and the changes applying to another stem,
be so oblivious to the massive redundancy among all the stems
undergoing a change? Also, how is the speaker supposed to
generalize the rules to new verbs if they are constrained to apply
only to the currently stipulated ones?

The obvious way to handle these families is to distill out their
common denominator and attach it as a condition to the rule. In
the *ĭ-ŭ* family the *ĭ* vowel tends to be preceded by a consonant
cluster followed by *ng*. The consonant *ng* can be further analysed
into the features *nasal* (pronounced through the nose) and *velar*
(pronounced with the tongue against the soft palate or velum).
Perhaps, then, the rule should be 'Lower the vowel from *ĭ* to
ŭ if the stem has the pattern *consonant-consonant-ĭ-velar nasal
consonant*.' Unfortunately, this rule would make errors both of
commission and omission. It would falsely include *bring–brought*
and *spring–sprang*, which do not change their vowels to *ŭ*, and it
would falsely exclude *stick–stuck* and *spin–spun*, which do. These
verbs obviously belong in the class, but each one violates the
condition by an eyelash. The *k* of *stick* is not nasal velar like *ng*,
but it is a velar, pronounced at the same place in the mouth. The *n*
of *spin* also is not a nasal velar, but it is a nasal, pronounced
through the nose, just as *ng* is.[23]

The problem, first pointed out by the linguist Joan Bybee and
the psychologist Dan Slobin, is that the irregular clusters are
family resemblance categories.[24] They don't have strict, all-or-none
definitions that specify which verbs are in and which verbs are
out. Instead they have fuzzy boundaries and members that are in
or out to various degrees depending on how many properties they
share with one another. *String* and *sling* are prototypical members
of the *ĭ-ŭ* class, packing into one word all the consonants that are
prevalent in the family. *Spin* and *stick* each misses by a different
feature; *dig–dug* and *win–won* are farther toward the periphery;
and *sneak–snuck, drag–drug, skin–skun*, and *climb–clumb* are in a
muzzy zone at the edge where speakers differ as to their
acceptability. No rule can cleanly pick out the *ĭ-ŭ* verbs, which is
why Chomsky, Halle, and Mohanan didn't bother looking for the

conditions that triggered each rule but resorted to listing the verbs individually.

The other irregular families work in the same way. For example, *blow–blew, grow–grew,* and *throw–threw* are stereotypical *ow–ew* verbs, but the rule for the class cannot demand that a word conform to the condition *consonant-consonant-ō. Know–knew* is in the family but misses the rule by one consonant, *draw–drew* and *fly–flew* miss by a vowel, and *slay–slew* and *crow–crew* are neither clearly in nor clearly out, but muzzy.

Membership in an irregular family is also probabilistic when it comes to people generalizing a pattern to new verbs. Dialectal irregular forms tend to be close in sound to many members of a family. For example, *bring–brang* is close to *sing–sang, ring–rang, spring–sprang, drink–drank,* and *shrink–shrank,* and *write–writ* is close to *bite–bit* and *light–lit.* Fei Xu and I found that children's creative irregulars work the same way. The childhood error *swing–swang* is close to *sing–sang* and all the rest; *sleep–slep* is close to *feed–fed, bleed–bled, meet–met,* and so on.[25] Bybee and Moder even quantified the effect by presenting their adult volunteers with nonsense words that varied in similarity to the typical members of the *ing–ung* family. *Spling* and *skring* fall smack in the middle of the family, and about 80 percent of the participants came up with forms like *splang, splung, skrang,* and *skrung. Krink, trig* and *pling* are less similar, and only about 50 percent of the people suggested *krunk, trug,* or *plang. Vin, sid,* and *kib* share only a vowel with the verbs in the family, and only about 20 percent of the people provided forms like *vun, sud,* or *kub.*[26]

Chomsky, Halle, and Mohanan have tweaked rules for maximum performance, but at a steep price. They were forced to make incredible claims about the mental entries of words, and their theory cannot handle the fuzzy and statistical – but psychologically active – patterns of similarity among the verbs undergoing a rule. The irregular patterns are just not very rulelike, and call out for something very different.

~

When the psychologists David Rumelhart and James McClelland announced their artificial neural network model of the past tense in 1986, the reaction was sensational.[27] Here was a model with none of the paraphernalia of linguistics – no words, no rules, no modules – but it acquired several hundred regular and irregular past-tense forms, generalized their patterns to new verbs, and made errors such as *breaked* and *comed*, just like children. COMPUTERS MIMIC BRAIN IN TEST, said a headline in the *Chicago Tribune*. A TURNING POINT IN LINGUISTICS, ran the title of a review in the *Times Literary Supplement*.[28] The implications were 'awesome,' said the reviewer, because 'to continue teaching [linguistics] in the orthodox style would be like keeping alchemy alive.' Rumelhart and McClelland's model helped to launch a new school of cognitive science known as connectionism or parallel distributed processing, which explains mental processes in terms of networks of interconnected simple units that vaguely resemble neurons (brain cells).[29] Many researchers saw connectionism as a paradigm shift or scientific revolution in the study of the mind.[30] Neural networks also became a fad in artificial intelligence and soon were put to use in picking stocks for mutual funds and controlling expensive Japanese appliances like rice cookers and washing machines.

No one doubts that language is computed by networks of neurons in the brain. Rules – even the pristine, logic-like rules of Chomsky and Halle – are intended as high-level descriptions of processes or structures that are implemented in some way in neural circuitry. The difference between connectionism and generative grammar lies in the *kinds* of mental operations that are thought to be implemented in neural networks. In particular, connectionism differs from generative grammar in the way that associationism differs from symbol manipulation. It lacks combinatorial rules organized into modules, and instead tries to accomplish intelligence using Hume's law of contiguity (if A appears with B, associate them) and his law of resemblance (if C looks like A, let it share A's associations). A neural network that works this way is called a *pattern associator memory* or a *perceptron*.

Here is how Rumelhart and McClelland's model of the past tense works. Despite my clash-of-the-Titans buildup, the model actually shares some important design features with the Chomsky-Halle theory. The input to the model is the sound of a verb stem, and the past tense is computed from it. That is different from a model that computes a past-tense form directly from the meaning of the verb and the concept of pastness. So Rumelhart and McClelland are committed to at least one module – a morphology box – sitting between meaning and sound. As with Chomsky and Halle, a single kind of machinery is charged with computing the past-tense forms of all verbs, regular, irregular, and suppletive (*go–went*); the verbs sit on a continuum of regularity from completely predictable to completely arbitrary. Past-tense forms are composed piecemeal out of miniregularities that are shared among verbs, so that *sleep–slept* combines the vowel change in *feed–fed* and the suffixation in *burn–burnt*. Rumelhart and McClelland also import the standard Chomskyan assumption that speech sounds are represented in the mind not as phonemes but as bundles of features such as 'voiced' and 'nasal.'

But everything else is different. Here is the heart of the model, the pattern associator memory:

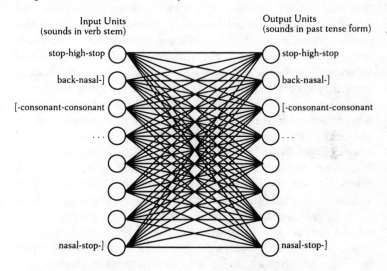

The left-hand column is the input layer, where the verb stem is entered. It contains 460 vaguely neuronlike units, each of which can be either on or off. Each unit represents a tiny stretch of sound that might appear in an English verb, such as a high vowel between two stop consonants, or a back vowel followed by nasal consonant at the end of the word. The beginning and end of a word are symbolized by open and close brackets ('[' and ']'). There are no units for individual verbs; a verb is entered by turning on the units for the sounds it contains. As a result, similar-sounding verbs share representational real estate. Most of the units that are turned on when *shrink* is fed in are also turned on when *drink* is fed in (consonant cluster at the beginning of the word, high vowel between two sonorant consonants, and so on). These units have no idea which word they are currently representing.

The right-hand column has an identical bank of units, and they represent the output of the model: the sound of the past-tense form. Every input is connected to every output by a synapselike connection that can vary in strength, from strongly excitatory (an input signal tends to turn the unit on), to neutral (an input signal has no effect), to strongly inhibitory (an input signal tends to turn the unit off). In effect each connection is a probabilistic microrule that states something like, 'If the stem contains a stop consonant followed by a high vowel, the past-tense form is likely to contain a nasal consonant at the end.' With 460 input units connected to 460 output units, we have $460 \times 460 = 211,600$ microrules in all. When an input unit is turned on, it sends a signal down all its lines to the output layer, where the signal is multiplied by the strength of each connection and fed to that output unit. Whether a given output unit turns on depends, in a probabilistic way, on the sum of the signals that feed into it and on its own level of triggerhappiness or *threshold*. The higher the summed signal is above the threshold, the more likely the unit is to turn on; the lower the summed signal is below the threshold, the more likely the unit is to turn off.

In the neonate network the connections have strengths of zero, so the output layer is completely turned off, regardless of the input. The connections then are changed in a learning procedure, in which the model is 'taught' with a set of verbs and their correct past-tense forms. Of course, Rumelhart and McClelland do not actually believe that a schoolmarm has to drill children with verb conjugations. They assume that children, when hearing a past-tense form in their parents' speech, recognize that it *is* the past-tense form of a familiar verb, dredge the verb stem out of memory, feed it into their past tense network, and silently compare their network's output with what they just heard. Skeptics might wonder how a child is supposed to do all this without the benefit of the lexical and grammatical machinery that Rumelhart and McClelland claim to have made obsolete, but let's put that aside for now.

The learning procedure works like this. The correct form from the parents is displayed in a special layer of 'teacher' units. The model compares its output, unit by unit, with the correct output (*walked* for *walk, came* for *come*, and so on). The model then adjusts the connection strengths a tiny amount up or down depending on the difference (see the figure below).

If a unit is off (say, the unit for ă), and the teacher says it should be on (because the correct past-tense form is *rang*) the model has to make the input word (*ring*) more likely to turn on that unit in the future. All of the connections from incoming lines that are currently active are strengthened an iota, and the ă unit's threshold is lowered an iota, making it more triggerhappy. In

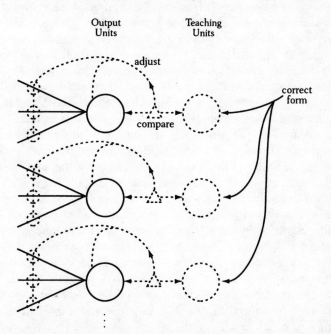

contrast, if a unit is on (for example, the unit for ĭ) and the teacher says it should be *off* (because the correct past-tense form is *rang*), the model has to make the input word *less* likely to turn on that unit in the future. All of the connections from incoming lines that are currently active are weakened an iota (possibly driving the connection down to a negative or inhibitory value), and the unit's threshold is raised an iota, making it less trigger-happy.

The model is trained on a list of verbs and their past-tense forms, presented over and over and over. A given connection will be buffeted up and down by successive verbs in a training run, but eventually it will settle on the strength value that does the best job, in combination with the other connections, of producing correct past-tense forms. The network's knowledge of the various verbs and their past-tense forms is smeared across the 211,600 connection strengths; one cannot point to a circumscribed part of the network that implements a particular word, a particular irregular family, or a regular rule.

Rumelhart and McClelland trained their network on a list of 420 verbs presented 200 times, for a total of 84,000 trials. To everyone's surprise, the model did quite well, computing most of the correct sound stretches for all 420 verbs. That means that a single set of connection strengths was able to convert *look* to *looked, seem* to *seemed, melt* to *melted, hit* to *hit, make* to *made, sing* to *sang,* even *go* to *went.* Then Rumelhart and McClelland challenged the network with 86 new verbs, which it had not been trained on: a test of generalization or productivity like the *wug*-test, the sine qua non of rules. The model offered the correct past-tense form with *-ed* for about three quarters of the new regular verbs, and made reasonable overgeneralization errors such as *catched* and *digged* for most of the new irregulars.

Even more impressively, the model mimicked some of the tendencies of children as they acquire English. At one point in training it produced errors such as *gived* for verbs that it had previously produced correctly. It also analogized new irregular verbs to families of similar-sounding old irregular verbs; for example, it guessed *cling–clung, sip–sept, slip–slept, bid–bid,* and *kid–kid.* It produced blends such as *gaved* and *stepped* that also occasionally come out of the mouths of children. It was less tempted to tack -*ed* onto an irregular verb from a large family, such as *feel*, than onto an irregular verb from a small family, such as *blow*. And it was bashful about sticking -*ed* onto verbs that already end in *t* or *d*, a common reluctance of human beings that we observed in chapter 2.

Rumelhart and McClelland's pattern associator memory is not made of some miraculous wonder tissue. It works by one trick: Rather than associating a word with a word, it associates the *properties* of a word – its phonological features – with the *properties* of another word, and thereby enjoys automatic generalization by similarity. That is, rather than associating *drink* with *drank*, it associates *dr* with *dr*, *dr* with *rang*, *ring* with *rang*, *ink* with *ank*, and so on. At the same time, it negatively associates *dr* with *nked*, *ink* with *nked*, and so on, inhibiting the incorrect regular form *drinked*.

Crucially, these associations are superimposed across the different words in the training set. When the model trained on *drink* is then trained on *shrink*, it strengthens many of the same connections, such as *ring* with *rang* and *ink* with *ank*. That makes *shrink* easier to learn – most of its connections have been prestrengthened – and it makes subsequent family members, such as *sink*, easier still. It's a short step to generalize to verbs that have not been trained at all, such as *stink* – the *ing–ang* connections have already been strengthened, and the *ing–inged* connections have already been weakened. The same trick works for the regular verbs: When the model is trained on *walk–walked*, it strengthens connections between *alk* and *alked*, restrengthens them when trained on *talk–talked*, and automatically generalizes them to *stalk–stalked*. The only difference between regular and irregular verbs is that the regulars are more plentiful, more diverse, and more consistent in the patterning of their past-tense forms. With thousands of strong connections conspiring to turn on a *t* or *d* at the end of a word, the model's first tendency will be to output a regular form.

The mainspring of the model, then, is forming associations between features and features, and that duplicates the human habit that embarrassed the words-and-rules theory: generalizing irregular patterns to similar words. The key idea is not original to Rumelhart and McClelland. Associating features with features is inherent to the design of many associationist theories, and goes back to the eighteenth-century English physician and philosopher David Hartley.[31] Hartley pointed out that if the brain represented the properties of an object individually, then Hume's two laws of association – contiguity and similarity – could be pared down to one law, contiguity. Similarity is nothing but shared properties, so associations among properties give you generalization-by-similarity for free. That is, if an association between bread and nourishment is in fact stored in the brain as a set of associations between beigeness, sponginess, savoriness, and nutrition, then when we encounter cake, which is also beige and spongy, 'nutritious' pops into mind automatically; no extra device in the brain has to register the fact that bread and cake are similar and make a point of transferring associations from one to the other.

~

So does traditional linguistics, with its words and rules, have the status of alchemy? Not yet. In 1988 the linguist Alan Prince and I published a paper in the journal *Cognition* that went after the pattern associator model hammer and tongs. We pointed out many facts about human language that the model, and the connectionist approach to language in general, ignored or mishandled.[32] Other trenchant critiques appeared around that time or in the years since.[33] In a recent book the mathematician and former *Scientific American* columnist A. K. Dewdney lumped connectionism with N-rays, cold fusion, and psychoanalysis as case studies of 'bad science' in need of debunking.[34] That is unfair, but connectionism *has* been overhyped and its problems as a theory of the mind are real.

Some of the problems might be obvious from the dissection of the language system in chapter 2. First, Rumelhart and McClelland's pattern associator memory is a device that only *produces* past-tense forms. You cannot turn the arrows around and get the model to run backward and *recognize* past-tense forms. Obviously people do both. Not only can we say *walked*, but when we hear *walked* we know it means walk in the past. Children are not separately trained to produce *-ed* and to understand *-ed*. The most straightforward explanation is that they learn rules and lexical entries, a database that can be accessed equally well by a module that sends commands to the tongue and a module that interprets sounds coming in from the ear.

Second, the model computes every detail of the pronunciation of the past-tense form. Yet we saw that many of these details, such as the choice among *-t*, *-d*, and *id*, are found in fifteen different parts of the language system. Surely they are computed by a single phonology module that is fed by the output of morphology and syntax, not duplicated by an amazing coincidence in fifteen different networks, one for the past tense, one for plurals, and so on.

Third, by forgoing the use of lexical entries and relying entirely on a word's sound to compute its past-tense form, the model cannot tell the difference between two words that have the same sound. It must give them the same past-tense form, and that won't work for soundalike verbs like *ring–rang* and *wring–wrung*, *break–broke* and *brake–braked*, or *meet–met* and *mete–meted*. One might reply that the problem could be fixed by adding a few units to the input that represented the meanings of words, in addition to their sounds. For example, a unit for 'striking' could turn on *ang* while a unit for 'squeezing' turned on *ung*, differentiating *ring* from *wring*. But as we saw in chapter 2, the meanings of words don't systematically predict their past-tense forms: *Hit*, *strike*, and *slap* are similar in meaning but have different past-tense forms; *take*, *undertake*, and *take a leak* are different in meaning but have the same past-tense forms. It's the raw fact that word 1 is not the same as word 2 which is not the same as word 3 that triggers the different idiosyncratic past-tense forms, and that is the distinction captured by lexical entries. Soundalike words with different plurals and pasts are widespread in English and give rise to many of the quirks that occupy letters to the language mavens – why a baseball player is said to have *flied out* to center field, why the hockey team in Toronto is called *The Maple Leafs*, why the plural of *Walkman* is often *Walkmans*. The answer involves a beautiful design feature of human language that we will explore in chapter 6 and that is quite unlike the knee-jerk associations that drive the Rumelhart-McClelland model.[35]

A fourth problem is that Rumelhart and McClelland had to use some jiggery-pokery to get the model to duplicate children's stages of language development. We will take a closer look at how children really do learn to use and misuse the past tense when we examine language acquisition in chapter 7.[36]

These troubles are all payback for the connectionists' distaste for carving a complex computational problem into a few simpler ones that can be farmed out to mental modules optimized for each. The problems could be solved by building separate networks for morphology, phonology, and the lexicon, much as in traditional linguistics but with the boxes fleshed out as neural networks.[37]

But there is one problem that cannot readily be solved by dividing up the computation into modules. It lies at the very core of the pattern associator model, and diagnoses the main flaw in the centuries-old theory of associationism. The problem could not be more basic: How do you represent an entity made of parts in a fixed arrangement, such as a word? Units can only be on or off; you can't inscribe them with symbols as if they were pads of paper or bytes in a computer. The first solution that comes to mind is to make the units into a phonetic alphabet. Assign one unit to $\bar{a}$, one unit to $\bar{a}$, one unit to b, one unit to d, and so on. Then simply turn on the units that spell out the word:

$$\begin{array}{l} \text{ă} \;\; \bigcirc \\ \text{ā} \;\; \bigcirc \\ \text{b} \;\; \bigcirc \\ \text{d} \;\; \bigcirc \\ \text{ĕ} \;\; \bigcirc \\ \vdots \end{array}$$

But this is a nonstarter. Information about the *order* of phonemes is lost: *pit* would be indistinguishable from *tip*, *Spiro Agnew* from *grow a penis*. If that's all there were to words, you would be solving anagrams every time you opened your mouth.

A better solution is to have an *array* of phoneme units, one bank for the first phoneme in a word, one for the second phoneme, and so on, up to the longest word that a person would ever be called on to remember:

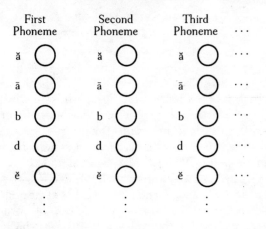

First Phoneme | Second Phoneme | Third Phoneme

ă · ā · b · d · ě

That solves the anagram problem, but it runs up against two new ones. First, how long *is* the longest word that the array should accommodate? Long enough for *antidisestablishmentarianism*, the longest word in standard dictionaries? For *floccinaucinihilipilification*, the longest word in the *Oxford English Dictionary*? What about *great-great-great-grandmother*, *great-great-great-great-great-grandmother*, and so on? There *is* no longest word, so something is wrong with a representation that forces us to decide what it is a priori.

The other problem is that the representation has a bank of units for the first phoneme in a word, a bank of units for the second phoneme, and so on, aligning the words in memory by their first phoneme, that is, left-justifying them. But the human mind does not count off phonemes from left to right when it perceives similarities among words and generalizes accordingly. The most tantalizing generalizations in the irregular past-tense system are in the *ing–ang–ung* family, with *ring–rang* and *drink–drank* and *spring–sprang* reinforcing each other and inspiring *fling–flang*, *bring–brang*, and *spling–splang*. But in a left-to-right array, the first three verbs do not overlap at all:

Positions:	1	2	3	4	5	6	7
ring:	r	i	ng				
drink:	d	r	i	ng	k		
spring:	s	p	r	i	ng		

Nothing that has been learned for *ring*, such as that it has a past tense with *ang*, will transfer to *drink* or *spring*; the two words are no more similar than *bird* and *clam*. But people *must* find them similar, for the irregular system is rife with families of words that don't align properly at their left edges, such as *dive, drive, strive*, and most obviously, prefixed forms such as *stand, withstand*, and *understand* or *come, become*, and *overcome*. The same problem fouls up any generalization that depends on the *ends* of words, and we know there is a huge one: the choice of *-t, -d*, or *id*, which hinges on whether the last phoneme is voiced, unvoiced, or a *t* or *d*. The last phoneme in a left-to-right representation can be position 2 (for the verb *add*), position 3 (for *ask*), position 4 (for *risk*), and so on, all the way up to position 23 for *floccinaucinihilipilify* and beyond. A left-to-right representation would have to learn how to pronounce the suffix separately for every length of word.

Rumelhart and McClelland must have recognized this problem, because they came up with a creative alternative: The units stand for things they called *Wickelphones*, named after the psychologist Wayne Wickelgren who first conceived them.[38] A Wickelphone is a sequence of three phonemes, like *ipt* or *str*. English has about forty phonemes, and if we add the special symbols '[' and ']' for the beginning and end of a word, there are about 67,000 possible Wickelphones, each needing a unit. By representing a word by its Wickelphones, one sidesteps both the anagram problem and the left-alignment problem. For example, *strip* contains the Wickelphones *ip], rip, str, tri*, and *[st*. You don't have to worry about their order, because they snap together in only one way: *[st* at the beginning, then *str*, then *tri*, and so on. And the Wickelphones for *rip* overlap the Wickelphones for *strip*, so their representations are similar, just as the human mind perceives them. (Rumelhart and McClelland in fact wanted to represent words in terms of their Chomsky-Halle-esque features, rather than their phonemes, so a unit actually stood for three *features* in a row, such as stop-high-stop or voice-unvoiced-voiced – a Wickelfeature.)

It may seem hard to believe that the simple act of registering a word in an unstructured bank of units is a near-insoluble problem, but it is: Wickelphones, though ingenious, don't work either. The human mind cares about *single* phonemes and the features that compose them, and the Wickelphone has submerged them into unbreakable chains of three-in-a-row. For example, *silt* and *slit* have no Wickelphones in common. The first dissolves into *[si, sil, ilt,* and *lt]*, the second into *[sl, sli, lit,* and *it]*. But people clearly hear them as similar, as we see in historical changes in English such as *brid* becoming *bird* and *thrid* becoming *third*. Worse, in some languages the Wickelphone cannot represent certain words at all. The Australian aboriginal language Oykangand has a word *algal* meaning 'straight' and a word *algalgal* meaning 'ramrod straight,' and they are made up of identical Wickelphones: *alg, al], gal, lga,* and *[al*:

Words:	[algal]	[algalgal]
Wickelphones:	[al	[al
	alg	alg
	lga	lga
	gal	gal
	al]	*alg* (already used)
		lga (already used)
		gal (already used)
		al]

Since units are either on or off, they have no way of representing *two* of something, and the Wickelphone theory therefore incorrectly predicts that Oykangand speakers should not exist (nor the speakers of many other languages; words like these are not uncommon).[39]

Also, a theory of how the mind represents things should predict what the mind finds easy and what it finds hard. The easy tasks should be computed by simple operations on the representation, the hard tasks by lengthy sequences of operations. Here, too, the Wickelphone makes the wrong prediction. When linguists explain to their classes how human languages use some kinds of rules and not others, they almost always use the same example: that no language has a rule that flips a word to its mirror-image, say, forming the plural by converting *tip* to *pit*, *gum* to *mug*, and *dog* to *god*. But a Wickelphone-to-Wickelphone network can do exactly that, and quite easily: strengthen every connection between a Wickelphone *ABC* in the input and its mirror-image Wickelphone *CBA* in the output, and weaken all the other connections.

Not only is mirror-reversal easy, but it is no harder to learn than the simplest conceivable relation between input and output: copying the stem verbatim, which involves strengthening the connections between *ABC* and *ABC*. The only difference between mirror-image reversal and verbatim copying is that we, the theorists peering into the model, can read the unit labels and see that *ABC* goes to *CBA* in one case and to *ABC* in the other. But the model cannot read its own node labels; all it cares about is the consistency of the input-output relations, and they are the same in both cases. Likewise, all kinds of crazy rules, such as replacing all *a*s with *b*s, all *b*s with *c*s, all *c*s with *d*s, and so on, are as easy to learn as copying the input to the output.

This is not just a quibble; it explains an embarrassing lapse in the performance of Rumelhart and McClelland's model. The model was mute when asked for the past tenses of simple but somewhat unusual-sounding words, like *jump*, *pump*, *warm*, and *trail*. And it garbled several others, turning *squat* into *squakt*, *tour* into *toureder*, and *mail* into *membled*. The lapses are puzzling to *us* because intuitively nothing could be simpler than copying a stem over to the past-tense form before adding *-ed*. But a pattern associator memory has no placeholder called 'stem' that *can* be copied, and no operation to do the copying. All it does is associate sounds with sounds, and if the training set happens to be missing words with certain combinations of sounds such as *-ump* or *-ail*, the model will be at a loss, and will sit in silence or cough up a hairball of bits and pieces that are vaguely associated with the sounds it *has* been trained on.[40]

All the problems go away if you bring back the rationalist theory that the mind manipulates symbols organized into hierarchical structures by rules. A verb such as *to outstrip* might be represented something like this:

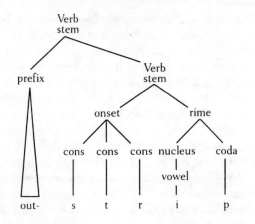

The phonemes are held in their correct order by a treelike scaffolding that embodies the morphological structure of the word (how it is built out of stems, prefixes, and suffixes) and the phonological structure of its parts (how they are built out of chunks like onsets, rimes, vowel nuclei, consonants and vowels, and ultimately, features). The similarity to other words such as *strip, restrip, trip, rip,* and *tip* falls mechanically out of the fact that they have identical subtrees, such as an identical 'stem' or an identical 'rime.' And computing the regular past-tense form is nothing but attaching a suffix next to the symbol 'verb stem':

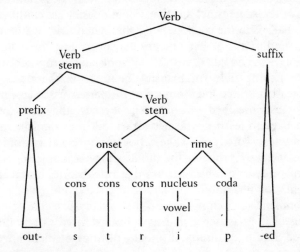

It doesn't matter whether the underbrush dangling beneath the 'verb stem' symbol is *walk, outstrip, jump, pump,* or *bftsplk* – if you have a mental symbol 'verb stem' and know how to put a suffix next to it, the entire vocabulary of verb stems lies waiting at your feet. Finally, since a tree structure is built out of recursive rules (for example, 'a stem can combine with a prefix to form a new stem'), no length limit needs to be set beforehand, and words of any length such as *re-outstrip* or *great-great-great-grandmother* can be represented.

Symbolic trees require fancier neural hardware than the smooth purée of units that is popular among connectionists, but those models hardly do justice to the brain anyway. Recently, a few neural network modelers have shown how hierarchical trees can be implemented in more organized neural networks.[41] One conjecture is that the periodic rhythms of neural firing, long downplayed in neuroscience, serve as the glue that binds together the units that represent an abstract *slot* in a tree and the units that represent its *content.* For example, when the units for the 'coda' slot fire at twenty times a second, and the units for *p* fire in synchrony with them also at twenty times a second, the system as a whole knows that the coda is *p.* Simultaneously, the units for 'nucleus' can be firing thirty times a second, and the units for *i* can be firing in synchrony thirty times a second, and the system knows that the nucleus is *i.* The units for 'coda,' 'nucleus,' *p,* and *i* are all active at the same time, but the system doesn't get confused and think that *i* is in the coda, because the shared firing patterns link each sound to its slot. That theory may or may not be right, but I mention it to show that abstract symbols and complex structure are not incompatible with plausible neural network models.

After Alan Prince and I took apart the pattern associator model, the linguists breathed a sigh of relief because they thought they didn't have to learn neural network modeling after all, and the connectionists dropped it like a hot potato. So it is ironic that Prince and I are probably the model's biggest fans. It does, after all, explain a major phenomenon that rule theories ignore, and it accounts for not one but several aspects of children's language development. In comparison, the twenty-five connectionist models of the past tense that have been devised in reply to our critique have been disappointments, not one of them anywhere

near as ambitious as the original.[42] Many sidestep the Wickelphone problem by using a Dick-and-Jane version of English that
contains only monosyllabic words made of a consonant, a vowel,
and a consonant, such as *walk* and *run*. Others implicitly concede
that words are composed of symbols for stems and symbols for
affixes and don't even bother computing a past-tense form. They
merely select from an innate menu of five or six units that stand
for the five or six suffixes or vowel changes in the language. Some
other mechanism then has to apply the suffix or vowel change to
the stem to get an actual past-tense form. That unmentioned
mechanism, of course, is what we call a *rule*. Many modelers beef
up the network with an intervening layer of units hidden between
the input and the output layers, but direct benchmark tests find
little or no improvement.[43] Each of the inventors has added a
different patch that narrowly fixes some problem that Prince and I
pointed out – what programmers call a hack or a kluge – but none
defends his brainchild as an actual theory of how that part of the
mind works. And no one has made an empirical prediction or
accounted for several kinds of data in the way that Rumelhart and
McClelland did.

~

One phenomenon, two models, both explaining too much to be
completely wrong, both too flawed to be completely right. Prince
and I have proposed a hybrid in which Chomsky and Halle
are basically right about regular inflection and Rumelhart and
McClelland are basically right about irregular inflection. Our
proposal is simply the traditional words-and-rules theory with a
twist. Regular verbs are computed by a rule that combines a
symbol for a verb stem with a symbol for the suffix. Irregular
verbs are pairs of words retrieved from the mental dictionary, a
part of memory. Here is the twist: Memory is not a list of
unrelated slots, like RAM in a computer, but is associative, a bit
like the Rumelhart-McClelland pattern associator memory. Not
only are words linked to words, but bits of words are linked to
bits of words. The bits are not Wickelphones, of course, but
substructures like stems, onsets, rimes, vowels, consonants, and
features, perhaps something like this:

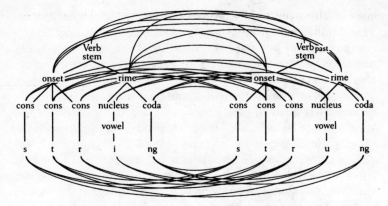

Furthermore, the nodes of one word (such as *string*) overlap the same nodes in other words (such as *sling, stick, stink*, and *swim*). As a result, irregular verbs show the kinds of associative effects found in a connectionist pattern associator. People find families of similar irregular verbs easier to store and recall because these verbs repeatedly strengthen their shared associations. And people occasionally generalize the irregular patterns to new, similar verbs, because the new verbs contain material that already had been associated with the pattern from the old verbs.

Prince and I were not the first to modify the words-and-rules theory in this way. Many generative linguists have been uncomfortable with the Chomsky-Halle ethos of using industrial-strength rules to account for everything that is systematic in language. Mark Aronoff, Joan Bresnan, Ray Jackendoff, Rochelle Lieber, Andrew Spencer, and others have suggested that language uses two kinds of rules: true rules that speakers generalize freely, and *lexical redundancy rules* that merely capture patterns of similarity among words stored in memory.[44] A memory system in which patterns of similarity are registered and occasionally generalized is simply a pattern associator memory, and Rumelhart and McClelland have given us a sketch of how one might work.

The modified words-and-rules theory may sound like a sappy attempt to get everyone to make nice and play together, but it makes a strong prediction. The prediction is that regular and irregular inflection are psychologically, and ultimately, neurologically distinguishable. But how could they be distinguished if both involve patterns that people can generalize? The answer is that irregular inflection depends on *memorized* words or forms *similar* to them, but regular inflection can apply to *any* word, regardless of whether the word is readily retrievable from memory. Regular inflection has that power because it is computed by a mental operation that does not *need* access to the contents of memory, namely, a symbol-processing operation or rule, which applies to any instance of the symbol 'verb.' The evidence will be woven through the rest of the book as we explore how words are used in conversation and in reading, how new words are created, how children learn their mother tongue, how language is organized in the brain, and whether the languages of the world conform to a universal design. We will see how in dozens of cases of language use that have nothing in common except a failure of access to memory, irregular patterns are disabled but the regular rule works fine.

If the modified words-and-rules theory is correct, it would have a pleasing implication for the centuries-old debate between associationism and rationalism: Both theories are right, but they are right about different parts of the mind.

5

WORD NERDS

When we use our native language, a torrent of words flows into and out of the brain. The occasional frustration of having a word stuck on the tip of the tongue, the slow ordeal of composing a passage in a foreign language, and the agony of a stroke victim struggling to answer a question remind us that our ordinary fluency with language is a precious gift.

This chapter looks at how words and rules pop into mind as we use language in real time. The topic offers a good test of the modified words-and-rules theory introduced at the end of chapter 4. According to the theory, regular forms are generated by rules, and irregular forms are retrieved from memory; the memory, however, is not a list of slots but is partly associative, linking patterns with patterns as well as words with words.

The theory differs from Chomsky and Halle's theory of generative phonology, with its battery of rules that generate both regular and irregular forms. It also differs from Rumelhart and McClelland's theory of connectionism, with its pattern associator memory that *stores* both regular and irregular forms. We saw how each model has problems with the facts of the English language. Chomsky and Halle, in leaching every bit of patterning out of memory and concentrating it in rules, had to propose implausible deep structures for words, and could not explain why irregular verbs come in families of similar forms. Rumelhart and McClelland, in dismantling every bit of structure in language, had to propose clumsy Wickelphones to represent words, and could not explain why regular verbs are copied so reliably into their past-tense forms.

In this chapter we will turn from the mechanics of these classic models to the general principles that power them. After all, some of their problems might come from details of implementation, which could always be improved in later models. Here we will see whether the intellectual core of each model – that memory is compressed to a minimum, in the case of Chomsky and Halle, and that generalization works by the laws of association, in the case of Rumelhart and McClelland – fits the facts of human language use.

The great challenge for any theory of language is productivity, the ability to generate and understand an unlimited number of new forms. One-piece words like *duck* and *walk* can always be listed in memory, but they are not the only words we trade in; we are constantly faced with new words formed by combining prefixes, stems, and suffixes. In a language such as Kivunjo or Turkish every word may come in a half a million to several million forms, and speakers could not possibly have memorized them all in childhood. Even in a morphologically challenged language like English we have to cope with new word forms every day. The psychologists Harald Baayen and Antoinette Renouf calculated that every time you open a newspaper you will be faced with at least one word with *un-* that you have never seen before, one with *-ness*, and one with *-ly*: words like *uncorkable, uncheesy, headmistressly, breathcatchingly, pinkness* and *outdoorsiness*.[1] And these are just three of the forty-odd common prefixes and suffixes in English.

Coping with new word forms is a problem for all language users, not just human language users. Computer programs that understand and produce conversational English need morphology modules to deal with novel forms of words.[2] Even my lowly spell-checker boasts one, according to its documentation:

> **spell** collects words from the named files, and looks them up in a hashed spelling list. Words that do not appear in the list, or cannot be derived from those that do appear by applying certain inflections, prefixes or suffixes, are displayed on the standard output.

The morphology module allows it to store just the stems of words and to compute the inflected and derived forms by rule. I asked the program to check the first two paragraphs in this chapter and report on its activity. About one out of every seven words was missing from its dictionary and had to be computed with rules:

According = accord + ing	looks = look + s
associative = associate – e + ion – ion + ive	modified = modify – y + ied
composing = compose – e + ing	occasional = occasion + al
flows = flow + s	partly = part + ly
forms = form + s	patterns = pattern + s
frustration = frustrate – e + ion	retrieved = retrieve + d
generated = generate + d	rules = rule + s
having = have – e + ing	slots = slot + s
introduced = introduce + d	struggling = struggle – e + ing
linking = link + ing	words = word + s

Similarly, the program can handle psycholinguistic jargon such as *overregularized* and *underlyingly* without complaining:

overregularized = over + regular + ize + d
underlyingly = under + lying + ly

If I add *mosh* and *Bork* to its dictionary, I don't have to add *moshed, moshes, moshing, Borked, Borks,* and *Borking* as well; the spell program passes the *wug*-test.

How does the human mind handle new inflected and derived words? Does it work like my spell-checker, with a minimal dictionary and a set of rules for carving unfamiliar words into prefixes, stems, and suffixes? Does it rely on an enormous dictionary that lists all the common forms of every word? According to the updated words-and-rules theory, the mind has rules for regular forms and relies on a pattern-associating memory for the irregular forms. The evidence for this hybrid model, we shall see, is that when people use an irregular form, they must have that form or similar forms in memory, whereas when they use a regular form, they don't need to access memory at all.

~

A simple property of memory is that the more often you hear something the better you remember it. Uncommon words, therefore, have weak memory entries and should be harder to retrieve. The words-and-rules theory predicts that rarity should hurt an irregular verb, but not a regular verb. The first place to look for this effect is in the statistics of the English language itself.

Here is a Top Ten list, the ten most common verbs in English:

Verb	Number of occurrences in a million words of text
1. *be*	39,175
2. *have*	12,458
3. *do*	4,367
4. *say*	2,765
5. *make*	2,312
6. *go*	1,844
7. *take*	1,575
8. *come*	1,561
9. *see*	1,513
10. *get*	1,486

The verbs are ranked by their frequencies in a million-word corpus of text assembled from newspapers, magazines, popular books, textbooks, and other sources, and analysed at Brown University by the computational linguists Nelson Francis and Henry Kučera.[3] You may have noticed something the verbs have in common: They are all irregular. Indeed, some are *really* irregular. The top four are the only verbs that are irregular not just in the past tense but also in the present tense: *be–is/are, have–has, do–does*, and *say–says*. And the #1 and #6 spot contain verbs that are so irregular that their past-tense forms are different words altogether: *be–was/were* and *go–went*.

There cannot be a Bottom Ten list for the least common verbs in English, since the truly rare ones won't turn up even in a million-word corpus. But we can look at the rarest verbs that do turn up, the 877 verbs tied for last place with a frequency of one in a million. Here are the first ten in alphabetical order:

Verb	Number of occurrences in a million words of text
abate	1
abbreviate	1
abhor	1
ablate	1
abridge	1
abrogate	1
acclimatize	1
acculturate	1
admix	1
adulterate	1

As you can see, they are all regular. To be exact, 860 of the one-in-a-million verbs, 98 percent, are regular. Another sixteen are prefixed irregulars that are parasites on their far more common roots: *bethink, forswear, inbreed, misread, outdraw, outfight, over-bear, overdrive, overlie, overwrite, presell, regrind, spellbind, un-bend, unbind, unwind.* Only one word among the rarities is an irregular root: *smite.*

Irregular verbs are the most common verbs and vice-versa, in English and in most other languages.[4] The explanation is simple. Irregular forms have to be memorized repeatedly, generation after generation, to survive in a language, and the commonly heard forms are the easiest to memorize. If an irregular verb slips in popularity, a generation of children will fail to hear its past tense often enough to remember it. Since the rule 'Add *-ed*' can apply to an item regardless of its frequency, the children will use the regular suffix, and that verb will be regular for them and for all subsequent generations. Those who cannot remember the past are condemned to compute it.

Joan Bybee did some historical digging to prove this conjecture. Remember that Old English had about three times as many strong irregular verbs as Modern English, including obsolete forms such as *cleave–clove, crow–crew, abide–abode, chide–chid*, and *geld–gelt.* Bybee looked at thirty-three strong verbs that survived in Modern English, and divided them into verbs that remained irregular and verbs that became regular. She then looked up their modern frequencies in the Brown corpus. The still-irregular verbs appear an average of 515 times per million words; the regular defectors appear an average of 21 times.[5]

We can feel this force of history acting today when we look at the past-tense forms that are low in frequency. *Smite* is the only one-in-a-million verb root clinging to an irregular form, *smote*. But no one can use it in conversation with a straight face; the form is sliding out of the language before our eyes. Likewise, *heave–hove, stave–stove, rend–rent, bid–bade, slay–slew, smell–smelt*, and *thrive–throve–thriven* are a bit peculiar, and one can predict they will go the way of *chid* and *crew*. Sometimes a form is familiar enough to block a regular version, but not quite familiar enough to sound natural, and speakers are left without any good past-tense form for it. Complete this sequence: *I stride, I strode, I have* ____. Most people grimace at this point, equally uncomfortable with *stridden* and *strided*. *Stridden* may be found in dictionaries and occasionally in prose (as in, 'where . . . a pinnacle of beauty had stridden the earth,' from Rebecca Goldstein's 1989 novel *The Late-Summer Passion of a Woman of Mind*), but it is not to be found in the million words of the Brown corpus. It hovers in the mists of memory, tainting *strided* without stepping onto the stage itself.

Crucially, this never happens with regular verbs. Complete this sequence: *I agglutinate, I* ____, *I have* ____. *Agglutinate*, like *smite*, appears once in a million words, and not in the past tense. Here, though, people have no reluctance to supply the elusive past and participle forms *agglutinated*. The other pastless verbs in the sample are just as easy to conjugate – *allure–allured, badger–badgered, carouse–caroused* – as are the countless verbs that are too rare to show up even in a million words: *to fleech, to fleer, to stint, to prescind, to anastomose*, and so on.

Does this difference really come from the faintness of a rare irregular past-tense form in memory, or does it come from a squeamishness about using the rare verb in *any* form? We may not use *smote* very often, but then we don't use *smite* every day, either. Are there any verbs that are fine in their bare, stem form and odd only in the past tense? That would be a good test of the theory that verbs and their past-tense forms are listed separately in memory: If so, one entry could be solid while the other is frail. The place to look is in idioms, clichés, and collocations, which often are used solely in the infinitive or the present tense.

Take the verb *forgo*. Though uncommon, it retains a certain liveliness, particularly in the sarcastic phrase *to forgo the pleasure of*, as in *You'll excuse me if I forgo the pleasure of watching the video of your wife giving birth.* Now try to put it in the past tense. The word is certainly not *forgoed*, but the alternative is not quite right either: *Last night I forwent the pleasure of watching Hank's vacation slides.*[6] Similarly, it is perfectly natural to say *I don't know how she can bear the guy*, but something is odd about *I don't know how she bore the guy.* Though one might say *I dig The Doors, man!*, it's much harder to say *In the '60s, your mother and I dug The Doors, son.* And while everyone knows what *That dress really becomes her* means, the past-tense version is almost unintelligible: *But her old dress became her even more.* Here is another example:

ZITS reprinted with special permission of King Features Syndicate.

Irregular stems and their past-tense forms can part company and accrue different degrees of familiarity – just what we would expect if they are separate entries in memory. The slippage can go the other way too, with an irregular past tense becoming more familiar than the verb itself. Recall from chapter 3 that people often are foggy about what verb stem goes with *smitten, rent, shod, hove*, and *wrought*. When that happens in the history of a language, a past-tense form can lose its moorings and drift over to some other verb. For example *went* originally went with *wend*, and now goes with *go*.

None of this happens to regular verbs. Some are used primarily in negations and hence appear in the infinitive, such as *He doesn't suffer fools gladly.* But when the cliché is coaxed into the past tense, nothing interesting happens: *None of them ever suffered fools gladly.* Other verbs that are common in general but rare in the past tense, such as *afford* and *cope*, also do just fine when plunked in the past tense, as in *I don't know how he afforded them* and *It's a miracle how she coped with him*. In the *Zits* cartoon, the joke would disappear if the irregular *bite* 'be bad' were replaced with its regular synonym *suck*. All this is what we would expect if people react to uncommon regular past-tense forms not by searching for them in memory but by analysing them into a stem and a suffix by rule. The past-tense form would inherit the familiarity of the verb as a whole, because the past-tense form simply *is* the verb (plus an ornament) as far as the mind is concerned.

We have seen that the distribution of verbs from the high end of the frequency range to the low end tells us something about the psychology of the people using the verbs. Equally informative are the clues found among the verbs with the lowest frequencies of all, those that appear exactly once. There is a lovely technical term for a word that appears once in a body of text: a *hapax legomenon*, plural *hapax legomena*, Greek for 'once said.' The term comes from philology, the study of old texts.

Hapax legomena can be a nuisance to scholars of ancient languages because with only one instance of a word one can never be sure what it means. But Harald Baayen has shown that they can be a gold mine for linguists interested in whether a prefix or suffix is truly regular and productive. Baayen devised a formula to capture in a single number the productivity of a suffix (or any other form): the number of hapax legomena, that is, the number of words with that suffix that appear exactly once in a corpus, divided by the number of times the suffix appears in the corpus summing over all words.[7] Here is an intuitive way to understand it.

Suppose you want to know how many fish are in a huge body of water. Obviously you can't count them all. You get your rod and reel and pull out a fish, tag it, release it, and give it time to swim away. Then you catch another fish, tag it and release it, then another, and so on, noting for each fish whether you had caught it before. If the body of water contains a small number of fish, at some point you'll keep catching the same few fish again and again. If it has an enormous number, most fish you pull out will not have been caught before. If the number is open-ended, or essentially infinite – if the fish breed faster than you can catch them – you will never catch the same fish twice. After a million tries, you can write down the number of fish that you have caught once, the number you have caught twice, and so on. The larger the proportion of the catch that consists of fish caught only once, you may conclude, the more fish are out there in the water.

The body of water is the English language. A fish is a suffixed word. A million casts of the fishing rod is a million-word corpus. A fish that has been caught ten times is a word with a frequency of ten per million. A fish that has been caught only once is a word with a frequency of one: a hapax legomenon. If the creel is filled with hapax legomena, the words must be breeding quickly. That is, the suffix must be productive and the set of words accepting it open-ended. The vague notion that a rule of language is 'productive' or 'open-ended' can therefore be translated into a number.

What happens when we cast our line into English and pull out regular past-tense forms? Of the 15,369 catches in the Brown corpus, 871 are hapax legomena, regular past-tense forms pulled out only once. That means that if one were to keep fishing (adding to the Brown corpus), 5.7 percent of the new regular past-tense forms would never have been seen in the past tense before. Is that a high rate or a low rate? The best comparison is to a class of words we know to be relatively unproductive, namely, verb stems such as *eat* and *walk*. New ones such as *mung* and *mosh* are created not by a rule but by occasional moments of inspiration in creative wordsmiths. When we bait our hooks for verb stems, we get 170,931 catches, of which 877 are hapax legomena, a rate of new word formation of only 0.5 percent. (Actually, this is an over-estimate, because it includes new words formed by derivational prefixes and suffixes such as *uncorkable* and *pinkness*.) That is a tenth of the value for the regular suffix, and shows that we breed new regular past-tense forms far more quickly than we invent verbs. The other interesting fishing expedition is for irregular past-tense forms. We land only 62 hapax legomena among 10,832 catches, a rate of 0.6 percent – essentially the same as the rate for verb stems.[8] The statistics confirm that regular forms are generated freely, presumably by a rule, whereas irregular past-tense forms are stored in the same manner as ordinary words.

So far I have been doing psychology in a roundabout way – by working backward from the statistics of English vocabulary, now and then asking for your gut reactions to verbs in various parts of the frequency range. But the same effects can be shown in systematic studies. Michael Ullman and I asked ninety-nine people to rate *their* gut reactions to several hundred verbs and their past-tense forms, each presented in a sentence, on a scale of 1 (unnatural) to 7 (natural).[9] We got their ratings of the bare verb stems, as well as their ratings of the past-tense forms, so that we could disentangle any queasiness about a past-tense form – say, *maimed* or *smote* – from queasiness about the verb itself, such as *to maim* or *to smite*.

Ullman found that ratings of irregular past-tense forms depended on their frequency in the language: The more common the verb form, the better people liked it. (This was true even when he controlled for the familiarity of the verb itself, using a standard statistical procedure.) Ratings of regular past-tense forms, in contrast, did not depend on their frequency in the language – a rare verb form such as *maimed* was just as natural sounding as a common one like *walked* (again, controlling for the fact that *maim* is less natural than *walk*). Instead, ratings of regular past-tense forms depended highly on the ratings of the *stems*: The more natural the stem, the more natural the past-tense form. That is just what we would expect if people saw the past-tense form as simply the stem itself, with a decoration added by a rule. People's ratings of the irregular past-tense forms correlated less well with their ratings of the stems, which is what we would expect if irregular past-tense forms are separate words, which are only *linked* to their stems.

The same patterns hold when people have to cough up past-tense forms under time pressure, as they do in rapid conversation. Sandeep Prasada, William Snyder, and I brought another group of volunteers into the lab, seated them in front of a computer screen that flashed verb stems at them, and asked them to blurt out the past-tense form as quickly as they could.[10] A microphone connected to a voice-operated trigger sent a signal to the computer. By timing the interval between stimulus and response, the computer could estimate how long it took people to read the verb, mentally compute the past-tense form, and begin to say it aloud.

We chose verbs in pairs, one used frequently in the past tense and one used less frequently, but both used equally often in the nonpast forms. (As before, we wanted to hold constant the familiarity of the verb, and compare only the past-tense forms.) For example, the stems *ring* and *strive* are both used about seven times per million words of text, but *rang* is used twenty-one times whereas *strove* is used only four times. Similar statistics hold for the regular items *pour* and *soak*, which are used equally often in the stem form, but *poured* is about twenty times as common as *soaked* among past-tense forms. All the verbs were presented in random order.

With irregular verbs, the more frequent past-tense forms came out of the people's mouths faster, as we would expect if they were stronger in memory and therefore quicker to retrieve. But with regular verbs, the more frequent past-tense forms were no faster than the less frequent ones, which suggests that people were not retrieving them preformed from memory but were assembling them on the fly. The effects were small – only a few hundredths of a second – but we found them in three experiments, and four other teams of experimenters have replicated the results.[11]

When you produce an irregular form, you not only have to dredge it out of memory but also must repress the 'Add *-ed*' rule so you don't say *breaked* or *broked*. Linguists call this principle *blocking* – the irregular form blocks the rule – and the experiments help us understand how the mind implements it. One possibility is that when we need to utter a past-tense form we first scan our list of irregular verbs to see if it is there, and if it isn't, we turn on the rule. That predicts that the slowest irregular verb (the one at the end of the list) should be faster than the fastest regular verb. The prediction is wrong: Irregular forms usually are slower to produce than regular forms; they are never faster.[12]

A more likely possibility is that words and rules are accessed in parallel, that is, at the same time. As we plan to utter a verb in the past tense, we simultaneously look up the word in memory and activate the rule. An inhibitory link runs from the memory box to the rule box, which gradually slows down the rule as evidence for a match is found, and eventually turns it off.

Memory lookup is not an alphabetical search, of course, like finding a word in a real dictionary. The phonemes and syllables in

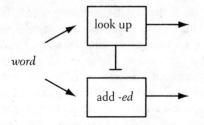

a word contact their counterparts in memory piecemeal, more and more of them finding a match as the milliseconds tick by. As soon as all the pieces match some entry, the irregular form linked to the entry is fetched and shunted to the vocal tract. While the lookup is in progress, the inhibitory signal sent to the rule box gets stronger and stronger, and when all goes well, the rule is braked to a halt. Occasionally the matching or fetching goes awry and the rule runs to completion, producing a speech error, such as *I carefully looked at 'em and choosed that one.*[13]

If a memory entry is faint or blurry because the word is uncommon, the matching and fetching will be especially erratic, and often the rule may not be braked in time. For example, many people may produce *slayed* as a speech error (it should be *slew*), and their listeners may tuck it away in memory as a genuine past-tense form for the verb. If that happens often enough, doublets such as *slew/slayed* and *strove/strived* will come into common parlance. Eventually the irregular form may disappear outright, like *chid, dempt,* and *abode.* Ullman found that verbs with doublet past-tense forms (*strived* and *strove, dreamed* and *dreamt, dived* and *dove*) have lower frequencies than verbs with just one past-tense form. Ullman also got snapshots of the weakening grip of these irregular forms and the rising strength of their regular counterparts: Among the doublets, the lower the frequency of the irregular form, the better-sounding the regular form.[14]

The flip side of failing to recall an irregular form is tripping a false alarm for one when the verb is in fact regular. Word lookup is not instantaneous, and as it proceeds a few irregular verbs in memory might crudely match a regular probe. That could temporarily slow down the rule until the last jots and tittles of the word are properly matched and the false matches have petered out; only then will the rule be allowed to proceed unhindered. This predicts that regular verbs that are similar to irregulars, inviting temporary false matches, should be slower to produce in the past tense. The psychologists Mark Seidenberg and Maggie Bruck found exactly that: Regular verbs such as *smell*, *greet*, and *bake* (which rhyme with irregular *tell*, *meet*, and *make*) are slower for people to utter in the past tense than regular verbs such as *walk* and *look*, which don't rhyme with any irregular and therefore have an unimpeded run through the rule box.[15] Incidentally, there is no contradiction between saying that regular past-tense forms don't depend on their memory entries and that they can be slowed down by temporary false matches with *other* verbs' memory entries. From your brain's point of view, no verb is either regular *or* irregular until it has been looked up in memory and discovered to have, or to lack, a special past-tense form.

~

The mental machinery for words also can be studied in the laboratory by asking people to recognize words, as they would when listening or reading. The problem with studying word recognition is that it happens privately inside the skull, and it's not clear at what point in this process people may be said to have 'recognized' the word. Is it when they know they have seen the word before? When they know what it means? When they know how to use it in a sentence? The experimental psycholinguists who call themselves 'word nerds' usually try to tap the moment at which people are willing to say that the word *is* a word and not just something that looks like a word. Volunteers see or hear a mixed-up series of real words and fake words, such as *narse* or *bluck*, and have to press one button if the item is a word and another button if it is not. This task, called *lexical decision*, doesn't correspond to anything that people do outside the lab, and no one really knows what goes on in people's heads as they do it, but hundreds of experiments have used the procedure.[16] The task is popular because it tells us something about how the mental dictionary is organized.

If people are given a word and then given it again, they are faster at recognizing it – that is, discriminating it from a nonword – the second time. Apparently a mental dictionary entry can be primed (sensitized or prepared) to match an instance of a word; the effect is called *repetition priming*. Interestingly, a word also can be primed, though not as strongly and not for as long, by a word that is associated or similar in meaning to it, such as *doctor* for *nurse* or *duck* for *goose*. Words must be hot-linked in memory, like pages on the World Wide Web, so that when one word is turned on, it becomes easier to turn on related words.

What happens if the priming word is in an inflected form and the target word (the one the person has to recognize) is its stem? That is, what happens when *walked* is presented as the priming word and then *walk* is presented as the target? According to the words-and-rules theory, if the prime is a regular form, the mind should analyse it as a stem plus a suffix, and the stem should prime the mental dictionary entry just as if the stem had been presented by itself. That is, *walked* should have the same effect on recognizing *walk* as *walk* itself does, because as far as the mind is concerned, the word *walked* just *is* the word *walk* with a suffix. In contrast, an irregular form like *swept* should be seen as a word that is distinct from its stem *sweep*, though hot-linked to it. Therefore, while *swept* might prime *sweep*, it should do so less effectively than *sweep* itself, because as far as the mind is concerned *swept* is its own word, not *sweep* with a vowel change and a *-t*. The psychologist Robert Stanners and his colleagues found exactly that, and four other laboratories have replicated the finding, including one that bypassed the button-push and measured the brain's response to the words directly via electrodes pasted to the scalp. (We will return to this technique in chapter 9.)[17]

You may be objecting that *walked* and *walk* overlap in letters more than *swept* and *sweep* do, and perhaps it's just the brute repetition of black lines in the shape w-a-l-k or a repetition of the sounds of *wŏk* that primes the word *walked*, and not anything so airy as a putative entry in a mental dictionary. But these studies include controls such as *market* as a prime for *mark* or *gravy* as a prime for *grave* – words that overlap as much as regular forms and their stems do but are not related in meaning or grammar.

Nothing much happens; the mind is not impressed by mere overlap in letters or sounds.

Another way to show that priming occurs deep in the mind and not in the senses is to play a tape of the priming word spoken aloud, and then flash a printed word on a computer screen. If a noise coming in from the ears helps a person to recognize a squiggle coming in from the eyes, what must have come between them was a representation of the word in the mind:

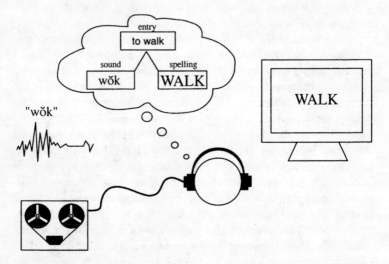

The psychologists William Marslen-Wilson and Lorraine Tyler did the experiment.[18] They found that when people heard *asked*, it was easier to read *ask*, but when they heard *gave*, it was no easier to read *give*. In a diabolical twist, the researchers ran the experiment again using a technique called *subliminal* priming. The priming word is flashed very quickly on the screen and is immediately covered up by a row of random squiggles. Subjects insist that they can't see a thing. But the word must have registered unconsciously in their brains, because invisible *filled* made them faster at recognizing *fill*. In contrast, invisible *rode* did not make them any faster at recognizing *ride*. As usual, mere overlap – *fillet* and *fill*, *rude* and *ride* – had no effect.

Is an irregular form, then, simply related to its stem the way *duck* is related to *goose* – as two words that just happen to be similar in meaning? Apparently not. A stem and its irregular past, even if they are stored separately, have to be treated as members of the conjugation of a single word; otherwise there is no reason that *rode* would block the generation of *rided*. Marslen-Wilson and Tyler in fact showed that irregular forms are tightly linked to their stems, not just loosely related like *duck* and *goose*. A word can prime a word with a similar meaning only if the second one appears immediately after the first; the priming by shared meaning dies down quickly. A word can prime *itself*, however, over many seconds, minutes, sometimes even hours or days. Marslen-Wilson and Tyler tried a variant of the priming experiment – a spoken word followed by another spoken word, rather than by a word flashed on a screen – and found that regular and irregular forms were equally effective in priming their stems. (No one knows why irregular forms can prime their stems in the all-sound version of the task, but not in the sound-then-screen version.) Words with similar meanings, such as *gold* and *silver*, primed each other as well. But grammar proved to be a stronger tie than meaning when the target did not follow the prime immediately but only after a delay of 35 seconds. By then, *gold* was a distant memory and an ineffective prime when *silver* came around, but *gave* continued to prime *give*, and of course *filled* continued to prime *fill*. That suggests that the mind represents *gave* not as a word that just happens to be similar in meaning to *give* but as a separate morpheme that represents the verb *give* in the past tense.[19]

~

These differences between regular and irregular verbs were nicely predicted by the words-and-rules theory, but do they rule out the alternatives?

Mark Seidenberg has suggested that one piece of evidence for rule use – the finding that uncommon regular past tense verbs are no slower to produce than common ones – is in fact a signature of connectionist pattern associator memories.[20] Pattern associators are designed to ignore words and remember patterns. Regular verbs are so plentiful that an uncommon one like *stalk* is bound to share patterns with many other regulars, such as *stop* and *walk*, so its rarity should not hurt it. Irregulars are few in number, so if one is uncommon, it will have no similar irregular forms to lean on, and it will be harder.

Seidenberg and the computer scientist Kim Daugherty devised a simulation much like the Rumelhart-McClelland model, but with an extra layer of units hidden between the input and output, a modification that can make a pattern associator more powerful:

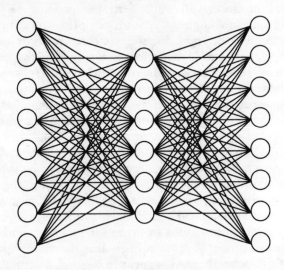

An immediate problem in comparing this model to people is that pattern associators compute everything in a constant number of steps, in this case two, and therefore take the same amount of time for all verbs, regular and irregular, common and uncommon – unlike people. So Daugherty and Seidenberg interpreted the error score for the pattern associator (the discrepancy between its guess and the correct answer) as an indicator of a verb's difficulty, and assumed that some black box fed by the associator would take longer to assemble the 'harder' verbs. Unfortunately, though the designers can see the model's error score, the model itself cannot. Once the training sequence is over, the model has no way of knowing whether its guesses are right or wrong, unless it memorizes all the words in the teaching sequence to use as a crib sheet. But the whole point of a pattern associator is to do away with entries for words!

In any event, the model did not act like people: To the modelers' disappointment, it found uncommon verbs harder than common verbs, whether they were irregular or regular. (Recall that people find the uncommon regular verbs no harder than the common ones.) The problem was that the *irregular* verbs, such as in *stick–stuck*, were interfering with the regular patterns that uncommon verbs such as *stalk* depend on. So Daugherty and Seidenberg retrained the model with only 24 irregular verbs and 309 regulars, giving the uncommon regulars plenty of similar neighbors to lean on and keeping the pesky irregulars in check. That did the trick – the model now found the uncommon regular verbs to be not much harder than common ones, mimicking the human being.

This decimation of their irregular verbs is questionable. Adults know about 165 irregulars, preschool children about 80,[21] and as we shall see in chapter 8, the ratio of regular to irregular words varies wildly from language to language and should not be precariously balanced on a narrow ledge of values just to get the model to behave properly with English. There is a more serious problem, however: The model's success hinges on its artificiality. The model is an idiot savant tailored to do one task: generate the sound of a past-tense form. It can be made insensitive to the frequency of regular words because it is insensitive to words, period; all it knows is the *sounds* of words. Unfamiliar *stalk* gets dissolved into *st-* and *-alk*, and both sounds can be made familiar if there are enough regulars in the training set and not too many irregulars. But real human beings do not live by sound alone; they need to know what a word means. The human head must contain something that differentiates *stalk* from *stop* and *walk.* And that something – a lexical entry, or some link between the sound and meaning of *stalk* – is bumped up and down in strength as the person encounters a word more or less often. If a pattern associator were to model people's actual knowledge of words, not just their sounds, it too would be accumulating information about the word as a whole, including its past-tense associations. That would make it unlikely to mimic the human pattern of generating uncommon regular past-tense forms as easily as common ones.

What about the opposite theory, Chomsky and Halle's all-rule model from generative phonology? According to their theory, regular verbs are not tagged for the *-ed* rule (they undergo it by default), whereas irregular verbs *are* tagged for their vowel-change rules. If the tags are strengthened by practice, then the uncommon irregulars should be slower and less certain, which squares with the facts.

But another prediction of the theory does not square with the facts. The theory's main precept is that any trace of redundant information is pried out of memory and computed by rule; the theory assumes that in the human brain, memory is expensive but computation is cheap. The ultimate redundancy is in the regular verbs, where storage of the past-tense forms would be unconscionably wasteful. If predictable forms are never stored, then surely regular verbs are not stored.

In contrast, the words-and-rules theory assumes that memory is constantly working alongside rules – that's how irregular verbs arise to begin with – and it would be a strange mental block indeed that would force the memory system to be amnesic for all the regular past-tense forms it hears. (After all, we remember them just fine in quotations such as *All men are created equal* and *The quick brown fox jumped over the lazy dog*.) The words-and-rules theory predicts only that people don't *depend* on stored past-tense forms, not that they are *incapable* of storing them. People use a rule to generate and judge past-tense forms when they need to, and if some regular forms have been stored in memory, they are available but not indispensable (indeed, the stored versions may even get in the way).[22]

Ullman's experiment uncovered two cases in which people do store and retrieve regular forms. In words that have two acceptable past-tense forms such as the doublets *dived/dove* and *dreamed/dreamt*, people's judgments about the regular versions, such as *dived* and *dreamed*, depended heavily on how often those forms are used in the language. This makes perfect sense, even within the Chomsky-Halle theory. Suppose the gatekeeper to memory follows the principle, 'Store anything that is unpredictable.' Irregular forms, of course, are unpredictable, so they (or their tags to the vowel-change rules) are always stored. But the regular member of a doublet is unpredictable, too. If you already know *dreamt*, the blocking principle rules out *dreamed*, just as *came* rules out *comed*. Suppose now that you hear someone say *dreamed*. If your language processing could be displayed as a readout on your forehead, it would say, '*Dreamt* plus the blocking principle says that *dreamed* shouldn't be in the language. But someone just said *dreamed*, so it must be in the language after all. Since my rule system won't generate it, I had better remember it.'

A similar explanation works for another case in which frequency makes a difference: regular verbs that rhyme with irregulars, such as *blink* (which rhymes with *stink* and *slink*) and *glide* (which rhymes with *ride* and *stride*). Human memory always holds out a temptation to generalize irregular patterns. So when a person contemplates the past tense of *blink* and *glide*, a little voice in the head will be whispering *blunk!* and *glode!* When the person hears someone say *blinked* or *glided*, she makes a mental note of the form, because its existence comes as a bit of a surprise.[23]

So the generative phonology theory could handle the data by allowing less predictable regular forms like *dreamed* and *blinked* to be marked for the regular rule. What about the other regular forms – ordinary past tenses and plurals that don't resemble irregular forms and are therefore 100 percent predictable? The answer is that some of them appear to be stored in memory, too. Four teams of psychologists have asked people to decide whether a printed string of letters is a word or not, the 'lexical decision' task described earlier.[24] All the studies found that people take a few hundredths of a second longer to recognize rare regular forms than common ones (holding constant the frequencies of the verbs or nouns themselves). That suggests that people do build up a memory trace for common regular forms, even though logically they don't have to. Human memory is not a scarce resource reserved for the incompressible nuggets that cannot be generated by rules. If a word form is common enough, we can look it up directly, rather than breaking it into parts and looking up the parts.

Why do people use memorized regular forms in these experiments but not in the experiments on rating and producing past-tense forms described earlier? Harald Baayen and the psychologist Robert Schreuder propose that the rule system and the word system process a word form in parallel (at the same time), as in the diagram on page 145, and that the two systems race against each other to produce or analyse a form.[25] Since each system will be faster with some words and slower with others, people will rely more on one system or the other in different tasks to optimize their ability to speak or understand quickly and without error. Whether past-tense forms will be looked up or generated by rule depends on the nature of the forms and on what the person needs to do with them. With irregular words there is no choice; only the word system can handle them. But with regular words, it depends on the task and on the word.

The tasks fall along a continuum. At one extreme is the leisurely process of rating or reflecting on the naturalness of forms, or choosing the most natural one. Since people need only determine whether a verb *could* have a well-formed regular past tense in the language, and the rule is always there to provide one, they don't care whether they have seen the past-tense form often, seldom, or not at all. In the middle of the continuum is the task of producing a form under time pressure. If many of the verbs are regular it will be fastest to let the rule generate the forms, in which case the frequency of the stored regular past will make no difference. But if the list contains many irregulars, speakers are being forced to go to memory so often that they might as well use any regular form they come across, and the fainter ones will take longer to retrieve than the stronger ones.[26] At the other end of the continuum is the task of deciding whether a string of letters is a word. Here people must say 'no' to forms that well might be words but that they have never seen before, such as *refeamed*, and 'yes' to forms that they indeed *have* seen before. The task discourages the use of rules and encourages matches against memory, and thus is a sensitive assay for any trace of a word lingering in memory. That is why in this task, common regular forms are faster to recognize than uncommon ones.

It depends on the words as well. A regular form stored in memory is useful only if it is strong enough to be retrieved quickly. Otherwise, the rule system will strip off the suffix and retrieve the stem (especially if the stem is common) before the inflected form can be retrieved, and the rule system will win the race.[27] The winner depends on other factors, too, such as how easy the suffix is to strip from the stem and whether the suffix is ambiguous because the language has a lot of syncretism (suffixes found in many conjugational slots).

Baayen and Schreuder have formalized the parallel-race theory as a mathematical model that predicts how long people should take to recognize stems, plurals, and past-tense forms with different frequencies in a language. The predictions are an excellent match to the behavior of real humans across a large set of experiments. When Baayen and Schreuder removed the rule pathway, making it a memory-only theory, or removed the word pathway, making it a rules-only theory, the model's resemblance to humans plummeted.[28] The two routes of the words-and-rules theory appear to be just the right number to account for actual human performance.

~

Frequent pairing is one of the engines of associationism, and similarity is the other. Hume enshrined it as his second law of association, and behaviorism depended on it in a version called stimulus generalization. Train a pigeon to peck at a green key, and it will peck furiously when the key stays green, vigorously when the key is yellowish green or bluish green, languorously when it is yellow or blue, and barely at all when it is red or violet. The pattern is called a generalization gradient, after the shape of the curve that emerges on a graph when the pecking rate is plotted against the wavelength of the key. Connectionism, too, runs on similarity. Here is how Rumelhart, McClelland, and their collaborator Geoffrey Hinton once explained the appeal of pattern associators: 'People are good at generalizing newly acquired knowledge. If, for example, you learn that chimpanzees like onions you will probably raise your estimate of the probability that gorillas like onions. In a network that uses distributed representations, this kind of generalization is automatic.'[29]

According to the words-and-rules theory, similarity is indeed important when people generalize an irregular pattern to a new verb, and for exactly the reason connectionists invoke: Similar forms overlap in memory, and the associations of one automatically transfer to the others. But similarity is not important when people generalize a regular pattern to a new verb, because that can be done by grammatical combination, which simply joins a suffix to any stem labeled as a verb.

Here is a way of visualizing the difference. Imagine that words fill a space with many dimensions, each corresponding to the presence of some sound, such as a consonant cluster at the beginning of a word or a vowel at the end. Irregular verbs fall into neighborhoods of similar forms that carve out regions of this space: *sing–sang, ring–rang,* and *drink–drank* cluster in one region; *hide–hid, slide–slid,* and *bite–bit* cluster in a second; *blow–blew, know–knew,* and *grow–grew* cluster in a third. When a verb falls in a neighborhood or in the halo surrounding it, people are tempted to generalize the local irregular pattern to it, as in *bring–brang, fight–fit,* and *snow–snew.* Here is a simplified picture, in which each dot is a word, clusters of dots are words with similar sounds, and the shaded areas are the sounds that tempt people to generalize an irregular pattern:

Now, how do *regular* words fit into this space of sounds? Do they fall into neighborhoods of their own – neighborhoods that just happen to be bigger, more populous, and more evenly sprinkled with words than the irregular neighborhoods?

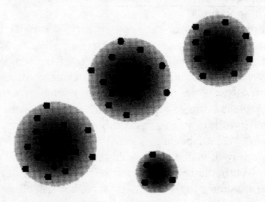

That is the connectionists' explanation of why the regular pattern is more easily generalized. But if regular forms are generated by a rule that merely states, 'Join a suffix to a verb,' the sounds of the verb should make no difference: The rule is triggered by a symbol, 'verb,' not by the sounds of particular verbs. The potential regular territory should not be an archipelago of neighborhoods in the territories unoccupied by irregulars. They should suffuse the entire space of sounds, with every potential word sound as eligible for the regular suffix as every other one. Graphically, the result would not look like a big set of clusters but like a uniform blanket:

The words-and-rules theory predicts that a verb with any sound at all can get the regular suffix, not only those in neighborhoods of familiar regular words.

As with the effects of frequency, the English language itself provides us with clues. First, regular verbs aren't intimidated by a gang of irregulars occupying a neighborhood, but tramp right through and pitch their tents. Every irregular family in English tolerates regular interlopers:[30]

Irregular Family	Regular Interloper
hit–hit, split–split	*pit–pitted*
cut–cut, shut–shut	*jut–jutted*
bleed–bled, feed–fed	*need–needed*
bend–bent, send–sent	*mend–mended*
sleep–slept, keep–kept	*seep–seeped*
sell–sold, tell–told	*spell–spelled*
bind–bound, find–found	*mind–minded*
grow–grew, blow–blew	*glow–glowed*
take–took, shake–shook	*fake–faked*
drink–drank, shrink–shrank	*blink–blinked*

   The most blatant sign of regular imperialism is the ability to
occupy the *same* patch of territory as an irregular verb – that is,
regular verbs that are homophonous (identical in sound) with
irregular families:

Irregular Verb	Regular Homophone
The dentures *fit*.	He *fitted* the dentures.
She *rang* the bell.	They *ringed* the city.
He *hung* the painting.	He *hanged* the rustler.
We *met* the passenger.	We *meted* out the punishment.
They *lay* on the couch.	They *lied* about what happened.

The regular carpetbaggers are not just a handful of tough verbs
that have clung to their territories against steep odds; they can be
created at the drop of a hat. Here is a do-it-yourself guide for
creating a regular verb with any sound you want. English has a
rule that creates a verb by prefixing *out-* to a person's name, as in
It out-herods Herod (from *Hamlet*) or *Clinton is trying to out-
Kennedy Kennedy*. Make your favorite sound into someone's
name. Presto! A regular verb. In 1983 everyone heard about Sally
Ride, America's first woman in space. But a few years later Mae
Jemison got even more publicity for being the first African
American woman in space – she *out-Sally-Rided Sally Ride* (not
out-Sally-Rode Sally Ride). For many years the most infamous
prison in New York was Sing Sing. But in 1971 a riot at the Attica
Correctional Facility propelled it into the limelight: Attica *out-
Sing-Singed Sing Sing* (not *out-Sing-Sang Sing Sing, out-Sang-Sang
Sing Sing, out-Sing-Sung Sing Sing,* or *out Sung-Sung Sing Sing*).
There is a popular board game in Japan called 'Go.' Suppose that
the Japanese continue to emulate American culture and that
Monopoly becomes even more popular. It would have *out-Go'd
Go* (not *out-Gone Go*). In the next chapter we will see experiments
that catch people in the act of creating these forms.[31]
   Finally, we find regular verbs boldly going where no verb has
gone before, into the nooks and crannies of phonological space
that rarely or never see verbs. Few if any English verbs end in *-ach,
-ev,* or *-a,* but we can say *Handel out-Bached Bach; Yeltsin out-
Gorbachev'd Gorbachev;* and *We rhumba'd and cha-cha'd all night.*
Nothing in English sounds anything like the word *oink,* but since

it found its way into print in 1969, no one has had any compunction about saying that a pig *oinked*.[32]

Michael Ullman has documented the wanderlust of the regular suffix in his study in which people judged the naturalness of hundreds of verbs and their past-tense forms.[33] He found that irregular forms that belonged to large families of similar forms (*ring–rang, sing–sang, drink–drank*, and so on) were judged as more natural than irregular forms with small or nonexistent families, such as *stand–stood*, holding all else constant. But with regular forms, it made no difference: verbs from large families, like *walk, talk, stalk*, and *balk*, were no more acceptable than verbs from tiny ones, such as *scour*.[34]

Can the act of slapping regular suffixes on strange-sounding verbs be reproduced in the lab? Sandeep Prasada and I replicated the experiment by Joan Bybee and Carol Moder that asked people to guess the past-tense forms of novel verbs like *spling*.[35] Recall that Bybee and Moder got a pigeonlike generalization gradient: The more similar a novel word was to existing irregular verbs, the more likely the subjects were to generalize an irregular pattern to it. For example, the pseudo-verb *spling* is similar to many real irregular verbs such as *sling, slink, string*, and *shrink*, and subjects inflected it as *splang* or *splung* 80 percent of the time. *Krink* is less similar, and it came back as *krank* or *krunk* only about 50 percent of the time. *Vin*, which is barely similar, came back as *van* or *vun* only about 10 percent of the time.

Prasada and I tested nonsense verbs like those, but we also tested nonsense verbs that varied in their similarity to *regular* verbs. A verb like *plip* sounds like many English regular verbs, such as *chip, clip, dip, drip, flip, grip, nip, pip, quip, rip, ship, sip, slip, skip, snip, strip, tip, trip, whip*, and *zip*. Verbs like *glinth* and *smaig* don't rhyme with any English verb roots. And verbs like *ploamph* and *smeerg* are very much unlike English verbs because they violate the sound pattern of the English language. English phonology doesn't allow a long vowel to precede a consonant cluster at the end of a syllable unless all the consonants are produced by the tip of the tongue. For example, *toast* is a possible English word, but *toask* and *toasp* are not.[36]

We gave these six kinds of verbs – three varying in similarity to irregular families, three varying in similarity to regular families –

to university students and to the trained Rumelhart-McClelland pattern associator model. For the irregular verbs the model gave a reasonably good impersonation of a human being, showing a generalization gradient in which only the verbs that sounded a lot like irregular verbs were readily given irregular forms. But for the regular verbs the model and the human diverged. People put *-ed* on strange-sounding *ploamph* at virtually the same rate as they put it on familiar-sounding *plip*. The pattern associator also had little trouble with *plip*, but with *ploamph* it could muster a regular form only about 10 percent of the time. Sometimes it offered hopeful guesses loosely based on irregular forms:

> *greem–grame*
> *proke–prokt*
> *brilth–brilt*

Sometimes it offered a form that belonged to some other verb it had been trained on:

> *brilth–prevailed*
> *ploag–pleaded*
> *proke–trusted*
> *krilg–brewed*

And sometimes it concocted bizarre blends:

> *slace–fraced*
> *smeeb–imin*
> *ploanth–bro*
> *smairf–spurice*
> *trilb–treelilt*
> *smeej–leefloag*
> *frilg–freezled*

This sorry performance confirmed our diagnosis that the model's errors in Rumelhart and McClelland's own tests – *mail–membled, tour–tourder*, and silence for *jump, pump, warm,* and *glare* – were not random noise. They were a symptom of the model's inability to generalize its training to words that did not sound like the words it had been trained on.

The failure is instructive. Pattern associator memories, unlike symbol crunchers, cannot exploit the basic gadget of computation called a *variable*. A variable such as 'verb' can stand for an entire class of items, regardless of their phonological content. That allows a rule to copy over the material of a stem and simply hang a suffix on it, whatever it is. A pattern associator, in contrast, has to be painstakingly trained with items bearing every input feature in the class. If a new item bearing a novel combination of features is presented, the model cannot automatically copy over the combination; it activates bits and pieces that are vaguely associated with the features and coughs them up in a hairball.

Connectionists were quick to blame these problems on the 1960s-era technology used in the original Rumelhart-McClelland model. *Forget* the Wickelphones, they said; we never meant for them to be taken seriously.[37] And forget the direct stimulus-response connections; the state of the art in connectionist models has three layers of units (as in the Daugherty-Seidenberg model described earlier), not two. The new layer, hidden between input and output, has been proven to allow a pattern associator to solve problems that a two-layer model cannot. For example, it can decide whether a number is odd or even, and can apply the formula 'A or B but not both,' neither of which the old models could do. In principle, this deluxe kind of model can be trained out of its slavery to similarity and create new families that don't necessarily reflect the shared features of the input.

The computational linguists Richard Sproat and Dana Egedi put these upgrades to the test.[38] They replaced the Wickelphones with input and output layers that any linguist would be proud of, and added a sophisticated decoder that turned the output layer into a sensible string of vowels and consonants. They also retrofitted the model with a hidden layer of units, and trained it using a state-of-the-art learning procedure. Like the original model, and like a human being, it generalized irregular patterns to verbs that were similar to the irregulars it had been trained on. But also like the original model, and unlike a human being, it generalized poorly to novel regular verbs. Some were left unchanged. Some were confused with other verbs in the training set, such as *train–trailed, spoke–smoked*, and *glow–glanced*. And fully a quarter were bizarre distortions, such as *conflict–conflafted, wink–wok, yield–rilt, satisfy–sedderded*, and *quiver–quess*.

A pattern associator's ineptitude with novel combinations appears to be deeply rooted in its design, not just a failing of a first-generation implementation. Many connectionists have gone back to the drawing board, but none has been able to get a pattern associator memory to generate new regular forms properly. Several modelers, stymied by the models' habit of outputting gibberish, have hardwired various patches into their model that are tailor-made for regular verbs. One team of modelers included a second pathway of connections that linked every input unit to its twin in the output, implementing by brute force the copying operation of a rule.[39] Another team added an innate clean-up network in which the units for *-ed* strengthen the units for an unchanged stem vowel and inhibit the units for a changed vowel, shamelessly wiring in the English past-tense rule.[40] And as mentioned in chapter 4, many connectionist modelers have given up on trying to generate past-tense forms altogether. Their output layer contains exactly one unit for every past-tense suffix or vowel change, turning inflection into a multiple-choice test among a few innate possibilities.[41] To turn the choice into an actual past-tense form, some *other* mechanism, hidden in the wings, would have to copy over the stem, find the pattern corresponding to the chosen unit, and apply the pattern to the stem. That mechanism, of course, is called a rule, just what connectionists claim to be doing without.

When it comes to generalizing regular inflection to novel words, pattern associators are simply the wrong tool for the job. The problem is that a single mechanism is being asked to do several jobs with contradictory demands. To discriminate among similar irregular sounds with different outputs, such as *drink–drank, slink–slunk, think–thought*, and *blink–blinked*, a pattern associator has to cultivate an ear for the tiniest nuances of the sound of the input and confine its associations to them. That is exactly the opposite of what it has to do to generalize the regular pattern to novel words, where it must be oblivious to differences in their sounds and plaster them all with the suffix. Moreover, with the same pathway it uses to discriminate and generalize among verbs, a pattern associator also must record all the sounds of the stem and try to copy them to the past-tense form. In a standard grammar these demands are handled by three parts that stay out of each other's way: a lexical entry (which marks a verb as irregular), a category symbol such as 'verb' (which can be joined to a suffix by a rule), and a phonological representation (which comes through in the output untouched). The pattern associator has to do the jobs of all three, none satisfactorily.

Perhaps the most important lesson of the chapter is that the mind, like any complex device, is a system of mechanisms optimized for different jobs. Any theory that has one mechanism doing all the work is proposing a kind of crippleware that the human brain is bound to outperform. If the mechanism is a set of rules, it loses the advantage of cacheing the results of frequently performed computations so that it can look them up quickly rather than recomputing them every time. If the mechanism is a set of associations, it loses the advantage of variables and the rules that combine them. As the psychologist William James wrote, 'Thought is . . . a kind of algebra . . . in which, though a particular quantity be marked by each letter, . . . it is not requisite that in every step each letter suggest to your thoughts that particular quantity it was appointed to stand for.'[42]

6

OF MICE AND MEN

Irregularity in language, the quintessence of illogic and caprice, often inspires bouts of idle curiosity. The *Boston Globe* columnist John Powers wonders:

> Why do artists show Adam with a belly-button? Why is Germany the Fatherland but Russia the Motherland? Who made 'impact' a verb? Why does Queen Elizabeth wear that kerchief? How can there be a Miss Universe pageant without Miss Pluto? Why is Peg short for Margaret? If tin whistles are made of tin, what are foghorns made of? Why do dilemmas have horns? If 'mice' is the plural of 'mouse,' why isn't 'hice' the plural of 'house'?[1]

Sometimes the only answer to such questions is, That's just the way it is. I have no idea why in Old English *mus* had the plural *mys* but *hus* had the plural *husas*, to say nothing of the Queen and her kerchief. But sometimes idle questions do have answers. In the comic strip 'Funky Winkerbean,' a Little Leaguer uses his time on the bench to ponder the mysteries of baseball (see the following page), but we can do much better than his teammate at resolving them.

The primary meaning of *plate* is 'a smooth, flat, relatively thin, rigid body of uniform thickness,' and according to the Rules of Major League Baseball, 'Home base shall be marked by a five sided slab of whitened rubber fixed in the ground level with the ground surface,' which fits the definition nicely. In the early days of baseball every pitch led to a fair strike, a foul strike, or a ball. A foul strike was any attempt to strike at the ball that was not fair (in bounds).

FUNKY WINKERBEAN by Tom Batiuk

The term *fair strike* fell into disuse, and *foul strike* was shortened to *strike*, with *foul* reserved for striking the ball out of bounds. Baseball managers, unlike the coaches of other sports, sometimes have to run onto the playing surface, to confer with the pitcher or kick dirt on the umpire's shoes. As for why a batter is said to have *flied out* – no mere mortal has even flown out to center field – there is an even more satisfying answer, and it is the topic of this chapter.

To begin with, yes, it really is *flied out*. In *The Careful Writer*, Theodore Bernstein notes,

> You won't find it in most dictionaries, but *flied* is the past tense of *fly* in one specialized field: baseball. You could not say of the batter who hoisted a can of corn to the center fielder that he 'flew out'; you must say he 'flied out.'[2]

Since we are discussing baseball slang, I can't resist explaining the lovely phrase *a can of corn*, an old term for a high, lazy fly ball. It goes back to the early decades of the twentieth century, when grocery stores stacked canned goods on a high shelf and the grocer would retrieve a can by tipping it with a pole or grabber and catching it as it fell, much as an outfielder catches a fly ball. The language columnist Jan Freeman adds, 'The fact that it's corn in the can . . . is probably based on euphony – *can of corn*, with its neat near-rhyme, sounds a lot snappier than *can of peaches* or *can of beans*.'[3]

But back to flying out. *Flied out* is one of many irregular forms that mysteriously turn up in regular garb when used in certain ways. For example, one might say *All my daughter's friends are lowlifes*, rather than *All my daughter's friends are lowlives*, even though the usual plural of *life* is irregular *lives*. One might say *I'm sick of dealing with all the Mickey Mouses in this administration*, not the *Mickey Mice*. Toronto has a hockey team called the *Maple Leafs*, not the *Maple Leaves*, and when their goon tries to decapitate an opposing sniper and is sent to the penalty box for high-sticking, we say that he *high-sticked* his opponent, not that he *high-stuck* him.

Flying out and other systematic regularizations, we shall see, offer an elegant corroboration of the theory that language in general, and the regular-irregular contrast in particular, may be explained as an interaction between words and rules.

In the preceding chapter we saw how people are happy to use a rule like 'Add -*ed*' whenever their memory does not supply an inflected form. That can happen for many reasons. With new coinages such as *to wug, to fax, to Bork,* and *to mosh* there *is* no past-tense form in memory. With rare verbs such as *to allure, to badger,* and *to carouse* the form may be too faint to retrieve reliably. With strange-sounding words such as *to ploamph* and *to frilg* there is no *similar* form in memory either, preventing the use of analogy. With words with irregular homophones such as *mete* and *lie*, or words with irregular neighbors such as *wink* and *blink*, there may be competing forms in memory. But in all these failures of memory, people are not left speechless; their rule can step into the breach and generate a regular past-tense form.

In this chapter we examine cases where memory is useless not for quantitative reasons (as when a word is relatively strong or strange) but for qualitative reasons. The forms that surprise us, such as *flied out* and *lowlifes*, either violate the standard format for a word stored in memory or skirt the mechanism that funnels information from memory to the rules that compute the word's form. The regular suffix rises to the occasion, just as it does when a word is rare or strange. That underscores the power of a rule: It can apply whenever memory fails, regardless of the reason for the failure.

These examples also add to the debate on whether rule processing or memory associations are the main motor of productivity in language. In the preceding chapter we saw that the chief rival to rules, the pattern associator memory, has trouble generating regular forms for novel and unusual words. Yet the connectionists who defend these memory models have not conceded. The behavior of pattern associators and other artificial neural networks depends on dozens of settings, such as the number of hidden layers, how many units are in each one, and the nature of the training set. Tweaking these networks has become part of the neural network modeler's art, and any failure of a model is taken as a challenge to squeeze out more performance by souping it up or by adjusting the richness or leanness of the mixture of regular and irregular forms in the input.

I suspect that ultimately little will come of this trial and error, because the problem with pattern associators lies in their very design. In the next two chapters we will see that no magical combination of settings is likely to work for all inflections in all languages. But in this chapter, I set aside questions about numbers. Rules will show their worth, and pattern associators their limitations, because of the *kind* of information captured in words, not *how many* words of various types there are.

~

Systematic regularization immediately proves that sound alone cannot be the input to the device that computes inflected forms, as it is in most pattern associator memories. A given sound such as *fly* can come out the other end of the device as *flew* and *flown* when referring to birds, but as *flied* when referring to ballplayers. The question is: What is that extra input, and why does it make a difference?

Many language mavens, psychologists, and connectionists have come up with the same explanation, which can be called the semantic stretch theory.[4] It comes from the intuition that language is a direct conversion from meaning to sound, and states that the extra input features are *semantic*. When a verb is given an extended or metaphorical meaning, the new sense is felt to be dissimilar from the original, and this inhibits the speaker from using the original's irregular form. People sense that *flying out* is different in meaning from *flying*, and that a *lowlife* is different in meaning from a *life*, so they are inhibited from borrowing the irregular forms for those words. Perhaps they house a pattern associator augmented with units for bits of meaning (a unit for 'wing-flapping,' a unit for 'sleaziness,' and so on), and the pattern of a new word with a stretched meaning does not overlap enough with the pattern of the word with the original meaning to parasitize its associations. Or perhaps people are trying to make themselves clear and worry that the irregular form will give their listeners the wrong idea, such as that a ballplayer has acquired superhuman powers.

The main problem with the semantic stretch theory is that its basic premise is wrong: Semantic stretching in itself has no effect on a word's past tense or plural. There are hundreds, perhaps thousands, of examples in which the meaning of an irregular word is stretched, sometimes to the breaking point, and people do not abandon its irregular forms:[5]

- If a new word is formed from an old irregular word by *prefixing*, the new word stays irregular. When *eat* begets *overeat*, the new past tense is *overate*, not *overeated*. Similarly, we get *overshot* (not *overshooted*), *undid*, *preshrank*, *remade*, *outsold*, and so on.
- New nouns constantly are being formed by *compounding* a word onto an existing noun. When the input is irregular, so is the output: *bogeymen*, not *bogeymans*; *superwomen*, not *superwomans*; also *muskoxen*, *stepchildren*, *milkteeth*.
- Using a noun as a *metaphor* also does nothing to its irregularity. Misguided reviewers of my books attack *straw men*, not *straw mans*, and we speak of *chessmen*, *snowmen*, *sawteeth*, *children of a lesser god*, and being *six feet under*. The petroleum industry refers to freeloaders who tap into wells and pipelines as *oil mice*.[6] A species of wasp that hunts for bees is called a *beewolf*; several of them are known as *beewolves*. My computer, alluding to a spawned program called a 'child process,' recently spat out the eloquent message, `sendmail [95]: NOQUEUE: SYSERR: getrequests: accept: No children.`
- English has hundreds of *idioms* based on irregular verbs, as we saw in chapter 2, and they steadfastly retain the irregularity of the originals, no matter how strained or opaque the metaphor: *cut a deal*, not *cutted*; *took a leak*, *bought the farm*, *caught a cold*, *hit the fan*, *blew him off*, *lost his marbles*, *put him down*, *came off well*, *went bananas*, *threw up*.

These irregular loyalists also falsify the suggestion that people regularize words to avoid ambiguity and make themselves clear.[7] Many of these idioms are ambiguous between literal and idiomatic senses, such as *bought the farm* and *threw up*, and some are ambiguous with other idioms as well: *blew away*, for example, could mean 'wafted,' 'impressed,' or 'assassinated'; *put him down* could mean 'lower,' 'insult,' or 'euthanize.' But that doesn't tempt anyone to single out one of the meanings in each set by saying *buyed the farm, throwed up, blowed him away,* or *putted him down.* Conversely, the past tense of *to grandstand* is *grandstanded,* not *grandstood,* but *grandstood* would be perfectly unambiguous if anyone said it. The same is true of *Mickey Mice, highstuck,* and *lowlives,* which would be perfectly clear, especially in context. But with these *un*ambiguous words people *are* tempted, even compelled, to use a regular past-tense form.[8]

It's not that meaning is irrelevant to the abandonment of irregular forms; it *is* relevant, but only sometimes and in certain ways, and meaning is not the only thing relevant. A theory that does predict when a word will lose its irregularity has been developed by the linguists Paul Kiparsky, Edwin Williams, Rochelle Lieber, and Elizabeth Selkirk, with amendments from me and my collaborators.[9] It comes right out of the words-and-rules theory, once the words part and the rules part have been fleshed out in more detail than I have given you so far. To preview:

- Words are stored in the mental dictionary not as haphazard bundles of information but in a standard format called a *root.*
- Rules don't just throw words or parts of words together; they provide a scheme in which the properties of the new combination can be computed from the properties of the parts and the way they are arranged. A combination that obeys this scheme is said to have a *head.*

We shall see that a word that conforms to these standards – a word with a root and a head – is well-behaved: If it looks like it should be irregular, it *is* irregular. A word that violates the standards has to give up its irregular form; these are the words, such as *flied out* and *lowlifes,* that arouse the curiosity of language-lovers. This explanation may be called the word structure theory, because it says that the structure of a word – in particular, whether it has a root and a head – determines whether it gets to keep an irregular form.

This chapter will show how the word structure theory explains dozens of puzzles about English words that fill the language columns and cartoon pages. Since it appeals to the essence of words (the root) and to the essence of rules (the head), the success of the theory will stand as a confirmation of the words-and-rules theory more generally. We begin with words that cannot find their roots; then we will turn to words that have lost their heads.

~

People are infinitely creative with the sounds they use in conversation. They salt their speech with gestures, sound effects, foreignisms, names, and quotations, all used as if they were actual words:

> So he starts to argue with me, and I just went [*rolls eyes*].
> When I hit the rock, the tire made a pfffffffffffffft sound.
> This townhouse has that *je ne sais quoi.*
> I've been Norman Mailer'd, Maxwell Taylor'd. I've been Rolling Stoned and Beatled till I'm blind.[10]
> If he 'Yes, Dear's me one more time, I'll scream.

Yet we all sense that these quasi-words are special. Most speech is filled with ordinary words like *dog* and *walk* that feel as if they conform to a set of standards for a basic word in English. These words, when unadorned by prefixes and suffixes, can be called canonical roots (roots for short), and they are stored in a particular way in memory.

A root occupies a distinct entry in the mental dictionary, like an entry in a real dictionary. It specifies the word's part-of-speech category, such as 'noun' or 'verb.' It specifies the word's meaning. And it specifies the word's sound, which conforms to a regulation template for standard words in the language.[11] In English the template for a canonical root is a monosyllable, or a monosyllable with an unstressed bit dangling off the end. Other languages have different templates; Italian, for example, does not allow monosyllables for canonical nouns and verbs. (A rough and ready test for a standard word sound in a language is the sound of its nicknames: In English, uncanonical *Bartholomew* becomes canonical *Bart*; *Elizabeth* becomes *Lisa, Liza, Libby, Liddy, Lizzy, Liz, Betty, Betsy, Beth* or *Bess.*) A canonical root also embodies

Ferdinand de Saussure's conception of the linguistic sign as an arbitrary pairing between a meaning and a sound, one of the foundations of modern linguistics discussed in chapter 1.[12] Speakers tacitly sense that a canonical root doesn't have to sound like its referent, as does *oink* or *pfffffffffffft*; it *symbolizes* the referent by a conventional pairing they have learned.

And here is the key to irregularity. An irregular plural or past-tense form is a root linked to another root: *sank* to *sink, feet* to *foot*:

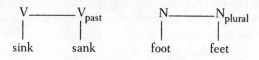

Irregulars by definition are arbitrary, and as we saw in chapter 3, they are canonical English sounds: monosyllables such as *stuck* and *mice*, monosyllabic roots adorned with prefixes such as *became* and *understood*, or words with insubstantial second syllables such as *children* and *oxen*. The fact that irregulars are tied to roots, not words, explains six cases in which words cannot have irregular forms, even if their sound calls for one. The explanation is that the words are not represented in the mind as canonical roots, the only legitimate anchors for irregularity, but as stretches of sound pressed into service as a word; the difference in mental representation from the roots shown above might be as follows:

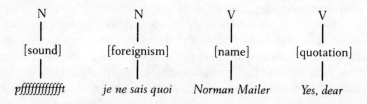

When a word is rootless and thereby disconnected from inflected forms stored in memory, however, it is not left without a past tense or plural; the rule rushes in and turns it into a regular form by adding a suffix.

The first example of this effect is onomatopoeia, where a sequence of vowels and consonants is construed not as a sound arbitrarily paired with a meaning but as a direct rendering of a sound in the world. No one of course thinks that onomatopoeic forms are particularly accurate, as in a tape recording; they vary from language to language, have conventional forms, and usually are compatible with the language's sound pattern. But they *can* violate canonical sound patterns – no word in English has a sequence like *oink* – and crucially, people *perceive* them to resemble sounds. Onomatopoeic verbs and nouns need past tense and plural forms, but because they are not canonical roots, they cannot tap into the lexicon of roots and linked irregular forms that encourage irregular analogies. Onomatopoeic forms therefore are regular, even when their sound would otherwise tempt people to borrow an irregular pattern, *spling–splang–splung* style:

> The engine *pinged* [not *pang* or *pung*].
> My grant got *dinged* [not *dang* or *dung*].
> That presentation really *zinged* [not *zang* or *zung*].
> The canary *peeped* [not *pept*].
> Her computer *beeped* [not *bept*].

A second kind of sound that lacks a canonical root is a quotation. A quotation does not have to use canonical words; it reflects a stretch of sound that someone else has said, as in *Elmer shouted, 'Dwat!'*. They may happen to be real words, of course, as in *I hate how he begins every sentence with 'actually,'* but that is just a coincidence; you can quote any sound anyone else has made. As with onomatopoeia, quotations are not perceived as roots and fail to link to roots and their associations in memory. Stored irregular plurals are not tapped, and the regular applies, as in *While checking for sexist writing, I found three 'man's on page 1* (not *three 'men'*). In Jane Austen's *Mansfield Park*, a character says of his drama-loving father, 'How many a time have we mourned over the dead body of Julius Caesar, and *to be'd* and *not to be'd*, in this very room, for his amusement?'

A third way a word can be rootless is to be based on a name. In modern English, names are meaningless noises. Many happen to sound like roots, like *Shepherd* and *Green*, because surnames originally were based on a person's residence, occupation, father, or distinguishing features. But these names have long since lost their meanings; no one expects Professor Shepherd to be a shepherd or Mrs Green to be green. Other names may have nothing to do with English words, such as *Dweezil Zappa, Carl Yastrzemski,* or *Seamus McGillicuddy.* As with onomatopoeia and quotations, names are mentally registered as stretches of sound, not canonical roots, and hence do not hook up with roots with the same or similar sounds in memory – and that forces people to regularize them:

> We're having Julia Child and her husband over for dinner. You
> 　　know, the *Childs* are really great cooks [not *the Children*].
> Why hasn't the German literary world seen any more Thomas
> 　　*Manns* [not *Menn*]?
> All the producers are looking for likeable historians, but there
> 　　aren't many Shelby *Footes* out there [not *Feete*].

A fourth class of rootless words are foreign borrowings such as *latke* and *cappuccino,* which are patently not canonical English words but sounds taken from other languages. These words, despite their alienation from the lexicon of English roots, happily receive regular inflection, as in *latkes* and *cappuccinos.* They do so even when irregular patterns beckon. Despite the widespread irregular pattern in *thief–thieves, leaf–leaves, shelf–shelves,* and *life–lives,* nouns originally borrowed from French or German have regular plurals instead. From French we have *beefs, chiefs,* and *gulfs,* not *beeves, chieves,* or *gulves*; from German we have *fifes,* not the *fives and dra* (fifes and drums) of Richard Lederer's irregular-loving Farmer Pluribus whom we met in chapter 3.[13] If your pet mongoose gives birth, you have *mongooses,* not *mongeese,* because the word comes from *mangus* in Marathi, a language of south India. More than one *talisman* is a bunch of *talismans,* not *talismen,* because the source is the Arabic *tilasm.* Almost all of the thousands of French and Latin verbs that were loaned to English since 1066 are regular: *derided,* not *derode*; *succumbed,* not *succame.*

Modern English speakers, of course, do not have a collective memory of the cadences of an ancient Saxon fatherland. There must be a source in a speaker's own experience for the inkling that a word is not of native stock. Multilingual or cosmopolitan speakers may literally recognize a foreign word, which is probably how *chiefs, succumbed,* and *mongooses* began their English lives with regular plurals in centuries past. And if the first speakers of a new word use a regular plural, other speakers generally follow suit, because people pay attention to any irregular-sounding word that appears in a regular form (as we saw on pages 152–3). But even monolinguals can recognize loan words when they violate the canonical English sound pattern. Recent immigrant words like *cappuccino* immediately give themselves away, and even long-established French and Latin words have a distinctive sound: They tend to be bisyllabic with stress on the second syllable, such as *deride.* Even when speakers are unaware that a word at some earlier point was a foreign borrowing, they may sense that the words sound fancy or stuffy and not good old everyday English. (Several experiments have found that people prefer native-sounding words to Latinate-sounding words in a variety of everyday English constructions, even if they cannot put their finger on the difference.[14]) Conversely, when speakers have no sense at all that a word has been borrowed they treat it as a standard root, and in those cases they feel free to make it irregular if it resembles other irregular roots. *Quit* and *cost*, both imported from French but assimilated as standard English monosyllables, are examples. Their past-tense forms are irregular no-changers, analogous to *hit* and *cast.*

A fifth class consists of words that are recognized as rootless because they were concocted by artificial means. *To synch* is a truncation of *to synchronize*, and its past tense is generally *synched* (as in *lip-synched*), not *sanch* or *sunch*. The system manager of a computer installation is sometimes called a *sysman*, plural *sysmans*. Acronyms are other examples of ham-fisted wordsmithing, and they easily undergo regular suffixation, as in *PCs, TVs, SOBs*, and the much maligned *RBIs*. Even when an acronym matches an irregular sound, the irregular form is unavailable and regular suffixation applies. If a container with a mixture of oxygen and xenon were labeled with the acronym OX, it is doubtful that several tanksful would be called *OXen*. The example is a bit contrived, but in other languages they are plentiful, as we shall see in chapter 8.

The sixth and final example consists of words that lack their own roots because they are converted from a root of a different part-of-speech category. English is notorious for converting roots from other categories into verbs. Columnist John Powers wondered at the beginning of this chapter who made *impact* a verb, and Calvin tells Hobbes:

CALVIN AND HOBBES © 1993 Watterson.

Far from weirding the language, verbed nouns are punctiliously lawful. Verbs that are recognized as thinly disguised nouns or adjectives don't accept irregular forms, even when they sound like an irregular verb:

Boom-Boom Geoffrion got *high-sticked!* [not *high-stuck*]	hit with a high *stick*
Powell *ringed* the city with artillery.	formed a *ring* around
I *steeled* myself for a visit with my dentist, Dr de Sade.	made like *steel*
Harvey *bared* his soul on Oprah.	laid *bare*
Mongo *spitted* the pig.	put on a *spit*
Vernon *braked* for the moose.	applied the *brakes*
Mae Jemison *out-Sally-Rided* Sally Ride.	outdid *Sally Ride*
Swans are *dark-meated* fowl.	having dark *meat*
Poor Bowser had to be *de-flea'd*.	had *fleas* removed
Babs quickly *righted* the canoe.	set *right*
We *sleighed* over the river and through the wood.	went by *sleigh*
Martina *two-setted* Chris.	beat in two *sets*
After you've *meaned* both columns, you can do the t-test.	computed the *mean* (average)[15]

Mom was flying home. In a box. To be *waked* and buried.	given a *wake*[16]
Most snow or sugar snap peas need to be *stringed*.	have the *string* removed[17]

The explanation is that a noun root like *stick* cannot have an irregular past tense associated with it because the concept of past tense makes no sense for a noun and hence cannot be listed with it. (What could it possibly mean for *hockey stick* to have a 'past tense'?) The irregular past-tense form *stuck* that is stored with the verb root *stick* is not treated as relevant, because *to high-stick* doesn't have that root. The regular rule is not restricted to verb roots or to anything else, but applies by default. The rule therefore is fully available – indeed is the only way – to inflect verbs without verb roots. The same thing happens with nouns (as we shall see in chapter 8 [pages 248–9]).

We almost have an explanation for *flied out*. Baseball fans recognize that the verb *to fly out* is based on the noun *a fly*, a.k.a. a can of corn, namely a high, arcing ball. *To fly out* means to make an out by hitting a fly that gets caught. One might conclude that *flew out* is avoided because it is based on a noun, just as *high-stuck* and *de-fled* are avoided because *they* are based on nouns. But we have a problem. The noun *a fly* was itself converted from the everyday verb *to fly*, meaning 'to slip the surly bonds of earth and dance the skies on laughter-silvered wings.'[18] Since the baseball *fly* is a double convert – a verb from a noun from a verb – it does have a verb root, and that root comes with *flew* and *flown*. Something is missing from the explanation. The missing piece comes not from the nature of words but from the nature of rules.

∼

The point of grammatical rules is to define new combinations in which the meaning of the whole can be computed from the meanings of the parts and the way they are arranged. Some rules (the rules of syntax) build sentences and phrases out of words; others (the rules of morphology) build complex words out of simple words and bits of words such as prefixes and suffixes. When a new word surfaces, like *weirding* or *long-billed thrasher*, grammatical rules allow the speaker to coin it and the listener to understand it.

Take the verb *overeat*. It is based on the verb root *eat*:

The root is then encrusted with a prefix, yielding the following mental structure, in which the V at the top stands for the whole word *overeat*:

How do we know how to use the new word *overeat*? Easy – we give it the properties of the rightmost thing inside it, *eat*. What part of speech is *overeat*? It is a verb, just as *eat* is a verb. What does *overeat* mean? It refers to a kind of eating – eating too much – just as *eat* refers to eating. And what is its past-tense form? *Overate*, not *overeated*, just as the past-tense form of *eat* is *ate*, not *eated*.

A new complex word inherits its traits (including any irregular forms) from the memory entry for the rightmost word inside it, the *head* of the word. The pipeline of information from the head at the bottom of the tree to a new, complex word formed from it, and then to an even bigger word formed from that one, can be depicted like this:

The percolation of information up from the head can explain the examples that disconfirm the semantic stretch theory (the theory that a change of meaning poisons an irregular). For example, the compound *workman* is formed by prefixing the noun *man* with the verb *work*:

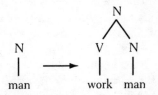

The traits of the whole word come from the traits of the rightmost word inside it, the head, in this case *man*. *Workman* is a noun, just as *man* is a noun. It refers to a kind of man, just as *man* refers to a man. And its plural is *workmen*, because the plural of *man* is *men*. A similar explanation works for other compounds and metaphors, such as *bogeymen, superwomen,* and *sawteeth,* and for prefixed verbs such as *understand–understood* and *become–became.* It also works for idioms like *took a nap* and *threw up,* as soon as we remember that these are phrases, not words, and that in English the head of a phrase is on the left, not the right (hence the great *mothers-in-law/mother-in-laws* debate in chapter 2).

All of which brings us to lowlifes, flying out, and Mickey Mouse. A few complex words are *headless*: They can't get their properties from their rightmost element if they are to work the way the speaker wants them to work. The information percolator must be turned off for the word to be interpreted and used properly. As a result, the pipeline that carries stored information from the word's root is clogged, and any irregular form stored with the root is imprisoned in memory, unable to percolate up to apply to the whole word. The regular rule, acting as the default, steps in to supply the word with a past-tense or plural form, undeterred by the fact that the sound of the word smacks of irregularity.

How does a word lose its head? One way is to be a compound that doesn't refer to the kind of thing indicated by its rightmost word. Instead it refers to something else, which merely *has* or *does something to* the kind of thing indicated by the rightmost word. Though a *workman* is a kind of man and a *bluebird* is a kind of bird, a *cutthroat* is not a kind of throat, nor is a *lazybones* a kind of bones. Linguists call these *bahuvrihi* compounds, from the Sanskrit expression 'having much rice.'[19] (The term comes from a school of Indian linguists working 2500 years ago who left us a remarkably sophisticated analysis of the grammar of Sanskrit.) Similarly, a *lowlife* is not a kind of life, but a kind of person: a person who *has* (or leads) a low life. For it to have that meaning, the percolation pipeline, which would ordinarily make *lowlife* mean a kind of life, must be plugged up. With the data pipeline to memory disabled, there is no way for the other information stored with *life* to be passed upward either, such as the linked plural form *lives*. With the irregular plural unavailable, regular -*s* gets the call, and we have *lowlifes*.

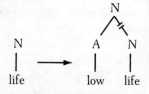

Headlessness explains at least four kinds of plurals and past-tense forms that have puzzled word-watchers for decades. Bernstein complains that 'a few plurals seem almost unreasonable: *talismans, mongooses, still lifes*.'[20] We have already demystified the first two, and now we see that a *still life* is not a kind of life but a kind of painting. To have that meaning it must be percolation-proof, sealing off *lives* in the mental lexicon and defaulting to *lifes*. Similarly, a *flatfoot* is not a kind of foot but a neighborhood policeman (his feet flattened from so much walking), and several of them are called *flatfoots*, not *flatfeet*. An inexperienced woodsman or cub scout is called a *tenderfoot*, and dictionaries give *tenderfoots* as one of its plurals.

If I were Professor Kugelmass in the Woody Allen story about the machine that could project people into the novel of their choice, I would ask to be projected into Richard Russo's *Nobody's Fool* so I could resolve a controversy at the local high school:

> . . . A controversy had erupted on the editorial page of the *North Bath Weekly Journal* over whether the plural of Sabertooth

should be Sabertooths or Saberteeth. When the cheerleaders led
the spell cheer, how should it go? The principal said Saberteeth
sounded elitist and silly and dental. The chair of the high
school's English department disagreed, claiming this latest
outrage was yet another symptom of the erosion of the English
language, and he threatened to resign if he and his staff were
expected to sanction tooths as the plural of tooth. Why not? the
public librarian had asked in the next letter to the editor. Wasn't
this, after all, the same English department that had sanctioned
'antelopes' as the plural of 'antelope'? The letters continued to
pour in for weeks. Beryl Peoples, who'd nursed a twenty-year
grudge against the principal for caving in and allowing history
courses in the junior and senior high school to be redesignated
'social studies,' had the last editorial word, reminding her fellow
citizens that the sabertooth tiger was an extinct animal. Food, she
suggested, for thought.[21]

Fortunately, they worked it out without my help: The new banner
read 'GO SABERTOOTHS! TROUNCE SCHUYLER SPRINGS!' It was
the right decision, because *sabertooth* is a bahuvrihi: It refers not
to a kind of tooth but to a kind of cat.

I would also defend Bilbo Baggins in J. R. R. Tolkien's *The
Fellowship of the Ring*:

> 'My dear Bagginses and Boffins,' he began again; 'and my dear
> Tooks and Brandybucks, and Grubbs, and Chubbs, and
> Burrowses, and Hornblowers, and Bolgers, Gracegirdles, Good-
> bodies, Brockhouses and Proudfoots.' 'ProudFEET!' shouted an
> elderly hobbit from the back of the pavilion. His name, of
> course, was Proudfoot, and well merited; his feet were large,
> exceptionally furry, and both were on the table.
>   'Proudfoots,' repeated Bilbo.[22]

And I wish I could have put my two cents into this conversation
between a know-it-all narrator and a character called Buffalo Gal
in a short story by Alison Baker:

> 'They come and they go,' Buffalo Gal said. 'They might as well be
> bigfoots.'

'Bigfeet,' I said.
'Whatever,' Buffalo Gal said.[23]

Here is another mystery solved by the word structure theory. A
1989 article in *Newsweek* began:

> It's been ten years since the Sony Walkman was born. Fifty
> million of the machines have been sold. Yet nobody knows the
> correct plural form of Walkman. Is it Walkmans? Is it Walkmen?
> We can only guess. Nonsexists might suggest a new name:
> Walkperson. But then, would the plural be Walkpersons or
> Walkpeople? Sony Corp. avoids the issue entirely by using
> Walkman only as an adjective. In the interest of consistent
> usage – and trademark protection – Sony talks about 'Walkman®
> personal stereos.' Everyone else, when talking about personal
> stereos – whether Panasonic, Toshiba or Aiwa – calls them
> 'Walkmans.' Or 'Walkmen.' Never 'personal stereos.'[24]

Some people, at least, are completely confident that the plural
should be regular. The owner of a San Francisco store had it bent
into the tubing of a neon sign:

A walkman, of course, is not a kind of man, so many of us, like
the sign maker, interpret it as headless and hence bereft of the
irregular plural *men*. It's not exactly a bahuvrihi either, because
while a *lowlife* has a low life and a *flatfoot* has flat feet, in no sense
does a *walkman* have a man. The closest gloss might be 'that

which allows a man to walk (while listening to music).' Parsing it may be futile, however, because Japanese companies often use meaningless English names and slogans just for the cachet, such as 'Supreme Liberal,' 'For vibratory refreshment,' and 'Love the earth with honest poverty.'

Headlessness also explains a second class of regularizations, eponyms. An eponym is a word that comes from a name, such as *atlas, boycott, bowdlerize, cinderella, maverick, quixotic, sandwich, scrooge, shylock, tantalize,* and, according to legend, *crap*, after Thomas Crapper, a nineteenth-century British inventor who improved the flush toilet. (Crapper really existed, but the noun, originally 'chaff,' dates back to Middle English.)[25]

A *Mickey Mouse*, in the sense of a simpleton, is an eponym. It began with the ordinary noun *mouse*. Walt Disney made it a name when he christened his diminutive hero *Mickey Mouse*. Names are somewhat like nouns, but in English they are not the same thing (you don't ordinarily say *She's talking to the Mildred* or *I left work because sick Jason came home early*). Then in colloquial speech the name was converted back into a common noun, *a Mickey Mouse*:

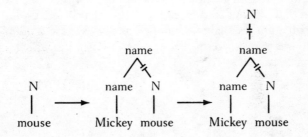

The new noun is headless, because the percolation pipeline had to be blocked twice: once to convert the noun *mouse* into a name, and then to convert the name back into a noun. (It also had to be blocked to get the meaning to come out right: A Mickey Mouse is not a kind of mouse in the sense that a workman is a kind of man.) With percolation turned off, *mice* is trapped in the lexicon, and the plural rule gives us *Mickey Mouses*. Note that *Mickey Mouses*, a double convert, gets a slightly different explanation from *the Childs*, a single convert. With Mickey, but not with Julia, the surname has a visible connection with the ordinary noun, so the problem is not the lack of a root; it is the inaccessibility of a root.

Nouns can be based on other kinds of names, such as works of art, products, or teams. We might say that Michael Keaton starred in the first two *Batmans*, not the first two *Batmen*. I can imagine someone arguing that Roy Orbison's original recording is the best of all the *Pretty Womans*, and Bobby Darin's is the best of all the *Mack the Knifes*. I have seen or heard mentions of *Spectrums* and *Quantums* (bicycles), *Elfs* (cars), *John Deeres* (tractors), *Top Shelfs* (frozen dinners), *Sea Wolfs* (navy Aircraft), *Supermans* (comic books), and *Maple Leafs* (gold coins).[26] In *Popular Photography* a journalist wrote of a new camera, 'As Canon squeezes out more production, ELPHs keep selling out. (What is the plural of ELPH, anyway? ELPHS? ELVES? ELVIS?)'[27] As for sports teams, we have the *Maple Leafs* high-sticking in Toronto and the *Marlins* (not *Marlin*) hoisting cans of corn in Florida. At this point killjoys will bring up the *Timberwolves*, who shoot hoops in Minnesota; I will get to them, and to other apparent counter-examples, in the next section.

At long last, the complete explanation for why no mere mortal has ever flown out to center field. Recall that in the evolution of baseball argot the plain verb *to fly* was converted to a noun, *a fly*, which was then converted back to the verb *to fly*, meaning 'to hit a fly that is caught':

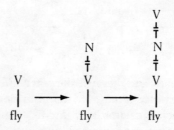

The new verb (top V) is sealed off from the root verb (bottom V) at two layers, the one that converted the verb to a noun, and the one that converted the noun back into a verb. Percolation had to be blocked both times to allow the verb to change categories rather than blindly receiving the category from one layer down.

Baseball cognoscenti can hear the *fly ball* in *flying out*, so for them the forms *flew* and *flown* are unable to climb out of the lexical entry for *fly*. The word turns to *-ed* as the last resort and becomes *flied out*.

Here are some other verbs-from-nouns-from-verbs that shook off the irregularity of their roots:

> Once again, Perot *grandstanded* to the audience.
>   (to stand → a grandstand → to play to the grandstand)
> Vera *costed* out the grant.
>   (to cost → the cost → to ascertain the cost)
> Doctor Crunch *encasted* my leg.
>   (to cast → a cast → to put in a cast)
> She threw out all her *runned* nylons.
>   (to run → a run → having a run)
> A doctor who slided a sample.[28]
>   (to slide → a slide → to place on a slide)
> Many projects could be *offshooted* from television in the classroom.[29]
>   (to shoot → an offshoot → to make an offshoot)
> In each of the past two seasons, Cleveland State guard William Stanley has sported a self-styled, one-of-a-kind hairdo. In 1987–88 it was a half-foot-high flattop. Last season he went to a bilevel box cut. This season, as a senior, Stanley has *outdo'ed* himself.[30]
>   (to do hair → a hairdo → a 'do → to have a more impressive 'do than)
> The preterite of *to joyride* is not *joyrode*, nor even *joyridden*, but *joyrided*.[31]
>   (to ride → a joyride → to take a joyride)

Headlessness explains a fourth curiosity, this one a quirk of spelling. English spelling is a rule system that connects the sounds of words with their written forms. As with grammar, spelling is rife with irregularities and complications, especially in frequent words such as *eye, of, have,* and *would.* One complication is that when a noun ending in *y* gets a suffix, the *y* becomes *ie*, as in the plurals *army–armies, body–bodies,* and *cherry–cherries,* and the derived nouns in *happy–happiness, pretty–prettiness,* and *ugly–ugliness.* Yet the regular spelling rule, in which an $\bar{e}$ sound at the end of a stem is spelled *y*, sometimes reasserts itself. Here is an example from an article on fashions in women's names:

> Bettys abound, past and present – from Crocker to Boop, from Grable to Friedan to Ford. . . . If you're looking for a Betty under 40, though, good luck. . . . Bettys are so endangered that they've formed a club – lots of clubs, actually. In fact, the Bettys of Nebraska just held a convention simply to rejoice in their Bettyness.[32]

Betty, of course, is not just any old noun but a noun that comes from a name. And pluralized names shed their irregular spellings, just as pluralized names shed their irregular plurals: *the Kennedys, the Fogartys, the Kansas Citys, the Germanys, the Emmys, the Tonys; I'll have two Bloody Marys.* So do nouns converted from other grammatical categories: A sign outside an apartment-motel advertized *Dailys Weeklys Monthlys Yearlys*, all nouns derived from adverbs.[33] When a noun is based on a nonnoun, people seal off their associations to irregularities, even in spelling.

The spelling effect doesn't always work. I have seen *Dollies* (the eponymous cloned sheep), *dailies* (newspapers), *onlies* (only children), and *goody-goodies*, not to mention *the Alleghenies* and *the Rockies*. Nevertheless it is striking that spelling, which most of the time is a clumsy afterthought pasted on to the standard equipment of language, should often showcase a deep principle of grammatical organization.

~

An irony of systematic regularization is that ordinary speakers apply abstract grammatical principles instinctively, while many style manual authors and language mavens are oblivious to them and hector people into sticking with the irregular. Here is Theodore Bernstein contrasting *flied* (which he explains with the semantic stretch theory) with *broadcasted*:

> If you think you have correctly forecasted the immediate future of English and have casted your lot with the permissivists, you may be receptive to *broadcasted*, at least in radio usage, as are

some dictionaries. The rest of us, however, will decide that no matter how desirable it may be to convert all irregular verbs into regular ones, this cannot be done by ukase, nor can it be accomplished overnight. We shall continue to use *broadcast* as the past tense and participle, feeling that there is no reason for *broadcasted* other than one of analogy or consistency or logic, which the permissivists themselves so often scorn. Nor is this position inconsistent with our position on *flied*, the baseball term, which has a real reason for being. The fact – the inescapable fact – is that there are some irregular verbs.[34]

This had long been a losing battle. In *The American Language* Mencken notes that 'The effort of purists to establish *broadcast* as the preterite has had some success on higher levels, but very little on lower. "Ed Wynn *broadcasted* last night" is what one commonly hears.'[35] Henry Fowler, like Bernstein, was scornful, though his rationale is closer to the linguistic truth:

> If etymology is to be our guide, the question whether we are to say *forecast* or *forecasted* in the past tense and participle depends on whether we regard the verb or the noun as the original from which the other is formed. If the verb is original (= to guess beforehand) the past and p.p. [perfect participle] will be *cast* as it is in that verb uncompounded; if the verb is derived (= to make a forecast) they will be *forecasted*, the ordinary inflexion of a verb. The verb is in fact recorded 150 years earlier than the noun, and we may therefore thankfully rid ourselves of the ugly *forecasted*; it may be hoped that we should do so even if history were against us, but this time it is kind. The same is true of *broadcast*; and *broadcasted*, though dubiously recognized in the OED Supp., may be allowed to die.[36]

If only Fowler had couched his explanation in terms of speakers' intuitions about whether the verb was based on the noun or vice-versa, rather than on whether the verb or noun appeared in the language first (a bit of history lost on modern speakers), he would have been less offended. People may very well interpret the verb *to broadcast* as 'to do a broadcast,' rather than interpreting *a broadcast* as 'an act of broadcasting'; in contemporary English the noun is far more frequent than the verb and may feel more basic.[37]

Other indignant critics also forget to ask whether a noun may be lurking behind a supposedly slovenly use of a regularized verb. The *Boston Globe* ombudsman spinelessly agreed with a reader's complaint about one of his colleagues:

> A woman wrote: 'I join other readers in lamenting the lack of attention given to good writing, spelling, and grammar these days.' One article she sent left out a key comma and contained the phrase 'he may of been.' Another article read, 'Martyny subletted a Kenmore square apartment.' It's sublet.[38]

Not necessarily! In American English hardly anyone uses the verb *to let,* meaning 'to lease,' but many speakers use the noun *a sublet.* They might analyse *to sublet* as a verb from a noun (to arrange a sublet) rather than as the verb *to let* with the prefix *sub-*. And that would lead to a regular past-tense form. Similarly ashamed, *Scientific American* published this letter under the heading 'Nobody's Perfekt':

> Have you actually used 'inputted' as the past tense of a verb? Yes, in the caption of the figure on page 150 of the January issue. I am upset. I am appalled. I am horrified. I am out putted.[39]

The nouns *input* and *output* are about sixty times more common than the verbs *to input* and *to output*. If the caption writer interpreted *to input* as 'to produce input,' the regular past-tense form would follow as night follows day.

～

Before we congratulate the word structure theory for explaining so many kinks and vagaries, we need to worry about the examples it doesn't explain. I've already mentioned the *Timberwolves* and the factions that like *Bigfeet* and *Saberteeth,* and I confess to having heard *flew out* and *flown out* a number of times. A skeptic might wonder whether regularization is really that systematic after all. Perhaps people carelessly turn irregulars into regulars, just as they have been doing with simple verbs for hundreds of years (*chid* becoming *chided*, *holp* becoming *helped*, and so on). Perhaps I have cherry-picked a few examples that are consistent with a needlessly fancy theory.

How can we prove that these effects are alive in the minds of speakers? By bringing the speakers into the lab, presenting them with *new* headless or rootless verbs, and seeing whether they regularize them more than they regularize pure verb roots. John Kim, Alan Prince, Sandeep Prasada, and I gave people a questionnaire with three dozen irregular-sounding verbs.[40] Half the sentences used the verb root metaphorically:

> When guests come, if they arrive with slides my hopes for a lively evening quickly sink.
>
> When I saw Bob and Margaret carrying six boxes, my hopes <u>sinked</u> instantly.
>
> sounds bad |_|_|_|_|_|_|_| sounds good
>   1 2 3 4 5 6 7
>
> When I saw Bob and Margaret carrying six boxes, my hopes <u>sank</u> instantly.
>
> sounds bad |_|_|_|_|_|_|_| sounds good
>   1 2 3 4 5 6 7

These sentences served as a control group. They contained a verb in a metaphorical or extended sense, so the semantic stretch theory predicted that they would scare people away from the irregular forms. We were confident, though, that people would stick with the irregular forms for these items, because they have the same root as the standard irregular verbs. So we used the ratings of these items as a baseline measure of people's fondness for the various irregular past-tense forms.

The sentences we really were interested in contained verbs based on nouns:

> When guests come, I hide the dirty dishes by putting them in boxes or in the empty sink.
>
> Bob and Margaret were early so I quickly boxed the plates and <u>sinked</u> the glasses.

sounds bad |__|__|__|__|__|__|__| sounds good
          1   2   3   4   5   6   7

Bob and Margaret were early so I quickly boxed the plates and
<u>sank</u> the glasses.

sounds bad |__|__|__|__|__|__|__| sounds good
          1   2   3   4   5   6   7

If the word structure theory is true and people spontaneously
regularize headless and rootless words, these verbs-from-nouns
should be rated differently from the controls: People should flip
to regular *sinked*, or at least should dilute their preference for
irregular *sank*. We matched up the items in pairs, every stretched
verb having a verb-from-noun with the same sound and vice-
versa. Across the experiment the matched sentences were
presented equally often so that any fondness or distaste for an
irregular sound would cancel out. Every subject, though, saw a
particular verb such as *sink* only as a stretched verb or only as a
verb-from-noun, so they wouldn't make side-by-side comparisons and concoct their own theories of what we were after. We
also tested a few verbs-from-nouns that are already in the
language: *flied out, grandstanded, ringed the city*, and so on.

The results were gratifying. With every verb, the subjects liked
the regular past tense (such as *sinked*) better when presented as a
verb-from-noun (*sinked the dishes*) than when presented as a
stretched verb (*my hopes sinked*). For 90 percent of the verbs they
liked the regular form so much that they rated it higher than the
irregular form – including, to our relief, the item *flied out*. We
also tested the semantic stretch theory, asking a second group of
subjects to rate the degree of stretching in the meanings of the
verbs (for example, how much the verb *to sink* has been stretched
in *our hopes sank*). If the theory is correct, those ratings should
predict the preferences of the original group of subjects for
regular and irregular forms. Using a standard statistical procedure, we pitted semantic stretching against word structure to see
whether one, the other, or both were necessary to explain the
preferences for regularity or irregularity. Word structure was
necessary to explain the preferences; semantic stretching was
unnecessary.

We also wondered whether the effect might be a fussy affectation of pointy-headed, Volvo-driving, endive-nibbling, chablis-sipping young urban professionals. It seemed unlikely, given that most language mavens fail to grasp the principle, but still worth ruling out. So we gave the questionnaire to a new sample of people without a college education, recruited through an ad in the local tabloid paper. The results were the same.

~

Why then would Chris Berman of ESPN say that 'Jose Offerman flew out to center field in the ninth inning'?[41] Why would William Safire, of all people, refer to the president's spin doctors as 'the bigfeet of the Opinion Mafia'?[42] Why is there a Native American nation called *The Blackfeet,* and a species of goose called *pinkfeet,* when they should behave like *Proudfoots* and *flatfoots*? And what about those damn Timberwolves?

Now that we have the results of the experiment we needn't worry about the phenomenon itself; people do in fact regularize rootless and headless words. At worst, the exceptions force us to say that regularization is a real effect but only a statistical one. It tips the odds away from the massive tendency to avoid *flied* and *foots* and *wolfs* in ordinary speech, even if it doesn't flip the preference all the way to 100 percent use of the regular form all of the time.

But the word structure theory is in even better shape than that. Like its parent, the words-and-rules theory, it is about the psychology of flesh-and-blood speakers rather than some scholar informed by the best etymologies philology has to offer. People should regularize headless forms only when they *perceive* the words to be headless. They may not be conscious of a word's derivation, or be able to explain it to others, but they should have a sense that the word is based on another word (for example, that *to fly out* is based on *a fly*). When they don't – when they are oblivious to the noun in a verb-from-a-noun and imagine that it is just a stretched verb root – the theory predicts that they *should* stick with the irregular. First let's consider some ways in which people *might* misanalyse headless words; then I'll present evidence that whenever people do so, they stick with irregular forms.

Sometimes there is a way to stretch a verb root to refer directly to an action, skipping the middleman noun. In sportscasting it's common to personify the ball and describe it with the name of the player who propelled it and whose fate is tied to it. Baseketball commentators often say *Jordan got blocked* and *Larry is rejected*, where it is the ball that is blocked or rejected, not the man. If an announcer similarly personifies a flying baseball as its hitter, the hitter could be said to have *flown out*; the noun *fly* need never have entered his mind.

A bahuvrihi noun also can invite misanalysis, because people may think not of the possessor of the rightmost word but of the rightmost word itself. In the rhetorical device called synecdoche, a part symbolizes the whole, as in *The Celtics need fresh legs*; *She got a new set of wheels*; *I counted heads*; and *He is chasing skirts*. Perhaps *Blackfeet, pinkfeet,* and the occasional *bigfeet* and *tenderfeet* are being used synecdochically as standard compounds like *fresh legs* rather than literally as bahuvrihi compounds like *lowlife*.

In naming a sports team (or understanding its name), usually its members are metaphorically identified with some referent (a Lion, a Tiger, a Bear), and the name of the referent is then pluralized (The Lions, The Tigers, The Bears). Pluralized names, as we saw with the Childs, the Manns, and the Mickey Mouses, lose their irregularity. So when a Maple Leaf joins his teammates on the ice, they are the Maple Leafs.

But there are other ways to name a team. Instead of metaphorically naming a member and pluralizing the name, one can name the entire team at once. That's what gave us those tacky mass-noun teams: The Utah Jazz, Miami Heat, Orlando Magic, Colorado Avalanche, Tampa Bay Lightning, Dallas Burn, San Jose Clash, and Kansas City Wiz. Entire teams also have been given x-rated plural names such as the Boston Red Sox, Chicago White Sox, Everett Aquasox, and West Tenn Diamond Jaxx. No singular was ever pluralized; no player identifies himself as a 'Red Sock.'

Now, if entire teams can be identified with mass or plural referents, it's easy to imagine that the roundball team in the Land of Lakes is being identified with a *pack* of wolves – wolves do, after all, fast-break in packs. No one ever had to figure out how to pluralize a Timberwolf because the whole team was named after the plural *timberwolves* to start with. (Once the team is named, it's easy to work backward to the name of a single player, a *Timberwolf.*) The situation is different north of the border, where Torontonians began with the symbol of Canada, the proudly singular Maple Leaf.

So there are many ways in which people *may* at times shortcut a derivation. But without independent evidence that people have taken the shortcuts exactly when they fail to regularize, the explanations would merely be escape hatches for the word structure theory, making it unfalsifiable. So Kim, Prince, Prasada, and I went back to the lab to break the logical circle. We wanted to measure people's tendency to shortcut a derivation, or to encourage them to do so, and then see if that increased the appeal of irregular forms.[43]

First we dusted off our materials from the earlier experiment and asked a new group of subjects to rate, on a scale of 1 to 7, how similar in meaning the noun was to the similar-sounding root verb. We asked our subjects, 'How similar is the *fly* in *fly ball* to the *fly* in *birds fly south*? How similar is the *sink* in *kitchen sink* to the *sink* in *sink the ship*?' and so on. Their responses gave us a measure of how vulnerable each verb-from-noun was to being misanalysed: With nouns that feel similar to the soundalike verbs, people might be more tempted to misanalyse the verb-from-noun as a simple variant of the verb. (For example, if a fly ball was perceived as being similar to what birds do, people might misanalyse *fly out* as a symbolic version of *to fly* rather than as a conversion from *a fly.*) That in turn should lead to seemingly embarrassing irregular forms such as *flew* and *flown*. Note the contrast with the semantic stretch theory: We are predicting that a metaphoric interpretation should make the irregular *more* appealing, not less so.

With these ratings in hand we went back to the data from the original experiment. As predicted, the verbs-from-nouns that were rated as most shortcuttable were the ones for which the earlier subjects had been least attracted to a regular form such as *flied out*.

In another experiment we gave people irregular-sounding verbs and nouns that were clearly related to one another, and tried to manipulate whether the verb was perceived to be based on the noun or the noun was perceived to be based on the verb. Compare these two *kleeds*:

> Mary got a brand new <u>kleed</u> for her birthday.
> She liked it so much, she <u>kleeded/kled</u> for a week.

> It has been a long time since I have had a nice, long <u>kleed.</u>
> I <u>kleeded/kled</u> quite often in the old days.

In both cases people see a noun and then a related verb. In one, the noun refers to a physical object and the verb is based on it: *to kleed* means to use a kleed. In the other, the verb is an action and the noun is based on it; *a kleed* is an interlude of kleeding. The stimuli were identical; only the readers' analysis of the verb – as a verb-from-a-noun or as a root verb – varied. As we predicted, that difference affected their ratings of the past-tense forms. With a verb-from-a-noun (use a kleed), regular *kleeded* was as acceptable as irregular *kled*; with a simple verb that just happened to be accompanied by a noun-from-a-verb (have a long kleed), the regular form was much less acceptable than the irregular.[44]

Both experiments show that the exceptions to the regularization effect are exceptions that prove the rule: When people don't perceive a word as headless, they don't plug the pipeline that sends irregular forms up from memory either. That secures a major kind of evidence for the word structure theory and for its parent, the words-and-rules theory. Irregular forms are word roots stored in memory; regular forms are computed when memory fails to cough up a form, for any reason. That in turn shows that a rule is a mental operation that manipulates variables, such as 'verb' or 'noun,' rather than an association to concrete memories of particular words and their sound patterns. It also shows that people erect an abstract mental scaffolding around words. The memory blockage in the examples in this chapter come from the nature of the mental scaffolding: People sense whether a word is stored as a root, and whether a word has a structure that allows information about the root to percolate up from memory.

~

When I lecture on regularity and irregularity in language, the question I am asked most often is, 'What's the deal with the plural of computer *mouse*?' Here, as a public service, is my best guess.

The use of *mouse* for 'pointing device' goes back to 1965.[45] The first mouse had its wire coming out the front toward the user, and it reminded the inventor, the computer scientist Douglas Engelbardt, of *Mus musculus*. Decades later the wireless pointing device was introduced; it is sometimes called a *hamster*.

Many people are squeamish about referring to more than one of them as *mice*. In 1992 I surveyed several dozen mail-order ads and found that many used plural headings for every category of hardware but the mouse, like this: Desktops–Notebooks–Monitors–Printers–Keyboards–Mouse. A few others played it safe by advertising Pointing Devices or Input Devices. More than half did use *mice*; none used *mouses*. *Mice*phobia is beginning to abate, and today *mice* is common in stores, magazines, and web pages (for example, my local CompUSA has an aisle labeled 'KEYBOARDS/MICE'). Many people, though, still wince at *mice*, though not to the point of using *mouses*, which remains rare among native speakers.

I would love to chalk up *mice* avoidance as another case cracked by the word structure theory, but in this case it is of no help. The root of *mouse* the pointing device is indubitably *mouse* the rodent, and the word is based on a transparent metaphor that should allow the irregular plural to bubble up unscathed. Thankfully, the facts don't call for such heavy-duty machinery in any case. People are a bit skittish about *mice*; they don't shun it entirely or switch to *mouses*. We need a different kind of explanation.

The explanation comes in two parts. One is familiar: Irregular forms are stored in memory, regular forms don't have to be. The other concerns the nature of the concept 'plural,' which is not as straightforward as one might think. Plural means 'more than one of,' but there are lots of ways in which objects can come in multitudes.

We can behold a small number of objects, each apprehended as an individual:

Several of them can be parts of a larger object:

Each can be a part of a larger object, several of which are under consideration:

The objects can congregate in an amorphous swarm or mass:

Or they can be distributed diffusely and randomly in the surrounding environment:

Many languages don't even *have* a unified plural marker that treats all the more-than-ones the same; they use various constructions for pairs, swarms, herds, families, and so on.[46]

Suppose the regular plural suffix -*s* simply means 'more than one of,' so that *hands* = 'more than one' + 'hand' and *rats* = 'more than one' + 'rat.' But suppose that no single concept of plurality is shared by all the *irregular* plurals. They have to be stored separately in memory anyway because of their idiosyncratic sounds, and that means each can have its own meaning slot in which a unique, concrete representation of more-than-one-*of-that-kind-of-thing* can be entered. It could even be a mental image of a typical multitude of that kind of thing: a committee of men or women, a flock of geese, a pair of feet, a set of teeth, a brood of children, a team of oxen, an infestation of lice.

Consider now what happens when you are called on to refer to more than one pointing device. Pointing devices come one to a computer, and several of them would imply several computers. But several little rodents tend to scurry, unattached, throughout the house or in meadows and woods. The metaphorical aptness of '*mouse*: the single rodent' for '*mouse*: the single pointing device' evaporates when we now have to think of '*mice*: the scattered vermin' as a metaphor for '*mice*: the accessory attached to each of several computers.' And that, I submit, makes people uneasy about calling the pointing devices *mice*.

Evidence? First, the same thing happens with other nouns. Remember that irregular plurals happily lend themselves to metaphors such as *sawteeth, God's children, chessmen,* and *oil mice.* But that happens only when the kind of plurality of the original word (small set, swarm, pair, attachment, and so on) matches the kind of plurality of the metaphor. When it doesn't, we get the same feeling of queasiness that surrounds computer *mice*. For example, *foot* is often used as a metaphor for a nether extremity:

> There was a cottage at the foot of every mountain.
> An ambassador was seated at the foot of each table.
> The page number is printed at the foot of each page.

But feet come two to a body, and when a metaphorical foot comes one to an object, the plurals are tainted:

There were cottages at the feet of the mountains.
Ambassadors were seated at the feet of the tables.
Page numbers are printed at the feet of the pages.

Note that as with the pointing device, the odor surrounding the irregular is not bad enough to drive people to the regular *foots*, though it sometimes forces them to an awkward singular. According to the *Encyclopedia Britannica*, the painter Paul Klee described his boyhood education as 'mountains immeasurably high but with no foot.' ('Mountains with no feet,' though more accurate, does not sound right.) A similar mismatch explains the slight strangeness of *Parish and McHale had excellent first halves*, where each player had an excellent first half. A headline in the *New York Times* about tastes in classical music read, 'Classical Radio Plays Only to Sweet Tooths.' Presumably it is because every listener has a single sweet tooth, and *sweet teeth* would connote a mouthful of them. The plural-mismatch effect may even have contributed to the *Toronto Maple Leafs*, a collection of individuals quite unlike a mass of foliage.

A disfavored person is sometimes compared to an ignominious animal:

Silly goose!
Clumsy ox!
Filthy louse!

When the animals congregate in flocks, teams, or infestations, though, it's strange to refer to several such people with the irregular plurals:

Silly geese!
Clumsy oxen!
Filthy lice!

Hence the lyrics from *Gentlemen Prefer Blondes*:

He's your guy when stocks are high
But beware when they start to descend.
'Cause that's when those louses

Go back to their spouses.
Diamonds are a girl's best friend.[47]

The choice of plurals therefore depends on how the mind construes multitudes. That part of our cognitive machinery not only affects how we use the language today; it can shape a language over centuries.

Often it's unclear whether a multitude is best perceived as one big thing or many little things, as we see in expressions like 'He can't see the forest for the trees' and 'The whole is more than the sum of the parts.' Whenever a collection of individual things is reconceptualized as a single gestalt, the plural for that collection can cease to feel like a plural, and the language can change.

In his song *One Hippopotami*, the comedian Alan Sherman sang, 'The plural of "half" is "whole"; the plural of "two minks" is "one mink stole." ' It is an astute observation. The linguist Peter Tiersma has found that whenever a set of objects can easily be construed as a single assemblage, a regular plural is in danger of congealing into a mass noun or an irregular plural.[48] This is happening today to the noun *data*, which often refers to large quantities of information and which is easily conceived of as stuff rather than things; the word is turning from a plural (*many data*) to a mass noun (*much data*). The effect is widespread. In language after language things that come in groups, such as children, gregarious animals, and paired or clustered body parts, end up unmarked, irregular, or transformed into a singular, sometimes to get pluralized all over again by a subsequent generation of speakers. Nonstandard dialects are filled with double plurals such as *oxens, dices, lices,* and *feets,* and that is how we got the strangest plural in Standard English, *children.* Once it was *childer,* with the old plural suffix -*er* also seen in the German equivalent *Kinder.* But people stopped hearing it as a plural, and when they had to refer to more than one child, they added a second plural marker, -*en.* Today many rural and foreign speakers still don't think of *children* as plural, and have added a third suffix, yielding the triply plural *childrens.*

~

A regular rule is a powerful instrument, creating inflected forms for a motley collection of rare, strange, and eccentric words. Is there any place it *cannot* work? Indeed there is, and it is my final demonstration of the difference in kind between regular and irregular inflection.

Regular plurals don't like appearing inside compounds. We speak of *anteaters, bird-watchers, Beatle records, Yankee fans, two-pound bags, three-week vacations,* and *all-season tires,* even though it's ants that are eaten, birds that are watched, all four Beatles that played on *Sgt. Pepper's* and the white album, and so on.[49] The discomfort is not shared by irregular plurals, though, as we see when we lay compounds with regulars and irregulars side by side. An apartment infested with mice is *mice-infested,* but an apartment infested with rats is not *rats-infested*; it is *rat-infested* even though by definition a single rat is not an infestation. Mice and rats are similar creatures, so the effect is not caused by a difference in meaning; it is caused by sheer irregularity. We also have *teethmarks* but not *clawsmarks, men-bashing* but not *guys-bashing,* and a song about a *purple-people-eater,* but never a song about a *purple-babies-eater.*[50]

Here are some real-life examples in which *mice* rush in where *rats* (or trackballs) fear to tread:

Mice Bait (a sign outside a general store)
Mice Cube (a better mousetrap)
mice-drivers (Microsoft software)
I felt mice-feet of apprehension scurrying over my skin. (From *The Edible Woman*, by Margaret Atwood)
Bad maps, mice-infested lodgings, and strict rules. (Description of the Appalachian Trail in the *New York Times Book Review*)
Frozen Mice Sperm (headline)[51]
Cells Implanted in Mice Brains; Hope Is Voiced for Mental Ills (headline)[52]
Mobile Phone Radiation Mice Tumor Link Much Stronger Than Expected (headline)[53]
Mice Accessories (a sign in a computer store)

Not far from where I work a flock of geese has taken up residence, and the city thoughtfully put up a sign declaring that part of

Memorial Drive a *GEESE CROSSING*. Had it been a flock of ducks, I doubt the sign would have announced a *DUCKS CROSSING*. A store in Florida selling exotic leather fashion accessories had a sign for 'Chicken Feet' wallets (irregular and plural) and a sign next to it for 'Turkey Leg' wallets (regular and singular). A periodical called *Rural Heritage* describes itself as 'a bimonthly journal for small farmers and loggers who use draft horse, mule, and oxen power.'

More examples can be found in the two-headed compounds that Sanskrit grammarians called *dvandva* (two and two), in which two nouns apply equally to some chimerical or twice-described person.[54] Examples include *man-child, manfish, man Friday, manservant, man-woman, woman-doctor, girlfriend, boy-friend, boy-king, player-coach*, and *singer-songwriter*. Dvandva compounds can be doubly pluralized, but only when the first noun is irregular: *men-children, menfish, menservants, gentlemen-farmers, women writers*, and *women-doctors*, but not *boys-kings, girlsfriends*, or *players-coaches*.

Could the effect have a boring explanation, such as that it sounds funny to have an -*s* suffix sandwiched inside a compound? There is a kind of word that nicely rules out that possibility. Richard Lederer asks, 'Doesn't it seem just a little loopy that we can make amends but never just one amend; that no matter how carefully we comb through the annals of history, we can never discover just one annal; that we can never pull a shenanigan, be in a doldrum, or get a jitter, a willy, a delirium tremens, a jimjam, or a heebie-jeebie?'[55] Lederer is alluding to *pluralia tantum*: Nouns that are always plural. Because they are not the result of pluralizing a singular, the complete plural form, -*s* and all, has to be stored in memory. Pluralia tantum in a sense are irregular regulars, and indeed they are happy to appear inside compounds: *almsgiver* (not *almgiver*), *arms race* (not *arm race*), *blues rocker* (not *blue rocker*), *clothesbrush, Humanities department, jeans maker, newsmaker, oddsmaker, painstaking*.[56]

(Incidentally, do not be distracted by the inconsistent way compounds are spelled in English: sometimes as one word, as in *teethmarks*; sometimes with a hyphen, as in *mice-infested*; sometimes as two words, as in *geese crossing*. The way to recognize a compound is by its composition, such as being two nouns in a row, and by its stress pattern: Compounds usually have their main stress on the first syllable, whereas phrases have their main stress on the second. Compare *bláckboard* with *blàck bóard*, *dárkroom* with *dàrk roóm*.)

The linguist Paul Kiparsky explained this effect with an influential theory.[57] Words are built in several stages, like a product on an assembly line. First there is a lexicon of memorized roots, including, as we would expect, irregular forms. (Kiparsky actually proposed that this box had rules generating irregulars, as in Chomsky and Halle's rules-all-the-way-down theory discussed in chapter 4, but his explanation works the same way if we assume irregular forms are stored whole.) The lexicon provides the input to regular derivational morphology, the rules that create complex words out of simple words and morphemes like *learn + -able*, *dance + -er*, and *black + top*. The output of this box, a complex word or stem, is then inputted to a third box, regular inflection, which modifies the word for its syntactic role in the sentence: past or present, singular or plural. The schematic for morphology would look something like this:

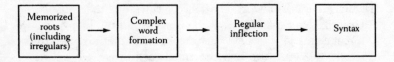

The word *mice*, stored as a root in the first box, is available as an input to the compounding rule in the second box, where it is joined to *infested* to yield *mice-infested*. *Rats*, however, is *not* stored as a memorized root in the first box; it is formed from *rat* by a regular inflectional rule in the third box, too late to feed the compounding rule in the second box. Hence we get *rat-infested* but not *rats-infested*.

Kiparsky's explanation is easy to understand and highlights a qualitative difference between irregular and regular forms: Irregulars are roots and can be the *input* to the process of word formation; regulars are the products of rules and have to be the *output* of the process of word formation. But we should not take the theory too literally and believe that words are always formed in conveyor-belt style or that anything stored in memory can be inserted into a compound. Remember from chapter 5 that some regular plurals are stored in memory, even though they don't have to be. Being stored, however, can't be an admission ticket to a compound: *cats* presumably is common enough to have a presence in memory, but *cats-infested* sounds bad nonetheless. Instead we should interpret Kiparsky's model as laying out the *logic* of word formation: what kinds of words may snap together in which ways to form bigger words. The restriction is actually that the *kind* of word that must be stored in memory – a root – is the input to complex word formation. Irregular *mice* is a root, just like a plain old noun such as *duck* or *rat*, and it may be entered into a compound; *cats* and *rats* are not roots but are syntax-ready words, and are the wrong kind of entities to enter into a compound.

There are counterexamples, as there always are in linguistics.[58] For a while Annie Senghas and I collected as many as we could find and put them in – what else? – a counterexamples list. Here are some compounds that contain regular plurals, contrary to everything I have told you so far:

admissions committee	MIT Innovative Structures Program
Boston Antiques Show	morphemes project
Celtics fan	personals ad
chemical weapons attack	publications catalogue
claims applications	ratings data
comics syndicate	records department
cuts package	repeated measures design
enemies list	singles bar
faces lab	skills gap
gimmicks war	top videos list
grades meeting	twins project
injuries report	unemployment benefits cut
landmarks commission	

What is going on? Could we have deluded ourselves by crowing about the examples that fit a false theory while ignoring the examples that contradict it? It seemed unlikely, given the obvious contrast of naturalness within such nicely matched pairs as *mice-infested* and *rats-infested* or *teethmarks* and *clawsmarks*. But the only way to know is to watch people deal with new examples. Senghas, Kim, and I made up a new questionnaire with items like these:[59]

> Hordes of rabid rats are swarming out of the Callahan Tunnel since construction began there. The governor has given a <u>rats-alert</u> advising people to stay in their homes.

> My cat Muffin left three dead mice on my doorstep this morning. She's a pretty good <u>mice-hunter</u>.

> The senior fraternity brothers just bought an awful contraption for hazing week. One by one, each pledge will put one of his feet into the small box while the fraternity president cranks the <u>feet-crusher</u> tight.

> At the ski lodge they have a huge central fireplace with a wooden rail around it that all the people rest their hands on when they get cold. It's the best <u>hands-warmer</u> I know.

People rated the naturalness of these compounds, which contain plurals, and rated the same compounds when they contained singulars: *rat-alert, mouse-hunter, foot-crusher, hand-warmer*. The questionnaire came in different versions, one with *rats-alert* and *mice-hunter*, the other with *mice-alert* and *rats-hunter*. This ensured that any differences in the plausibility of the items themselves would cancel out when we compared the average ratings of regular and irregular plurals.

The outcome was clear-cut. People liked compounds with irregular plurals, such as *feet-crusher*, significantly better than compounds with regular plurals, such as *hands-crusher*. They liked the singulars best of all – *foot-crusher, hand-crusher* – but when forced to consider the plurals, they liked the irregulars much better.

We were relieved, but still had a mystery to solve. Regular plurals never sound as good inside compounds as irregular plurals do, and they are usually reduced to their singular form. Yet sometimes they sound good enough for people to say *injuries list* and *landmarks commission*. Why is *rats-hunter* bad but *injuries list* good? Why are the boosters of one team called *Jets*

Fans and the boosters of another called *Raider Rooters*?[60] We had designed the questionnaire to test various explanations, but none worked.

The mystery was solved by the psychologists Maria Alegre and Peter Gordon, and the solution comes from the first law of language: Strings are nothing, trees are everything.[61] Alegre and Gordon began with a well-known phenomenon we encountered in chapter 2 when mulling over *mother-in-laws*: Words in English sometimes swallow entire phrases, not just other words. Here are some examples from the linguist Rochelle Lieber:[62]

> the Charles-and-Di syndrome
> a pipe-and-slipper husband
> over-the-fence gossip
> off-the-rack dresses
> God-is-dead theology
> a seat-of-the-pants executive
> a who's-the-boss wink
> a floor-of-the-birdcage taste

These compounds cannot possibly come off the conveyor-belt model of word building, because phrases like *off the rack* and sentences like *God is dead* are assembled by the rules of *syntax*, not the rules of morphology. The completed phrase has to be routed backward to the word-formation box, where it can be joined with *dress* or *theology* to form the compound:

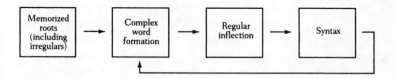

The good news is that this loop provides a route by which regular plurals can appear inside compounds. A plural can be created by a rule of regular inflection, grown into a one-word noun phrase (NP) in the syntax box, and sent back to the word-formation box to be inserted into a compound. The result is a recursive tree like this:[63]

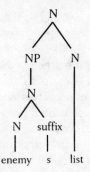

The problem now is that the theory is in danger of saying, Heads I win, tails you lose. When we find an irregular inside a compound, we call it a word; when we find a regular, we call it a one-word phrase. Without some independent way of telling a word-inside-a-compound from a phrase-inside-a-compound, we have sacrificed the original explanation for the difference between *mice-infested* and *rats-infested*, and the theory becomes useless. Alegre and Gordon knew, however, that there *are* ways to tell a word from a phrase. One is based on tree structure, the other on meaning.

What is a *red rat eater*? It could be a rat-eater that is red:

Or it could be an eater of red rats:

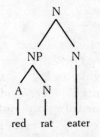

The second tree is made possible by the recursive loop that allows a phrase, in this case *red rat*, to be injected into the compound.

What then is a *red rats eater*? Alegre and Gordon's theory states it would have to be an eater of red rats (as in the second tree), not a red eater of rats (as in the first tree). That's because the *rat + s* combination *has* to be inside a phrase. It was born in the third box, too late to sneak directly into the compound, and could only have been admitted via the long loop through the syntax box and then backward to the word-formation box. Once you let the syntax build a phrase to accommodate the plural, it can put an adjective like *red* in the phrase too.

So ask yourself: What do *you* think a *red rats eater* is? If you can only imagine something that eats red rats, you have confirmed the explanation. You don't have to ask yourself, however; Senghas and I already have asked a group of adults, and Alegre and Gordon have asked a group of preschool children.[64] The adults were shown three-word compounds such as *torn receipts envelope*, sometimes with a singular noun (*receipt*), sometimes with a plural noun (*receipts*). They were asked to choose from a pair of descriptions: 'the envelope for torn receipts,' or 'the torn envelope for receipts.' The children had to pick the *green spiders eater* (or the *green spider eater*) from a pair of pictures, one with a green monster eating brown spiders, one with a brown monster eating green spiders. Everyone interpreted the compounds as Alegre and Gordon predicted: Adults interpreted a *torn receipts envelope* as an envelope for torn receipts; children interpreted a *green spiders eater* as an eater of green spiders.

This leaves us with a final question: Why do people sometimes use the loop and say *enemies list*, but sometimes shun it and say *rat-infested*? Perhaps there is a subtle difference in meaning, and if Alegre and Gordon's theory is on the right track, it should be predictable from the difference in meaning between a phrase and a word.

Words are generic: *dog* by itself refers to dogginess. A *dog hater* needn't hate any dog in particular; he may never even have met a dog. Phrases, though, are particular: *the dog, my dog*, and *a big dog* single out particular canines.[65] Alegre and Gordon noticed that in most of the items in our counterexamples list, the plural noun referred to a heterogeneous collection of individuals, each treated as a distinct entity. The whole point of an *enemies list* is to keep tabs on particular people, and a *publications catalogue* names publications any one of which a reader might want to order. But when your apartment is *rat-infested*, one rat is as good or bad as another, and when apprehending a set of *clawmarks*, few people ponder every claw that made a mark.

As usual, we need an experiment to elevate the convenient story into a genuine explanation. Alegre and Gordon asked one group of subjects to rate the heterogeneity of the referents of the first noun in each of a set of compounds, and asked another group to rate the naturalness of the compounds when the first noun was plural. The heterogeneous nouns, as they predicted, made better compounds-with-plurals-inside.

~

In science the pursuit of idle curiosity often pays off in deeper understanding. People's inquisitiveness about *flied out, talismans, sabertooths, still lifes, out-putted, rat-infested,* and other unexpected forms has enhanced our understanding of regularity and irregularity and provided an entirely new kind of evidence in favor of the words-and-rules theory.

The examples themselves show that regular and irregular forms are qualitatively different, not merely endpoints on a continuum of predictability. The regular pattern can apply to special words such as *fly out* and *still life*; the irregular patterns cannot. Irregular plurals can easily go inside compounds such as *teethmarks* and *mice-infested*; regular plurals cannot.

The examples also show that people consider more than sound when they form new words: An input sound like *fly* can emerge in the output as either *flew* or *flied*, depending on the person's analysis of the whole word. But people consider more than meaning, too; a word's meaning may be stretched to a wispy filament, as in *threw up* or *cut a deal*, and people will inflect it as if nothing had happened.

Instead, people are instinctive linguists, assigning a structure to every word. We all tacitly judge whether a word is a canonical root or some other kind of sound, and we analyse how the word may have been constructed from other words. The analyses that lie behind people's choice of regular and irregular forms have deepened our understanding of the nature of words and rules. The most basic kind of word is a root, with a canonical sound arbitrarily paired with a meaning and a part of speech. The most basic action of rules is to compute the properties of a complex form from the properties of its parts and way they are arranged, with a special role for one position in the arrangement, the head.

Like many psycholinguists I have always thought of language in terms of rules and structures, but I was never sure I could defend the attitude against a hardened skeptic. It was only when I learned about the phenomena in this chapter that I became convinced that rules are living things inhabiting people's minds. The theory that rules are a major ingredient of language offers a reasonable enough explanation for how we inflect new words, rare words, and unusual words. But when the theory also gives a click of insight into why we say *lowlifes* instead of *lowlives* and why teeth leave *teethmarks* but claws leave *clawmarks*, and ten other mysteries, it begins to take on the ring of truth. All the more so when it survives a wave of assaults by dangerous counter-examples.

The phenomena we have examined also provide us with a set of instruments to probe people's mental representation of words when we turn in the coming chapters to three great challenges for the words-and-rules theory. Can we catch children in the act of learning a rule as they master their mother tongue? Do rules work in all the world's languages the way they do in English? And can we distinguish words and rules in the human brain?

7
KIDS SAY THE
DARNEDEST THINGS

Moving its translucent mass through the watery shadows of the
dock and then, past the dock (something so real which now is
not), the jellyfish swam in its slow float while we (I and my
daughter, then just three) ran back and forth predicting that
limp pink gleam and each embodiment it would seem.

'A jello umbrello!' she began and turned to me expectantly.
Censoring (an afterbirth, broken veins, or Medusa's myth, the
monstrous queen made mortal and mother), I stood in silence
until it ended with a shout: the jellyfish glided out. Now months
have passed, but surprise!

'The jellyfish was in my eyes!' Caroline calls while caught
between depth and surface of a dream. 'It bleeded and it singed!'
Her conjugations soon will exact simple irregularities and
tensing will be not verbs, but time's tentacles untangling her
parachute, waving at me.[1]

—*Susan Kinsolving, 'The Jellyfish,' 1999*

Grammatical errors like *bleeded* and *singed* have long epitomized
the innocence and freshness of children's minds. The errors are
acts of creation, in which children lift a pattern from their brief
experience and apply it with impeccable logic to new words, unaware that the adult world treats them as arbitrary exceptions. In
A Dark-Adapted Eye, the novelist Barbara Vine introduces an
unlikable child by remarking, 'He would refer to "adults" instead
of "grown-ups," for instance, and get all his past tenses right,
never saying "rided" for "rode" or "eated" for "ate".[2]
Children's errors with irregular verbs also have been prominent

in debates on the nature of language and mind. The neurologist Eric Lenneberg pointed to the errors when he and Noam Chomsky first argued that language was innate; the psychologists David Rumelhart and James McClelland set them as a benchmark when they first argued that language could be acquired by generic neural networks. Psychology textbooks cite the errors to rhapsodize that children are lovers of cognitive tidiness and simplicity; researchers who study learning in adults cite the errors as a paradigm case of the human habit of overgeneralizing rules to exceptional cases.[3]

Nothing is more important to the theory of words and rules than an explanation of how children acquire rules and apply them – indeed overapply them – to words. The simplicity of these errors is deceptive. As we shall see, it is not easy to explain why children start making them, and it's even harder to explain why they stop.

～

Overgeneralization errors are a symptom of the open-ended productivity of language, which children indulge in as soon as they begin to put words together. At around eighteen months children start to utter two-word microsentences like *See baby* and *More cereal*.[4] Some are simply telegraphic renditions of their parents' speech, but many are original productions. 'More outside!' says a tot who wants to play in the park. 'Allgone sticky!' says another after his mother has washed jam off his fingers. My favorites in the data from my own lab are 'Small loud' after someone had turned down the stereo, and 'Circle toast!' shouted repeatedly to uncomprehending parents who couldn't figure out that the child wanted a bagel.[5]

By their twos, children produce longer and more complicated sentences, and begin to supply grammatical morphemes such as *-ing*, *-ed*, *-s*, and the auxiliaries.[6] Sometime between the end of the second year and the end of the third year, children begin to overgeneralize *-ed* to irregular verbs. All children do it, though parents don't always notice it. My sister told me that her son Carl never made this kind of error, and as if to contradict her, he said *sticked* in my presence a minute later. When children are old enough to sit still in experiments, they pass the *wug*-test: After hearing that a man knows how to *rick* or *bing*, they say that yesterday he *ricked* or *binged.*[7]

Children regularize almost anything they can. They put -*ed* not only on irregular stems, as in *breaked* and *eated*, but on irregular past-tense forms, as in *broked* and *ated*. They put it on their own neologisms, such as *poonked, lightninged,* and *spidered*. They put it on past-tense forms that already have a suffix, as in *sweepened, presseded,* and *My brother got sick and pukeded.*[8]

The past tense is not the only source of irregularity in English, and it is not the only regular pattern children overgeneralize. Alongside past-tense errors such as *breaked* and *putted* we find plural errors such as *mans, foots, tooths,* and *mouses.*[9] Three English verbs are visibly irregular in the third-person singular present tense, and children overgeneralize -*s* to all three:

He just haves a cold.
She do's what her mother tells her.
No, she be's bad, then she be's good, OK?[10]

The suffixes -*er* and -*est* turn many adjectives into comparative or superlative forms. It's easy to forget that the rule has exceptions until we hear children adding suffixes to them. They overgeneralize the suffixes to polysyllabic adjectives, as in *specialer* and *powerfullest*, and to a handful of suppletive irregulars:[11]

"Wow! I must've been
gooder than I
thought!"

THE FAMILY CIRCUS. *Reprinted with special permission of King Features Syndicate.*

Children often generalize from *fourth, fifth,* and *sixth* to *oneth, twoth,* and *threeth,* or sometimes *firstth, secondth,* and *thirdth.* They leap from *myself, yourself,* and *herself* to *hisself,* and from *ourselves* and *yourselves* to *theirselves.* I have heard of one child who used *its* rather than *them* as the plural of the pronoun *it,* and another who liked drawing *rectangles, triangles,* and *cirtangles* (circles).[12]

Children are overzealous grammarians not only in applying inflections in their own speech but also in analysing them in the speech of others. They have little choice. Children are never given grammar lessons presenting *-ed* or *-s* with lists of stems to conjugate or decline; they must mentally snip the suffixes out of the full, inflected words they hear in conversation. As they are figuring it out, they occasionally snip too eagerly and come out with strange back-formations:

FOR BETTER OR WORSE *reprinted by permission of United Feature Syndicate, Inc.*

I suspect that comic strips showing a child making a speech error are usually based on real-life instances known to the cartoonist; in almost every example I have seen, similar errors have been documented in the scientific literature. Alan Prince studied a girl who, like April, was delighted by her discovery that *eats* and *cats* were really *eat + -s* and *cat + -s.* She used her new suffix snipper to derive *mik* (mix), *upstair, downstair, clo* (clothes), *len* (lens), *sentent* (sentence), *bok* (box), *brefek* (from *brefeks,* her word for breakfast), *trappy* (trapeze), even *Santa Claw.*[13] Another child, overhearing his mother say they had booze in the house, asked what a 'boo' was. One seven-year-old said of a sports match, 'I don't care who they're going to verse,' from expressions like *the Red Sox versus the Yankees.*[14]

We laugh, but adults do the same thing, or at least our

ancestors did. *Cherry* is a back-formation from *cerise*, and *pea* is the invented singular of the mass noun *pease*, as in the nursery rhyme 'Pease porridge hot, pease porridge cold.' (Perhaps someday a grain of rice will be known as a *rouse*.) Many people have to be reminded that there is no such thing as a *kudo*: The noun *kudos* is singular, from the Greek word for glory.

A striking feature of children's past-tense errors is that they appear, sometimes suddenly, after long stretches in which the children use the past tense correctly when they use it at all. A child might say *sang, went,* and *heard* for many months before coming out with *singed, goed,* and *heared*.[15] In a sense, the child gets worse before getting better; if the percentage of past-tense forms of irregular verbs that are correct is plotted over time, the shape of the graph looks a bit like a U. 'U-shaped development' fascinates child psychologists because with almost anything else you measure, children get better as they get older.[16] No one considers childhood to be a period of decline (that comes later), so the newly appearing errors are taken as a sign of a *reorganization* in the child's mind. A laundry list of disconnected items suddenly reveals itself as having a pattern, and the child extracts the pattern and applies it across the board.

In the case of the past tense, children have a smattering of regular forms such as *played* and *used* before they make their first error with an irregular, and they use them correctly to talk about events in the past.[17] Presumably they have memorized those forms as indigestible chunks and use them like any other word, with the 'pastness' simply being part of their meanings.

At a certain point a child notices (not consciously, of course) that many words come in ever-so-slightly different versions: *walk* and *walked, use* and *used, play* and *played, push* and *pushed.* Logically speaking, these could be interpreted as meaningless variations in pronunciation or speaking style, yet something impels children to seek a principle behind the variation. By subtracting *walk* from *walked, push* from *pushed,* and so on, a child can isolate *-ed.* By correlating its use with its meaning – that is, noticing that Mom and Dad use *-ed* when describing events that are over and done with – the child can infer that *-ed* means 'past tense.' This synopsis brushes aside many complexities, such as how the child knows to look out for 'present–past' instead of 'hot–cold,' 'indoor–outdoor,' 'good mood–bad mood,' and hundreds of other interesting distinctions. It also sweeps aside how a child deduces that the rule is obligatory: You can't say *I already eat breakfast this morning,* even though the meaning would be clear. Yet children do succeed, and once the rule has been discovered they can feed any verb into it, regular or irregular. They now can say *goed* and *heared* and *bleeded* and *singed* in situations where earlier they might have said *went* and *heard* and *bled* and *sang.*[18]

~

Unfortunately, the rule-epiphany theory by itself cannot explain why children make errors like *bleeded* and *singed.* I have said that children start saying *bleeded* and *singed* because they have acquired an 'Add *-ed*' rule. But adults have an 'Add *-ed*' rule too, and we don't say *singed.* (If we did, we wouldn't call the child's form an error.) Something important is missing: the difference between children and adults, and how children overcome the difference as they grow up.

A first guess is that children become adults because language development is driven by communication: Children improve their language in directions that allow them to communicate their wishes more effectively. Wrong. There is nothing unclear about the meaning of *bleeded* or *singed.* In fact, as long as children make these errors, their language is more communicative than adults'. English has about twenty-five irregular verbs that don't change their forms in the past tense, such as *cut, set,* and *put.* These verbs are ambiguous between past and nonpast: *On Tuesday I put the trash out* could mean last Tuesday, next Tuesday, or every Tuesday. The childlike form *On Tuesday I putted the trash out* could mean only a preceding Tuesday. A language is certainly a powerful tool for communication, but children could not acquire its details by figuring out which ones help in communication; they learn the whole language, with all its strengths and weaknesses, because they just can't help it.

A second guess is that we adults don't say *bleeded* and *singed* because we never hear other adults say them. Wrong again. Adults say lots of things they never hear other adults say. New verbs constantly enter the language – *to diss, to snarf, to fax, to mung, to wild, to flame, to mosh* – and an adult who learns *diss* in the present tense does not have to wait to hear someone say *dissed* before using it in the past tense. If adults say *dissed* even though they have never heard it, they should be willing to say *singed* even though they have never heard it.

The reason adults avoid making regularization errors is not that the error has never been heard; it's that the irregular counterpart *has* been heard. There must be a component of adult psychology that causes the experience of hearing an irregular form such as *sang* to inhibit the application of the *-ed* rule to that item. As noted in chapter 5, this component is called blocking: A specific form in the mental lexicon blocks the application of a general rule that would express the same grammatical notion (past tense, plural, and so on), perhaps through an inhibitory link from the lexicon to the rule.[19] Thus *sang*, listed as a past tense of *sing*, blocks the past tense rule, preempting *singed*; *geese*, listed as the plural of *goose*, blocks *gooses*; *better*, listed as a comparative of *good*, blocks *gooder*.

Perhaps, then, children lack the blocking principle and have to learn it. But how? To learn the blocking principle children first would have to know that forms like *singed* are ungrammatical. Remember, not hearing other people say *singed* isn't enough, because other people don't say *wugged* either, and may not say *munged* or *flamed*, but people do not avoid the unheard past-tense forms.

The only way for children to know that *singed* is ungrammatical is to use it and get a negative feedback signal from their parents – a correction, a frown, a puzzled look, or a non sequitur as a response. Information about what is *not* in a language is called *negative evidence*, and it is one solution to what linguists call 'the logical problem of language acquisition': how a child could, in principle, learn an entire infinite language from a finite sample of the behavior of its speakers.[20]

Children almost certainly do not solve the language acquisition problem by depending on negative feedback from parents. For one thing, parents could not very well correct or disapprove of their young children every time they err. Most of toddlers' sentences are ungrammatical in some way, so parents would be chiding them all day long. Parents focus on the content of their children's sentences, not their form, and let most errors slip by:

> FATHER: Where is that big piece of paper I gave you yesterday?
> ABE: Remember? I writed on it.
> FATHER: Oh that's right don't you have any paper down here buddy?[21]

What happens when parents do correct their children? The cartoonist Bil Keane shows two of the results:

"Mommy, Dolly hitted me."
"Dolly HIT me."
"You too?! Boy, she's in trouble!"

THE FAMILY CIRCUS *Bil Keane, Inc. Dist. By Cowles Synd., Inc.*

THE FAMILY CIRCUS

GOIN' TO GRANDMA'S IS A LOT FUNNER THAN GOIN' TO SCHOOL!

NOT FUNNER, BILLY. YOU MEAN 'GOIN' TO GRANDMA'S IS A LOT MORE FUN THAN GOIN' TO SCHOOL.' NOW LET'S HEAR YOU SAY IT CORRECTLY.

By BIL KEANE

GOIN' TO GRANDMA'S WAS A LOT FUNNER BUT NOW IT SEEMS JUST LIKE SCHOOL.

THE FAMILY CIRCUS *Bil Keane, Inc. Dist. By Cowles Synd., Inc.*

Keane has a fine ear for children's language, and the dialogues are not fanciful. Here is a real one, transcribed by the psychologist Courtney Cazden: [22]

> CHILD: My teacher holded the baby rabbits and we patted them.
> ADULT: Did you say your teacher held the baby rabbits?
> CHILD: Yes.
> ADULT: What did you say she did?
> CHILD: She holded the baby rabbits and we patted them.
> ADULT: Did you say she held them tightly?
> CHILD: No, she holded them loosely.

Systematic studies bear out the anecdotes. The linguist Arnold Zwicky, observing his daughter's overgeneralization of participle endings, reported that 'six subsequent months of frequent corrections by her parents had no noticeable effect.'[23] The psychologists James Morgan and Lisa Travis looked at transcribed speech of three children and their parents, sampled biweekly for several years. They wanted to see whether the children's errors elicited any consistent pattern from their parents – not only overt corrections, but partial or full repetitions, requests for clarification, questions, attempts to move the conversation on, or silence. No consistent pattern was found. In a follow-up study, Morgan and Travis found a different kind of grammatical error, in which parents do sometimes recast a child's sentence in correct English. But they found that the recasting had no effect – if anything, it had an adverse effect – on the child's subsequent improvement.[24]

The psychologist Karin Stromswold has a particularly dramatic demonstration that parental feedback cannot be crucial. She studied a child who, for unknown neurological reasons, was unable to talk, but who was an avid listener and understood complex sentences. When the boy was four, Stromswold tested his knowledge of past-tense forms by asking the boy to teach a dog puppet to talk. She asked him to give the dog a bone when it spoke correctly and a rock when it made an error. The boy awarded bones for *heated, baked, showed,* and *sewed,* and rocks for *eated, taked,* and *knowed.* He made just one error, awarding a bone for *goed,* similar to the performance of normal children. Somehow the boy, and presumably other children, can come to recognize that overgeneralized forms are ungrammatical without first having to make the errors and note their parents' response.[25]

Children must solve the logical problem of language acquisition in a different way. Perhaps, rather than learning the blocking principle from evidence that *singed* is not English, they *begin* with the blocking principle and use it to *deduce* that *singed* is not English. That is, blocking might be built in to the circuitry that drives language acquisition – what Chomsky calls Universal Grammar and what I have called the language instinct. As with all sane proposals about innate structure, such an instinct would not be an alternative to learning but rather an explanation of how learning works. In this case, because children hear parents say *sang* in the course of ordinary conversation, they retain *sang* in memory, and the blocking mechanism represses their tendency to say *singed*, turning them into adults.[26]

We need one more assumption to get the theory to work. If children already have blocking, and all else is the same, they should never say *singed* to begin with! Having heard their parents say *sang* even once should be enough to block the rule from applying to it. Fortunately, the extra assumption is as parsimonious as a theory in child psychology can be.

What is the simplest conceivable hypothesis of how children differ from adults? Answer: They have not lived as long. That is what being a child *means*. Now, among the experiences we accumulate as we live our lives is hearing the past-tense forms of irregular verbs. Human memory profits from repetition. If children have heard *sang* less often than adults have, their memory trace for it will be weaker and their ability to retrieve it will be less reliable. Sometimes, when they are trying to express the thought 'singing in the past,' *sang* will not pop into mind (or at least not quickly enough to get put into the sentence). Before children acquire the *-ed* rule, when they fail to retrieve *sang* they have no choice but to use the bare stem *sing*, even for events that happened in the past. But once they have acquired the rule, they can apply it to *sing*, creating *singed*, thereby satisfying the syntactic constraint that tense be marked in every sentence.

This minimalist theory combines a simple idea from linguistics (blocking) with a simple idea from psychology (memory improves with repetition). It explains why children get worse before they get better, and solves the logical problem of how they exorcise their errors without parental feedback. Correct forms such as *sang* that a child used early on do not go anywhere once the child has acquired the rule, nor are they incapable of blocking errors: They simply must be retrieved from memory to do the blocking, and they are not always retrieved. The cure for overgeneralization is living longer, hearing irregulars more often, and consolidating them in memory, improving their retrievability.

Indeed this account, which posits that the mind of a child and the mind of an adult work the same way, is deducible from the very logic of irregularity, augmented only by the fact that memory is fallible. What is the past-tense form of the verb *to shend*, meaning 'to shame'? If you answered *shended* then you have overgeneralized; the correct form is *shent*. This 'error,' of course, is to be expected. Irregular forms, by definition, are not predictable, so the only way you could have produced *shent* is if you had previously heard and remembered it. But you have heard it zero times and can't have remembered it. If in two years you were asked the question and erred once more, it still would not be surprising, because you would have heard it only once. Now put yourself in the child's shoes. Many verbs will be like *shent* for you: never heard, or not heard enough times to be recallable on demand. The mystery of why children say *singed* and *bleeded* has been solved.

～

When children say *singed*, are they simply little adults with bad memories? Gary Marcus and I combed through computer files with transcripts of the spontaneous speech of 83 children and extracted 11,500 sentences with irregular past-tense forms. We wanted to figure out when and why children start making errors, how often they do it, and with which verbs. Most of what we found fit the simple theory.[27]

First we looked at the error rate. If a child's language system is basically like an adult's, it should be designed to suppress the regularization of verbs that the child remembers are irregular. The suppression cannot be perfect because memory is not perfect, but children's memory for words ought to be fairly good; the child is, after all, using thousands of words and acquiring a new one, on average, every two hours. Overgeneralization errors should be the exception, not the rule, coming from the occasional breakdown of a system built to prevent the error. In fact the average error rate across children is only 4 percent. More than 95 percent of the time a child utters the past-tense form of an irregular verb it is a correct form like *sang*, not an error like *singed*. (Adults tend to over-estimate the error rate because they remember the errors, which stick out like sore thumbs, and fail to notice the boring correct forms.) Once children begin to make the errors in their third year, they continue at this low rate until well into the school-age years.

No verb is immune to the errors, not even those a child used correctly before the error-making began. Nor is any verb consistently erred on. A child might use *felt* when young, then both *felt* and *feeled* when somewhat older. The errors are haphazard; children sometimes use correct and incorrect versions in quick succession, like this: 'Daddy comed and said "hey, what are you doing laying down?" And then a doctor came . . .'[28] The hit-or-miss nature of the errors suggests that children are not ignorant of the correct forms; they are fallible at retrieving them. Some verbs are more error-prone than others, and the simple theory predicts that these should be the verbs that the child has heard less often. So we counted how often the children's parents used each irregular verb in the past tense. If a parent used *told* and *brought* more often than, say, *froze* and *won*, the child should have a stronger memory trace for *told* and *brought* than for *froze* and *won*, and should say *telled* and *bringed* less often than *freezed* and *winned*. We examined ninety irregular verbs and found that with every child, the more often the child's parents used a verb in the past tense, the less often the child regularized it.

Could children on some level really know that their errors are errors? Sometimes they do. The psycholinguists Dan Slobin and Tom Bever tried using their children's errors in their own speech, just for fun.[29] The children were not amused:

TOM: Where's Mommy?
CHILD: Mommy goed to the store.

TOM: Mommy goed to the store?
CHILD: NO! (*annoyed*) Daddy, *I* say it that way, not you.

CHILD: You readed some of it too . . . she readed all the rest.
DAN: She read the whole thing to you, huh?
CHILD: Nu-uh, you read some.
DAN: Oh, that's right, yeah. I readed the beginning of it.
CHILD: Readed? (*annoyed surprise*) Read! (*pronounced* rĕd)
DAN: Oh, yeah, read.
CHILD: Will you stop that, Papa?

In more controlled studies children are asked to judge the past-tense forms of a language-impaired puppet. They let many errors slip by, but they object to errors more often than to correct forms. And when asked to choose, the children, on average, prefer the correct forms.[30] All this suggests that children really do know irregular past-tense forms like *went* and *read*; their errors must be slip-ups in which they cannot slot an irregular form into a sentence in real time.

If overgeneralizing children are not qualitatively different from adults, we should see adults making the errors, and indeed they do, approximately once in every 25,000 sentences in which they use an irregular past-tense form.[31] This figure is about a thousand times less frequent than children's errors, but the estimate includes common verbs like *came* and *went* and *told* that have been drilled into our heads tens or hundreds of thousands of times. With the less common irregulars adults make 'errors' quite often. It's hard to say *how* often, because we adults get to say what counts as 'correct,' and if we regularize an irregular often enough, we simply declare by fiat that it is not an error! These muzzy alternatives – *dreamed* and *dreamt, pleaded* and *pled, leaped* and *leapt, strided* and *strode* – are lower in frequency than pure irregular verbs like *went* and *came*, much as children's errors such as *winned* tend to occur with the verbs they hear less often. Even among pure irregular verbs, those used with lower frequency like *slew* and *strove* are judged to be somewhat unnatural, and their regular counterparts are judged to be relatively unobjectionable.[32]

Over the long run this psychology changes the composition of a language. Say you have heard *strode* only a few times in your life –

more often than *shent*, but far less often than *held*. You would
have a weak memory trace for *strode*, just strong enough for you
to recognize it and for a little voice in your mind's ear to whisper,
'strode!', but not strong enough to block the regular rule from
applying. You may very well say *strided*, just as a child would say
hided. If your neighbors are similarly ambivalent, the language
community may be divided, with some people saying *strided*,
others saying *strode*, and still others, hearing their neighbors using
both forms without rhyme or reason, memorizing both and using
them interchangeably.

With rarer verbs adults' 'errors' create a vicious circle: They use
an irregular form less and less, so their children and neighbors
hear it slightly less often, causing their memory traces in turn to
be weaker, causing them to use it less (and regularize it more), in
turn causing *their* children and neighbors to hear it less, and so
on. An irregular form that falls below a critical frequency could
disappear outright after a few generations. As we saw in chapter 3,
that is exactly what has occurred in the history of English: The
irregular forms of less common verbs such as *chide-chid, cleave-
clove,* and *geld-gelt* became extinct.[33] Verbs, like all bits of culture,
can rise or fall in popularity, and one can imagine a time when the
verb *to geld* had slipped so far that a majority of adults lived their
lives without having heard *gelt*. When pressed, they would have
used *gelded*; the verb had become regular for them, and for all
subsequent generations. That is why irregular verbs tend to be
high in frequency; the list has been filtered repeatedly through the
minds of children and adults, both of whom regularize uncommon irregular verbs.

~

What launches the transformation from regurgitating correct
forms to creating incorrect ones? Why does a child wake up one
morning and start to say *bleeded* and *singed?*

The simplest theory is that that is precisely the point at which
the child has acquired the past-tense rule, a result of the process
described on page 214. The rule must be acquired at some point; it
could not possibly be innate, because some languages don't mark
tense on their verbs, and those that do don't use the English *-ed*.

Prior to learning the rule, a child with an irregular form stuck on the tip of her tongue could do no better than to utter the bare stem *sing*; with the rule in hand she can fill the vacuum with *singed*.

One way to confirm the theory is to watch what happens to *regular* verbs when the child makes the first error with an irregular. Before the first error children leave regular verbs unmarked most of the time; they say *Yesterday we walk*. Then they begin to mark these verbs most of the time, as in *Yesterday we walked*. It is during this transition that the first error with an irregular form, like *singed* or *heared,* appears. We can interpret the tandem development of *walked* and *singed* as two signs of a single underlying process, the acquisition of the *-ed* rule: correct performance where the rule is called for, and errors where it is not.[34]

This idea that children add a rule onto their list of words is simpler than the suggestion that children radically reorganize their language, abandoning the list in favor of an imperialistic rule system and then slowly reacquiring the list. The simpler idea also fits the facts better. When children begin to make these errors, where do the four percentage points of errors come from? Are they produced in situations where previously the child would have produced a correct form such as *sang*? Or are they produced in situations where the child would have produced a bare stem like *sing*? That is, do the errors drive out correct forms, a mysterious step backward on the road to adult language, or is one kind of error driving out a different kind? The data say that the errors of commission (*singed*) are driving out other errors, errors of omission (*sing*), not correct forms (*sang*). For example, before making his first overt error, one boy we studied used correct forms 74 percent of the time and produced the bare verb 26 percent of the time. When the errors began, at a rate of 2 percent, did they come out of the 74 percent of his verb usages that were correct, driving performance down to 72 percent? No – the correct forms *increased*, to 89 percent; the two percentage points of new errors came at the expense of the errors of omission, which dropped to 9 percent. Children don't backslide; when they supplement *sang* with *singed,* they take a step forward, because the syntax of the sentence, which demands a past-tense form, is satisfied more often.[35]

What triggers the 'Eureka!' moment, when a child first discovers a rule? Why does it dawn on some children in their late ones, but on others not until their late twos? I suspect we will never understand what triggers the very first error. Two children we studied made no errors for seven or eight months, popped out a single error (*feeled* or *heared*) just before turning three, and then went another five months before doing it again. Why the false start? What were the children thinking in the months when they failed to act on their epiphany? One possibility is that the gap is an illusion of sampling. Perhaps a newborn rule is wobbly and unreliable, and there are only so many times a child has the urge to use an irregular verb in the past tense, fails to retrieve its stored form, runs the rule to completion, and the tape recorder is running. A steady low probability in the mind of the child may surface as sputtering fits and starts in a record of the child's speech.

Another possibility is that language development at times really is chaotic, because the child is trying to make sense of the language with a changing brain. Synapses, the connections between brain cells, sprout and die in large numbers in the first few years of life, and the churning may temporarily swamp or wash away the newly laid down trace of a rule. Also, countless random events affect the microscopic structure of the growing brain. The human genome does not have nearly enough information to specify the wiring of the brain down to the last connection. We see this in identical twins, who share all their genes and most of their experiences, and who have similar, but not identical, brains, intellects, and personalities.[36]

Jennifer Ganger and I suspected that at least some of the timing of language development, including the past-tense rule, is controlled by a maturational clock. Children may begin to acquire a rule at a certain age for the same reason they grow hair or teeth or breasts at certain ages. If the clock is partly under the control of the genes, then identical twins should develop language in tighter synchrony than fraternal twins, who share only half their genes. We have enlisted the help of hundreds of mothers of twins who send us daily lists of their children's new words and word combinations. The checklists show that vocabulary growth, the first word combinations, and the rate of making past-tense errors are all in tighter lockstep in identical twins than in fraternal twins. The results tell us that at least some of the mental events that make a child say *singed* are heritable. The very first past-tense error, though, is not. When one twin makes an error like *singed* for the first time, an identical twin is no quicker to follow suit than a fraternal twin. These gaps – an average of thirty-four days between the first past-tense errors of two children with the same genes exposed to the same speech – are a reminder of the importance of sheer chance in children's development.[37]

~

I have explained children's creative errors by crediting them with a rule, but there is an alternative: Children might *analogize* from words they already know. They might say *holded* because *hold* sounds like *fold, mold,* and *scold,* whose past-tense forms are *folded, molded, and scolded.* Even with verbs like *sing* and *ring,* which do not rhyme exactly with any common regular verb, children could be reminded of bits and pieces of similar verbs like *sipped, banged, rimmed,* and *rigged,* and cobble together analogous *singed* and *ringed.*

That of course is the basis of the pattern associator memories developed by Rumelhart, McClelland, and their connectionist followers. Rumelhart and McClelland's model acquired hundreds of regular and irregular verbs, generalized to dozens of new verbs, and strikingly, appeared to go through a U-shaped sequence, first producing correct past-tense forms for irregular verbs and later overgeneralizing *-ed* to them. Yet the model had nothing that looks like a word, a rule, or a distinction between regular and irregular systems. How did they get a memory model to learn in a way that everyone has always taken to be a hallmark of rules?[38]

They had an ingenious idea. Rumelhart and McClelland figured that children acquire common verbs first, rarer verbs later. Since common verbs tend to be irregular, and rare verbs regular, the mixture of irregular and regular verbs in children's vocabularies should shift toward the regulars as their vocabulary grows and they begin to run out of irregulars and encounter more and more regulars. Moreover, children's vocabulary growth shows a big spurt several months after they learn their first words. That spurt could cause a sudden influx of regular verbs.

Pattern associator memories are highly sensitive to changes in the statistics of their input. When given a small number of oddball items, they memorize their patterns individually; when given a torrent of items sharing a pattern, they go with the numbers, extract the pattern, and run roughshod over the individual items, gradually reacquiring them over many subsequent bouts of training. That sounds a lot like children.

Rumelhart and McClelland imagined an extreme case: A child first learns a few common verbs, mostly irregular, followed by a spurt of hundreds of verbs, mostly regular. They consulted a list of word frequencies in English, found the dividing line that gave the strongest contrast in the mixture of regulars and irregulars, and trained their network in two stages accordingly. First the model was fed the 10 most frequent verbs in the list, 80 percent of which were irregular, 10 times apiece; then it was fed the next 410 verbs, 80 percent of which were *regular*, 190 times apiece. The network learned the first ten easily. Then, when bombarded by regular verbs, it strengthened thousands of connections to *-ed*, which overwhelmed the connections to the irregular pasts and led the model to make errors such as *breaked*. Connectionist modelers following in their footsteps used more sophisticated networks, but they also induced child-like behavior by changing the models' diet of regular and irregular verbs over time.[39]

Did the computer really 'mimic the brain,' as the headlines put it? It all depends on whether children begin to say *breaked* in response to an influx of regular verbs. Michelle Hollander and I checked the transcripts of the speech of three children over several years to see whether parents at some point start using more regular verbs when talking to their children. They did not. The proportion of regular verbs in parents' speech – about 25 to 30 percent – is the same when their children are two as when they are five. At first that may seem odd: when children are young, parents should favor the common, irregular verbs such as *make* and *do*; only when their children are older should they dip into the lower frequencies and use regular verbs such as *abate*, *abbreviate*, and *abhor*. The reason that scenario doesn't appear in the statistics is that common irregulars like *make*, *do*, and *hold* are indispensable, general-purpose verbs that people of all ages depend on in every conversation. *Abate, abbreviate,* and so on compete *with one another* for air time, so even when the number of different regular verbs rotating in and out of conversation increases, the proportion of conversation filled with regular verbs remains constant.

Perhaps then we should be looking not at the number of *times* the verbs are used but rather at the number of *verbs* in the child's vocabulary, each counted once. There the proportion of regular verbs *must* increase, because there are only so many irregular verbs in the language, and when they begin to run out, the child has to acquire more and more regulars. That's how Rumelhart and McClelland derived their prediction. But counting vocabulary items is a bit odd if you think about the actual events that make up language acquisition. Children presumably learn as they listen to the speech coming out of the mouths of their parents, not by scanning their own mental dictionaries and feeding each verb into their network once per pass. We wanted to be charitable, though, so we checked the transcripts to see whether there really is a vocabulary spurt, and thus a richer mixture of regular verbs, when children begin to overapply *-ed.*

There was not. Children's vocabularies spurt in the mid-to-late ones, about a year too soon to trigger their past-tense errors, which begin in the mid-to-late twos. In the years in which children make the errors, regulars are coming in at a *lower* rate than they were earlier, when the children were using the irregulars correctly. The timing is not what we would need to get a pattern associator to overgeneralize after an early stage of correct performance.

The general problem is that Rumelhart and McClelland balanced their model on a knife-edge of assumptions about the statistics of the speech input to the child. But language acquisition is a robust process that does not live or die by the nuances of parental speech statistics. Throughout the world's cultures, children must learn the combinatorial tools of their language across a wide range of input mixtures, as we will see in the next chapter. Closer to home, even the English plural shows statistics unlike those of the past tense. The handful of irregular nouns known to children (*men, children, feet, teeth*) never could dominate their noun vocabularies the way that irregular verbs, at least in theory, could dominate their verb vocabularies. Yet children show the same U-shaped development with plurals as with past-tense forms. When they begin to speak, all of their plurals are correct, and then they begin to overgeneralize at a low rate for several years.[40]

Michael Ullman and I gave the pattern associator model two more chances to prove that it mimics the human brain. If children, like the model, learn by analogy, their irregular verbs should be lured into error by similar-sounding regular verbs and protected from error by similar-sounding irregular verbs. If *holded* is an analogy from soundalike *folded*, then the more soundalike regulars a verb has, and the more frequent they are, the more likely the verb will be regularized. *Holded* might be more common than *singed*, for example, because *holded* is strongly attracted to frequent *folded* and to a lesser extent *scolded* and *molded*, whereas *singed* is weakly attracted to low-frequency *blinked* and not much else. But when we correlated the number of potential seducers of a verb with its error rate in children's speech, we found little to no effect.[41]

The model did mimic the brain in one way. If *drank* owes its survival to similar *irregular* forms in memory such as *sank* and *rang*, then verbs with more irregular allies, and more common ones, should be erred on less often. And indeed they are. This difference – irregular forms need similar irregulars, but regular forms do not need similar regulars – parallels the findings from adults discussed in earlier chapters. It reinforces the compromise conclusion that pattern associators capture something about irregular forms and the memory in which they are stored, but fail to capture the nature of regular forms and the system in which they are computed.

~

What have children actually acquired when we say they have acquired a past-tense rule? Is it just one more noise they can make, or is it the powerful combinatorial tool that, in conjunction with the rest of grammar, gives rise to the vast expressive range of a language and the elegant logic behind its apparent quirks?

Children's past-tense and plural rules really do seem like wobbly versions of the adult's, with their sweeping power to inflect any verb or noun. Children apply their past-tense rule to almost all their irregular verbs, despite the strong associations to irregular past-tense forms. They apply it to unusual-sounding verbs of their own creation, such as *lightninged, smunched,* and *poonked.* They apply the rules to words built out of phrases, such as *eat lunched, cut-upped egg,* and *There is two Empire Strikes Backs.* Bilingual children sometimes apply a rule to words from their other language, as in *perachs* and *sefers,* Hebrew for flower and book.[42]

Children also apply the rule to rootless and headless words, the ones that lead to such curiosities as *lowlifes* and *flied out,* explained in the preceeding chapter. Kim, Marcus, and I gave children a *wug*-test with a twist. Half the new verbs were identical in sound to irregulars but obviously were based on nouns, like *to fly* meaning 'to cover a piece of paper with flies' and *to ring* meaning 'to put a ring on something.' These are precisely the circumstances that for adults turn irregular-sounding verbs into regulars – *flied out to right field; high-sticked the goalie; ringed the city with artillery* – because a verb based on a noun lacks a root or head and cannot tap into the system of irregular roots stored in memory. Children as young as four work the same way. They regularize verbs based on nouns (as in *She flied the paper*) more often than they regularize verbs with verb roots (as in *They are flying down the road*).[43]

In a similar experiment children saw objects labelled with irregular nouns. Some were simple noun roots, such as a *fuzzy mouse* and a *little goose*; some were based on names, such as a *Mickey Mouse* and a *Mother Goose*; and some were bahuvrihi compounds, such as a *snaggletooth* (a walruslike creature) and a *bigfoot.* When asked to describe collections of these toys, the children used regular plurals for names and headless compounds (*Mickey Mouses, Mother Gooses, snaggletooths, bigfoots*) more often than for the simple noun roots (*fuzzy mouses, little gooses*). Children, like adults, don't just listen to a word's sound when they compute its inflection; they also analyse its grammatical structure.

Children are also sensitive to the other curiosity of irregular nouns discussed in the preceding chapter: the contrast between *mice-infested,* where an irregular behaves like any other word and can be inserted into a compound, and *rat-infested*, where the regular plural *rats* is computed too late to be inserted into a compound. I know of one child who insisted to his father that a building with mice was a *mice building*, and another child who said, 'These aren't only handcuffs; they can be feetcuffs.' I have never heard of a child saying anything like *rats building* or *handscuffs*, though of course we need an experiment to show the difference is real.

Peter Gordon introduced a set of three- to five-year-olds to Cookie Monster and asked them, 'Here is a monster who likes to eat X. What would you call him?' while varying the X. First he trained them on mass nouns like *mud*, which don't take a plural, until the children would say *mud-eater*. That introduced them to the compound construction without biasing their subsequent answers. Then he asked the children what they would call a monster who likes to eat rats. The children virtually always said *rat-eater*, not *rats-eater*, even though they had just heard the experimenter say *rats*. In contrast, they often called a monster who likes to eat mice a *mice-eater* – and those children who occasionally said *mouses* never used it in compounds, as in *mouses-eater*. The avoidance of regular plurals was not simply an aversion to the sound *-s* inside a compound. As with adults, when the children were asked about pluralia tantum nouns such as *pants* and *clothes*, which sound regular but have to be stored like irregulars, they were happy to call the monster a *pants-eater* or a *clothes-eater*.[44]

Gordon then tested whether children could have learned the distinction by noticing irregular plural-containing compounds such as *teethmarks* in their parents' speech, while noticing the absence of regular-plural-containing compounds such as *claws-marks*. He examined all the compounds in standard frequency counts and discovered that *neither* kind of plural-containing compound is common; virtually all commonly used compounds take a singular first noun, such as *toothbrush* and *mousetrap*. Most of the children walked into the lab never having heard a compound containing a plural, but the first time they faced the temptation they used irregular plurals and avoided using regular plurals. Children's sensitivity to the distinction between *mice-infested* and *rats-infested*, Gordon concluded, is a product of the innately specified architecture of their language system, not a product of tabulating forms in parental speech.

Kim and I asked the same question of children's ability to distinguish *flied* meaning 'covered with flies' from *flew* meaning 'soared.' Children hear plenty of verbs-from-nouns, such as *to fish, to plug, to rain,* and *to screw in.* We discovered, however, that they do not hear any verbs-from-nouns that sound like an irregular verb, such as *flied out* or *high-sticked.*[45] That means that prior experience could not have told them what to do when a verb's sound calls for one past-tense form and its structure calls for another; they tend toward the correct answer on their own.

Of course, the speech heard by young children must contain information that tells them that an inflection is regular to begin with. What is that information? It cannot be simply the presence of added material on some words, because that would not distinguish the regular *-ed* in *pat–patted* from the irregular *-en* in *shake–shaken.* Nor can it be the sheer number of words bearing added material, because we saw that children's use of a rule has nothing to do with the proportion of regular verbs in their parents' speech or in their own vocabularies.

How can children recognize a regular inflection when they hear one? Suppose that children's language systems are prepared for words and for rules, and are always on the lookout for examples of each in parents' speech. Children listen for stretches of sound that fit the canonical pattern for a word in their language and that are arbitrarily paired with a meaning. Those are grabbed by the word system and stored as roots. Children also listen for words that might be modified versions of some other word after a rule has had its way with it. Those are snipped into stems and suffixes and the suffix is stored separately as rule-ready material. To distinguish words from rules, then, children must have antennae for signs that a pattern of vowels and consonants has been *added* to a word rather than being *part* of that word.

What might those signs be? They could be the kinds of information that linguists themselves use to determine whether an inflection is regular, the kind of evidence we explored in the preceding two chapters. If children hear a suffixed version of a verb that falls into a family of similar irregular verbs, such as *blinked* and *showed* (which sound like they should belong to the *drink-shrink-sink* family and the *blow-grow-throw* family), they can infer that the words have been modified by something strong enough to have nullified the pull toward the family. If they notice suffixed verbs based on nouns, such as *combed* and *fished,* or on onomatopoeia, such as *cracked* and *squeaked,* they could hear the noun or the environmental sound inside the word, and assume that the residue must have been added by a rule that is free to apply to words that aren't verb roots. If children hear suffixed verbs with nonbasic sounds such as *attached* and *exercised* (which are polysyllabic), they can guess that these forms are unlikely to be roots linked in memory to other roots; the extra bit is likely to have been added by a rule that doesn't care about sound. We don't know whether children rely on these telltale signs, but we do know that the signs are available if children knew what to listen for. All four kinds of verbs may be found in the vocabularies of young children before the stage at which they clearly apply rules in errors such as *singed.*[46] So children *could* use these signs of wordhood and rulehood if they had the mental apparatus of words and rules to interpret what they hear. Once a suffix has been identified as a rule product using the audible cues, it would be available for productive combination with new verbs.

~

Children's speech errors, which make such engaging anecdotes in poetry, novels, television features, and web sites for parents, may help us untangle one of the thickest knots in science, nature and nurture. When a child says *It bleeded and it singed,* the fingerprints of learning are all over the sentence. Every bit of every word has been learned, including the past-tense suffix *-ed.* The very existence of the error comes from a process of learning that is still incomplete: mastery of the irregular forms *bled* and *sang.*

But learning is impossible without innately organized circuitry to do the learning, and these errors give us hints of how it works. Children are born to attend to minor differences in the pronunciation of words, such as *walk* and *walked.* They seek a systematic basis for the difference in the meaning or form of the sentence, rather than dismissing it as haphazard variation in speech styles. They dichotomize time into past and nonpast, and correlate half the time line with the evanescent word ending. They must have a built-in tendency to block the rule when a competing form is found in memory, because there is no way they could learn the blocking principle in the absence of usable feedback from their parents. Their use of the rule (though perhaps not the moment when they first use it) is partly guided by their genes. They spontaneously deploy their new rule to a wide range of words coined by an experimenter or by themselves, and to verbs whose irregular forms are too faint to retrieve. Children fit the rule into its proper place in the logic of their grammatical system, keeping regular forms out of certain word structures and irregular forms out of others.

I suspect that in other parts of our psychology the interaction of nature and nurture has a similar flavor: Every bit of content is learned, but the system doing the learning works by a logic innately specified. Charles Darwin captured the interaction when he called human language 'an instinctive tendency to acquire an art.' 'It certainly is not a true instinct,' he noted, 'for every language has to be learned. It differs, however, widely from all ordinary arts, for man has an instinctive tendency to speak, as we see in the babble of our young children; while no child has an instinctive tendency to brew, bake, or write.'[47]

8

THE HORRORS OF THE
GERMAN LANGUAGE

Though it is sometimes easy for Americans to forget, English is not the only language spoken in the world. Humans babble in some six thousand languages falling into thirty-odd families. For many reasons, those mother tongues are a motherlode for the understanding of language and mind.

First, no one is biologically disposed to speak a particular language. The experiments called immigration and conquest, in which children master languages unknown to their ancestors, settled that question long ago. This means that if some feature of language is the handiwork of a fundamental mechanism of the human language faculty, it ought to be visible anywhere from Lapland to Lesotho, from Peru to Papua New Guinea.

Also, to understand language we have to test hypotheses about cause and effect, but linguists don't have the luxury of synthesizing a language in a test tube and seeing how it is spoken, learned, and changed. The differences among languages already out there make up the only laboratory apparatus that allows a linguist to vary one factor and see how it affects another.

Finally, no one supposes that language evolved six thousand times. We find different languages because people move apart and lose touch, or split into factions that hate each other's guts. People always tinker with the way they talk, and as the tinkerings accumulate on different sides of the river, mountain range, or no-man's-land, the original language slowly splits in two. To compare two languages is to behold the histories of two peoples: their migrations, conquests, innovations, and daily struggles to make themselves understood.

What do all languages have in common, and how do they

The Ancestry of Modern English

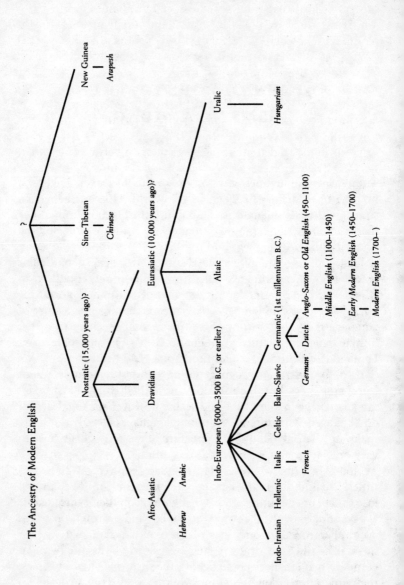

differ?[1] All languages have a stock of morphemes (word parts) and a set of conventions for assembling them into meaningful combinations such as complex words, phrases, and sentences. When words are assembled, they may accept suffixes, prefixes, and infixes (insertions), may undergo a change of vowels or consonants, or may be reduplicated: A part is repeated, as in Malay, where the plural of *orang* 'person' is *orang-orang*. Words are modified not only for tense and number but also for person, case (the role of a noun in the sentence), aspect (how an event unfolds in time), definiteness (the distinction between *the* and *a*), gender (kind), voice (active or passive), mood (indicative, imperative, subjunctive), polarity (true or false), and a handful of other distinctions. In all languages there are exceptions to some rules, that is, irregularity. As in English, irregular forms tend to belong to words that are used frequently.

This does not mean, however, that every inflection in every language has both regular and irregular words. Even in English we find all the possibilities. Though the plural and the past tense have mixtures of regular and irregular words, the progressive is completely regular: Even the most defiant verbs, *be, do,* and *have,* accept the *-ing* suffix without protest: *being, doing, having.* In contrast, the English word for the inhabitant of a city or state is completely irregular, with *Londoners* in London but *Bostonians* (not *Bostoners*) in Boston, *Louisianans* in Louisiana but *Hoosiers* in Indiana, and no one knowing what to call someone from Massachusetts, the Northwest Territories, or the United Kingdom. Other languages can be even more extreme. Turkish verb inflection is a combinatorial dream that cranks out millions of perfectly predictable forms; Russian noun inflection is a memorizational nightmare whose declensions are riddled with holes. The Russian humorist Mikhail Zoshchenko wrote a story about a night watchman who couldn't order a set of pokers because, like most Russians, he didn't know the genitive plural.[2]

Some languages, such as Chinese, don't inflect words; morphology consists of compounding and a few derivations. Others, such as those in the Bantu and Amerindian language families, assemble words in layer upon layer. When words are built in steps, regularity or irregularity can infect each of the steps separately. For example, French verbs fall into three families, of which the *-er* family is considered "regular" because it is the largest and absorbs most of the new words entering the language. Yet some *-er* words, such as *aller* (to go), have many irregularities; conversely, once you know that a verb belongs to one of the other two classes, you can apply a regular pattern of conjugation to many of the verbs within that class (though not all). In a language that builds words in stages, therefore, it isn't meaningful to say that a word is 'regular' or 'irregular' full stop; some parts of a word may be regular, others irregular, depending on whether the part was specified by a general rule or stored with the word's root.[3]

If you have studied a foreign language, you know about irregularity all too well. Everyone dreads the warnings in the tutorials: 'NOTE: NOT ALL VERBS ENDING IN *-RE* ARE REGULAR. YOU MUST LEARN WHICH VERBS ARE ACTUALLY REGULAR *-RE* VERBS AND WHICH FOLLOW AN IRREGULAR CONJUGATION PATTERN.' It also is easy to forget which nouns and verbs in a foreign language are regular and which are irregular. For example, one snatch of Chinese dialogue in a Hong Kong movie was rendered into the English subtitle: 'Greetings, large black person. Let us not forget to form a team up together and go into the country to inflict the pain of our karate feets on some ass of the giant lizard person.'[4]

Though every language textbook discusses 'regular' and 'irregular' forms, the concept of regularity they have in mind doesn't always match the one we have been examining here. 'Regular' almost always is equated with the pattern followed by a majority of words in the language, or with the pattern adopted by newly coined words. But I have been using it in a different sense, one that pertains to the mental processes of speakers rather than to numbers and citations in a dictionary: 'Regular' here refers to a rule that speakers treat as the *default*: an inflectional pattern they can apply to any word in a category, even if the word has never been stored with that pattern, or with any pattern, in memory.

According to this theory, a regular pattern could, in principle, apply to a *minority* of words in a language, with the majority having to be learned one by one. (As James Thurber said, 'There is no safety in numbers . . . or in anything else.') And the rule could fail to apply to a new word if the word is so similar to irregular words in memory that analogy is irresistible (as in to *spling*, which most people inflect as *splang* or *splung*). The only way to know whether an inflection is regular in the psychological sense is to see whether people apply it when their memory is blocked: when the word is new, rare, unusual, foreign, rootless, or headless (the circumstances encountered in chapters 5 and 6). The difference between the two senses of regular – 'majority of words' versus 'applied as the default' – uncovers intriguing aspects of the psychology of language and the history of language, and shows how one can affect the other.

~

Connectionist pattern associators love frequent patterns, and they have thrived from the fact that the regular past-tense suffix *-ed* is 'regular' in the textbooks' sense of applying to the majority of verbs in English. Pattern associator networks generalize well when a pattern is spread over many connections by a large and diverse set of items in the training set. It is no coincidence that people generalize *-ed* and that most verbs in English take *-ed*, the connectionists say; people, like pattern associators, go with the numbers.[5]

The argument has problems, as we saw in the last two chapters. Children presumably learn from hearing a word used, not from the mere existence of a word, and irregular verbs such as *did*, *made*, and *took* are heard more often than regular verbs; the high frequency with which each one is used makes up for the fact that there are fewer of them. Children fail to use the plural rule while their noun vocabulary is almost completely regular (since there are only a handful of irregular nouns in English), and they use the past-tense rule exuberantly some time after the vocabulary spurt in which the mixture of regular verbs increases most sharply. And the vocabulary numbers do nothing to help the pattern associator's problems with the regularization of homophones, headless words, and rootless words, because the models don't even register these distinctions and so must be color-blind to them.

Still, as long as the regular suffix is both the most common form *and* the most generalizable form, we can never tease them apart and definitively rule out the connectionists' suggestion that people are driven by the statistics. We need a language that breaks the confound by having a regular pattern found in a *minority* of words. If speakers still applied the pattern to words that lack associations in memory – that is, if they use it as the default with rare, novel, unusual, rootless, and headless words – then we would be sure that it is the *kind* of mental operation implementing the pattern, not the *prevalence* of the pattern, that makes the regular pattern special.

Note that this wish – for a language whose regular pattern is in the minority – is an oxymoron according to the textbook definition of 'regularity' as the majority pattern. It is unexceptional, though, according to the psychological definition as the product of a mental rule; the psychological definition says nothing about numbers. The question is: Does such a language exist? Could there be a language so perverse, so twisted, so sadistic, that it inflicts irregular forms on its speakers a majority of the time?

I quote from Mark Twain's essay, *Die Schrecken der Deutschen Sprache* (the horrors of the German language):

> A person who has not studied German can form no idea of what a perplexing language it is. Surely there is not another language that is so slipshod and systemless, and so slippery and elusive to the grasp. One is washed about in it, hither and thither, in the most helpless way; and when at last he thinks he has captured a rule which offers firm ground to take a rest on amid the general rage and turmoil of the ten parts of speech, he turns over the page and reads, 'Let the pupil make careful note of the following exceptions.' He runs his eye down and finds that *there are more exceptions to the rule than instances of it.*[6]

Perfect!

Twain knew, of course, that 'the awful German language' (as his title is usually translated) is no more awful than any other language for the children who acquire it as a mother tongue. But many foreigners, he noted, 'would rather decline two drinks than one German adjective.'

In standard (High) German, verbs have three forms: an infinitive, a preterite or simple past, and a participle: *kaufen–*

kaufte–gekauft 'to buy–bought–(has) bought.' (Mercifully, the simple past is seldom used in casual speech.) A participle has a prefix, usually *ge-*, the verb stem, and a suffix, either *-t* or *-en*. The verbs themselves come in three flavors. Weak verbs, such as *kaufen*, are regular; they are like English regular verbs such as *play–played*. Strong verbs are irregular: The stem usually changes unpredictably, as in *gehen–ging–gegangen* 'to go–went–(has) gone,' and they take the suffix *-en*. They are like English strong irregular verbs, such as *sing–sang–sung*. Mixed verbs also are irregular: They take the *-t* suffix, but their stem changes unpredictably, as in *rennen–rannte–gerannt* 'to run–ran–(has) run.' They are like English weak irregulars, such as *sleep–slept*.

The parallels are no coincidence. English and German evolved from a common ancestor, Proto-Germanic, spoken about two to three thousand years ago. The suffixes spelled *-ed* in English and *-t* in German are descendants of the dental suffix in Proto-Germanic (so called because it was pronounced with the tongue against the gum ridge behind the teeth). The vowel-change patterns, or ablaut, came from an even earlier ancestor, Proto-Indo-European, whose conjugations had long since decayed into irregularity. The parallels between English and German are visible in cognate verbs that have similar irregular forms, such as *singen–sang–gesungen* 'sing–sang–sung.'

In German, as in English, the irregular verbs have higher frequencies of use.[7] Among the thousand most common German verbs (which embrace about 96 percent of the verb uses in a large corpus), the irregulars are used an average of 640 times in every million words, and the regulars are used an average of 77 times. (The comparable figures for English are 684 and 73.) As in English, German irregular verbs come in families. For example, *singen–sang–gesungen* resembles *sinken–sank–gesunken* and *trinken–trank–getrunken; sehen–sah–gesehen* resembles *lesen–las–gelesen* and *geben–gab–gegeben*. Membership in the families, however, cannot be captured by a rule: The *singen* group has exceptions like *beginnen–begann–begonnen*, and the *sehen* group has exceptions like *gehen–ging–gegangen*. Furthermore, German speakers sometimes have muzzy judgments about their verbs: The preterite of *backen* can be either *buk* or *backte*, the participle either *gebacken* or *gebackt*.[8] All this suggests that in German, as in English, irregular forms are stored in an associative memory, which can encourage generalization of the patterns to similar new forms.

Yet the verbs of German and English do differ in one way: In German the irregular verbs are more plentiful. Of the thousand commonest verbs in English, a majority, 86 percent, are regular, but of the thousand commonest verbs in German, a minority, only 45 percent, are regular. Many connectionists have pointed to German as a troublesome case for any theory invoking rules, because it's not clear whether there *is* a regular class, at least by the traditional definition of 'regularity' as the majority pattern.[9]

But by the psychological definition of 'regularity' as the default, the German weak suffix passes with flying colors. The linguists Richard Wiese, Harald Clahsen, and Dieter Wunderlich have shown that German *-t* works just like English *-ed*: It goes on any verb, as long as the verb does not have an association with an irregular root in memory:[10]

- English speakers apply *-ed* to rare verbs, such as *ablate–ablated*; German speakers apply *-t* to rare verbs, such as *löten–gelötet* 'welded.'
- English speakers apply *-ed* to novel verbs, such as *wug–wugged*; German speakers apply *-t* to novel verbs, such as *faben–gefabt*.
- English speakers apply *-ed* to unusual-sounding verbs such as *to ploamph* and *to krilg*; German speakers apply *-t* to unusual-sounding verbs such as *quossen* and *rilken*.
- In English, *-ed* can apply to words with irregular homophones, such as *lie–lied* and *lie–lay* or *hang–hanged* and *hang–hung*. In German, *-t* can apply to words with irregular homophones, such as *malen–gemalt* 'to paint' and *mahlen–gemahlen* 'to grind,' or *schaffen–geschafft* 'to work' and *schaffen–geschaffen* 'to create.'
- In English, *-ed* can interlope into irregular-sounding territory, as in *winked* and *blinked*. In German, *-t* does the same. Weak *fehlen–gefehlt* 'to miss' rhymes with strong *stehlen–gestolen* 'to steal'; weak *kaufen–gekauft* 'buy' rhymes with strong *saufen–gesoffen* 'drink booze.'
- English speakers use *-ed* for onomatopoeia, as in *ping–pinged*, *ding–dinged*, and *peep–peeped*; German speakers use *-t* for

onomatopoeia, as in *brummen–gebrummt* 'growl,' *flüstern–geflüstert* 'whisper,' and *klatschen–geklatscht* 'clap.'

• As with English *out-Sally-Rided* and *high-sticked,* German uses -*t* for verbs that are derived from other categories and thus cannot have special past-tense roots listed in memory. These include verbs derived from nouns, such as *frühstücken–gefrühstückt* from *Frühstück* 'breakfast,' *baggern–gebaggert* 'to dredge' from *Bagger* 'excavator,' and *hausen–gehaust* 'to house' from *Haus* 'house.' The same is true for verbs derived from adjectives, such as *kürzen–gekürzt* 'shorten' from *kurz* 'short' and *saübern–gesäubert* 'to clean' from *sauber* 'clean.' It happens not only with existing verbs but with new ones made up on the spot: If someone were to coin a verb *gorbatschowen* 'to Gorbachev,' everyone would give it the participle *gegorbatschowt.*

• In both languages the weak suffix applies not only to verbs-from-nouns (which are rootless) but to verbs-from-nouns-from-verbs such as *flied out,* which have irregular roots but not in the head position from which an irregular form can percolate up. For example, the irregular verb *halten–hielt–gehalten* 'to hold' can be converted into the noun *Halt* 'a hold' which can be used in the compound *Haushalt* 'household.' The compound can be turned back into a verb 'to housekeep,' but the irregular forms are unavailable and the regular suffix applies: *haushalten–gehaushaltet.*

• German speakers, like English speakers, are prone to over-applying the weak suffix (-*t*) to irregular verbs, resulting in errors such as *gesingt* for *gesungen.* Adults make these errors occasionally; children make them more often, at rates comparable to English-speaking children. Despite the prevalence of vowel changes and -*en* among German participles, German-speaking children seldom overapply them; virtually all their errors consist of overapplying -*t*.[11]

To verify that flesh-and-blood German speakers, and not just dictionaries and linguists, apply -*t* to headless and rootless verbs, Marcus, Clahsen, Wiese, Ursula Brinkmann, and I ran an experiment in Germany parallel to the ones we have run in the United States.[12] German speakers were given a questionnaire that asked them to rate participle forms of novel verbs. Half the new verbs were based on nouns:

> Die kleinen dreieckigen Pfeifen für Yuppies sind bei der Kundschaft gut angekommen. Täglich muß Tabakhandler Meier die Regale auffüllen, auf denen die Pfeifen ausgestellt werden. Morgens ist daher immer seine erste Sorge: ['The little triangular pipes for yuppies are a success with the customers. Every day the tobacconist, Meier, has to fill the cabinets in which the pipes are exhibited. Therefore, his first concern every morning is:']

> Sind die Regale auch schon <u>bepfiffen</u>?
> ['Have the cabinets already been pippen?']
>   sounds bad |__|__|__|__|__|__|__| sounds good
>   1   2   3   4   5   6   7

> Sind die Regale auch schon <u>bepfeift</u>?
> ['Have the cabinets already been piped?']
>   sounds bad |__|__|__|__|__|__|__| sounds good
>   1   2   3   4   5   6   7

The other half were existing irregular verbs with a stretched meaning:

> Die schöne Ilse glaubt, mit ihrem Pfeifen Karriere beim Film machen zu können. Wenn sie beim Vorstellungsgespräch gefragt wird, was sie kann, fängt sie keineswegs an, aus Goethes Faust zu zitieren. Nein, nein, Ilse beginnt zu pfeifen. ['Pretty Ilse thinks she'll have a career in the movies by her whistling. When asked at the audition what she can do, she doesn't start reciting Goethe's Faust at all. No, Ilse starts to whistle.']

> Mittlerweile hat sie schon sieben fassungslose Regisseure <u>bepfiffen</u>.
> ['Meanwhile, she has already bewhustle seven speechless directors.']
>   sounds bad |__|__|__|__|__|__|__| sounds good
>   1   2   3   4   5   6   7

Mittlerweile hat sie schon sieben fassungslose Regisseure
  bepfeift.
['Meanwhile, she has already bewhistled seven speechless
  directors.']

sounds bad |__|__|__|__|__|__|__| sounds good
             1   2   3   4   5   6   7

These items provided a baseline for the popularity of the irregular
forms, and since their meaning has been stretched they bias the
experiment against us and toward the semantic stretch theory
(according to which any extension of meaning repels a speaker
from an irregular form).

As predicted, subjects liked the regular forms better than the
irregular forms when the verb was based on a noun, but preferred
the irregular forms when the verb was merely stretched in
meaning. Thus in German, as in English, irregular inflection is
confined to words that can be linked to roots in the mental
lexicon. Regular inflection applies to a 'verb,' period, and doesn't
care about the verb's memory status: It rushes in for words
without roots and for words whose roots are trapped in memory
by the absence of a percolation pipeline in the word's grammatical
structure.

So by nine of the tests that establish that *-ed* is the default past
tense in English, *-t* must be counted as the default participle in
German, despite the fact that *-t* applies to a minority of verbs in
German. The omnipotence of a regular rule does not seem to
depend on a person's previously having been swamped with
regular forms.

'Does not *seem* to depend,' I say, because the case is not
airtight. It depends on counting words in the two languages, and
counting words is always a tricky business. Often it is unclear
which things should count as a word, or how many times.

First, why stop at the commonest thousand words? Why not
count all of them? The reason is that a census of every word ever
spoken would turn up vast numbers of words that only a few
people know; certainly no one would know all of them. An
obscure bit of jargon in an anatomy journal can have no effect on
the reader of a fishing magazine and vice-versa. But if you do go
further down the German list, the regular words begin to gain on

the irregulars, because there are only so many irregulars and soon they become scarce and eventually run out. If you try to inventory every last German verb – by scanning a database pooled from unabridged dictionaries and millions of words of text – the regular verbs then become a *majority*, 78 percent. That's still lower than the proportion in a large database of English (95 percent), but the difference between the languages is no longer so clear-cut.

The other perennial problem in counting words is how to tally those that differ in meaning but share a root. German has many families of verbs such as *ankommen* 'arrive,' *aufkommen* 'blow up' (the onset of wind or a storm), and *bekommen* 'receive.' We counted them separately, just as we counted English *stand*, *withstand*, and *understand* separately, and *part*, *depart*, and *impart* separately. But that is debatable, because in German the prefix sometimes can be chopped off the verb and placed elsewhere in the sentence. As Twain explains:

> The German grammar is blistered all over with separable verbs; and the wider the two portions of one of them are spread apart, the better the author of the crime is pleased with his performance. A favorite one is *reiste ab* – which means departed. Here is an example which I culled from a novel and reduced to English:

> The trunks being now ready, he DE- after kissing his mother and sisters, and once more pressing to his bosom his adored Gretchen, who, dressed in simple white muslin, with a single tuberose in the ample folds of her rich brown hair, had tottered feebly down the stairs, still pale from the terror and excitement of the past evening, but longing to lay her poor aching head yet once again upon the breast of him whom she loved more dearly than life itself, PARTED.[13]

If you collapse all the verbs sharing a root, ignoring the differences in their meanings, then the regularity gap narrows even more: 83 percent of the German roots are regular, compared to 91 percent of the English roots.[14]

What we need is a pattern that applies to a minority of words regardless of how the words are counted. This time the German language, though awful to the language student, is helpful to the language scientist.

German has five plural suffixes: *-en, -s, -e, -er* and ∅ (zero, or no suffix at all). Just to keep learners on their toes, three of the suffixes are sometimes accompanied by a change in the vowel of the noun, the process called umlaut we met in chapter 3. Here are the eight varieties:

∅	der Daumen	die Daumen	'thumbs'
∅ with umlaut:	die Mutter	die Mütter	'mothers'
-e:	der Hund	die Hunde	'dogs'
-e with umlaut:	die Kuh	die Kühe	'cows'
-er:	das Kind	die Kinder	'children'
-er with umlaut:	der Wald	die Wälder	'forests'
-en:	die Straße	die Straßen	'streets'
-s:	das Auto	die Autos	'cars'

German textbook authors have made heroic efforts to impose order on this mess, but as Twain noted, the counterexamples outnumber the examples. One linguist eked out ten rules but tacked on seventeen lists of exceptions. There are some probabilities, such as that masculine and neuter nouns ending in *-er* and *-el* usually take zero, but nothing reliable.[15] Wiese and Wunderlich argue that seven of the eight plural classes are simply irregular. The reason that nouns with certain sounds get certain plurals is not that rules have put them there, but rather that irregular forms are stored in an associative memory, which makes families of similar forms easier to remember and encourages people to analogize the plural of one noun to a similar-sounding noun.

The eighth plural, *-s*, is different. It is by far the least common of the plural forms, no matter how you count them.[16] Among the two hundred most common nouns, only 1 percent, *Autos* and *Hobbys*, take *-s*. Among the 25,000 nouns in the largest database we could find, only 4 percent take *-s*. If you aggressively collapse nouns that share a root, and exclude obscure nouns that hardly anyone uses, you can boost the proportion to perhaps 9 percent. Even that liberal estimate doesn't come close to a conservative estimate of the percentage of nouns taking *-s* in English: 98

percent. No quibbles about word counting can alter the conclusion that nouns taking -s are a large majority in English and a small minority in German.

The -s plural is special for another reason: It is regular.[17] Decades ago a German linguist called it the *Notpluralendung* 'emergency plural ending,' which nicely captures the key trait of regularity in the psychological sense: It serves as the default, acting whenever memory retrieval comes up empty-handed.[18] The -s in High German is not a historical cousin of the -s in English (unlike -t and -ed), but its similarity to English -s is almost spooky:

- English -s applies to unusual-sounding words and to words borrowed from other languages; so does German -s. *Café*, borrowed from French, has a stress pattern unlike that of German roots, and *Kiosk*, from Turkish by way of French, is even stranger. They don't sound like anything in German, but they are not left without plurals: -s rushes in, giving *Cafés* and *Kiosks*.

- German -s also can set up camp in phonological territories that are tightly associated with other plurals. It attaches to nouns that rhyme with irregulars: *Schecks* 'cheques' despite *Flecken* 'spots,' *Labels* 'labels' despite *Kabel* 'cables,' *Relings* 'railings' despite *Ringe* 'rings.' That is, -s is the only plural that can appear with any noun, regardless of its sound.

- In English we talk about Julia and her husband, the *Childs* (not the *Children*). In German they talk about Thomas and his wife, *die Manns* (not *die Männ* or *die Männer*).

- We import Renault *Elfs* (not *Elves*); they export Opel *Kadetts* (not *Kadetten*).

- Eponymous titles work the same. We enjoy the films about the Caped Crusader, the three *Batmans* (not *Batmen*); they enjoy productions of the play about the omniscient alchemist, *Fausts* (not *Fäuste* 'fists').

- Both suffixes step in to pluralize quotations. 'While scanning for sexist writing, I found three "*man*"s on page 1' (not *men*). 'Nach Korrekturlesung für sexistische Wortwahl fand ich drei "*Mann*"s auf Seite 1' (not *Männer*).

- We use -s with nouns that were converted from other parts of speech. For example, the linguist George Curme quotes a complaint about workers loafing on the job: 'to obtain surreptitious smokes and loafs' (not *loaves*).[19] German speakers also use *-s* for nouns that were converted from other parts of speech, as in *Wenns und Abers* 'ifs and buts.' They even use it with nouns converted from entire verb phrases:

Rührmichnichtans	'touch me not ats'	touch-me-nots
Tunichtguts	'do no goods'	ne'er-do-wells
Dreikäsehochs	'three cheeses highs'	youngsters, squirts

• We would use *-s* for acronyms and truncations, such as *sysmans* and *OXes* (similar to our use of *-ed* for truncated verbs, as in *lip-synched*). That is one of its uses in German: *GmbHs* 'corporations'; *Wessis* 'West Germans,' from *Westdeutsche; Sozis* and *Nazis* 'socialists, National Socialists,' from *Sozialist.*

• English-speaking children say *mans*; German-speaking children say *Manns*. The little Dreikäsehochs overgeneralize *-en* more often than *-s*, but given how few nouns they hear with *-s*, it's remarkable that they overgeneralize it at all.[20]

• Perhaps the neatest corroboration of the special status of *-s* comes from a circumstance in which it *cannot* occur: inside compounds. Recall how English loves to form compounds as long as they don't have regular plurals inside them: *mice-infested* but not *rats-infested, teethmarks* but not *clawsmarks*. German is even more profligate with compounding, notoriously so. A dubious 'German Lesson' circulating on the Internet gives some examples:

Dog: Barkenpantensniffer
Dog Catcher: Barkenpantensniffersnatcher
Dog Catcher's Truck: Barkenpantensniffersnatcherwagen
Garage for Truck: Barkenpantensniffersnatcherwagenhaus
Truck Repairman: Barkenpantensniffersnatcherwagen-
    mechanikerwerker
Mechanics' Union: Barkenpantensniffersnatcherwagen-
    mechanikerwerkerfeatherbeddengefixengruppe

Piano: Plinkenplankenplunkenbox
Pianist: Plinkenplankenplunkenboxgepounder
Piano Stool: Plinkenplankenplunkenboxgepounderspinnenseat
Piano Recital: Plinkenplankenplunkenboxgepounderoffen-
    geshowenspelle

Fathers at the Recital: Plinkenplankenplunkenboxgepounder-
offengeshowenspellensnoozengruppe
Mothers at the Recital: Plinkenplankenplunkenboxgepounder-
offengeshowenspellensnoozengruppenuppenwakers

The *en* and *-er* sounds peppering the compounds are not entirely
fanciful, because in German, as in English, irregular plurals easily
appear in compounds:

Professor*en*kränzchen 'professors' circle'
Frau*en*laden 'women's center'
Schwein*e*stall 'pigsty'
Gäns*e*braten 'roast goose'
Büch*er*regal 'bookshelf'
Sozialist*en*treffen 'socialists' meeting'

This free and easy compounding does not extend to *-s* plurals,
however; *Sozi*streffen 'socialists' meeting' and *Autos*berg 'cars
heap' sound as awkward as our *rats-infested* and *clawsmarks*,
and children refuse to call a monster that eats cars an *Autos-
fresser*.[21]

The rulefulness of *-s* is not perfect. Headless 'has a' compounds
(bahuvrihis), which ought to be regularized, stay irregular in
German: *Groβmaul–Groβmäuler* 'braggarts,' literally 'bigmouths';
Geizhals–Geizhälse 'misers,' literally 'thrift necks.' And even
among the constructions that do regularize, we find sporadic
counterexamples. There are plausible explanations, but to allay all
fears about the reality of the *Notpluralendung* we wanted to show
it in action in the minds of live German speakers.[22]

We asked German speakers to rate the eight possible plural
forms for several kinds of made-up nouns. Half of them rhymed
with existing German irregular nouns and should be susceptible
to analogies. For example, *Pund* rhymes with *Hund–Hunde,
Pfund–Pfunde,* and *Grund–Gründe,* so people might well be
tempted to pluralize it as *Punde* or *Pünde.* The other half did
not sound like anything in German, such as *Fnöhk, Pröng,* and
Plaupf. (*Plaupf*, by the way, is German for 'ploamph.')

These sounds then were turned into three kinds of nouns
whose structures were made clear to the subjects by the

sentences in which we presented them. A third of the sounds were presented as roots, that is, ordinary German words. For example:

> Ich habe einen grünen KACH gegen meine Erkältung genommen.
> ['I have taken a green KACH for my cold.']

> Aber die weißen KACH sind oft billiger und helfen auch besser.
> Aber die weißen KÄCH sind oft billiger und helfen auch besser.
> Aber die weißen KACHE sind oft billiger und helfen auch besser.
> Aber die weißen KÄCHE sind oft billiger und helfen auch besser.
> Aber die weißen KACHEN sind oft billiger und helfen auch besser.
> Aber die weißen KACHER sind oft billiger und helfen auch besser.
> Aber die weißen KÄCHER sind oft billiger und helfen auch besser.
> Aber die weißen KACHS sind oft billiger und helfen auch besser.
> ['But the white kachs are often cheaper and work better.']

We predicted that these nouns would be eligible for regular *or* irregular inflection, and the choice should depend on similarity in sound: Nouns that rhyme with existing irregular nouns should tend to get their irregular plurals.

Another third of the sounds were presented as names, which should elicit the regular plural, *-s*.

> Mein Freund Hans KACH und seine Frau Helga KACH sind ein bißchen komisch.
> ['My friend Hans Kach and his wife Helga Kach are a bit strange.']

> Die KACH versuchen immer, ihre Schuhe anzuziehen, bevor sie die Socken anhaben.
> Die KÄCH versuchen immer, ihre Schuhe anzuziehen, bevor sie die Socken anhaben.
> [etc.]

['The Kachs always try to put on their shoes before they put on
their socks.']

The remaining third of the sounds were presented as if they had
been borrowed from a foreign language:

Die französische 'KACH' sieht schwarz am besten aus.
['The French "kach" looks best in black.']

Aber eigentlich sehen KACH in jeder Farbe gut aus.
['But actually kachs look good in any color.']

These nouns also should get -*s*, though the tendency should be
diluted when they rhyme with an existing German noun, because
that makes them easier to assimilate to the native stock (just as
native-sounding imports to English such as *quit* and *cost* sometimes become irregular).

What happened? As predicted, the subjects preferred irregular
plurals for the roots, and the preference shrank when the roots
didn't rhyme with existing irregular nouns – the classic associationist trend of generalization by similarity that we saw for
English in chapter 5. With the names, however, the preference
flipped: The -*s* plural sounded better across the board, whether
the name rhymed with an irregular noun or not. And with the
foreign borrowings, the -*s* plural also shot up compared with the
roots. With items that rhymed with German nouns and thus
could easily be assimilated, subjects slightly preferred an irregular;
with nouns that didn't rhyme, they slightly preferred -*s*.

So -*s* really is different from the other seven plural forms.
The other seven are irregular, and can be generalized only by
analogy to roots. The -*s* is regular, and is used as the default,
in all the 'emergencies' in which memory and analogy fail:
unusual roots, unassimilable borrowings, names, acronyms,
truncations, phrases, and quotations. These emergencies may
strike you as a motley collection of exotic constructions, but that
is exactly the point. The heterogeneity of the constructions, and
people's ability to apply the regular suffix to them the first time
they are faced with the choice, show that people do not have to be
trained to associate the suffix with each construction separately –

the constructions need only be given the mental label 'noun.' And the power of -*s* to serve as a default even though it is rare among words shows that regularity cannot depend on a pattern being stamped into a person's mind through exposure to a large number of regular words. Regularity comes instead from the mind's ability to acquire symbolic rules: operations that apply fully to any instance of a category.

～

The idea that a class of words can be both regular and in the minority upends the traditional concept of regularity found in every language textbook. How could the commonsense notion have been so wrong? The reason is that the notion is based on a correlation: The most generalizable inflection in English is also in the majority. The traditional view assumes a casual relation: Frequent experience leads to a greater tendency to generalize. But as with many correlations, the casual arrow can be flipped around. The 'English language' that provides the input to a child did not float down from the sky. A language is the product of generations of learners and could reflect, rather than shape, their tastes and propensities. English words may be mostly regular because they are the products of a rule, not the other way around.

In Proto-Germanic, the ancestor of English and German, a majority of verbs were strong; they were the forerunners of today's irregular verbs. There was also a precursor of the weak -*ed*/-*t* suffix, perhaps a reduced form of the verb *do*. Owing to its origin as a freestanding word shuffled around by the rules of syntax, the suffix kept its habit of promiscuously abutting against other words regardless of their sound or composition. It became the suffix of choice for new words that could not easily be associated with the existing strong classes: words borrowed from other languages, and words derived from other categories.[23]

As it happens, the major growth areas in English over subsequent centuries were precisely those kinds of words. In 1066 William the Conqueror invaded England. As noted in the satirical history *1066 and All That*, 'The Norman Conquest was a Good Thing, as from this time onwards England stopped being conquered and thus was able to become top nation.'[24] The conquest had another important consequence. Norman French became the language of the government and aristocracy for the ensuing 150 years, and during that interlude the English language was flooded with French words, including a large number of verbs. In the following centuries, English also absorbed numerous verbs from Latin due to the influences of the Church and of Renaissance scholars, who needed many words for abstract concepts. By sampling from an electronic dictionary I have estimated that about 60 percent of English verb roots came from French or Latin. English also is notorious for verbing nouns: Another 20 percent of our verbs are converts from the noun category. Both kinds of verbs, once introduced, *had* to be regular for grammatical reasons: They are rootless or headless.

A reliance on these lazy ways of forming new words is not all that surprising. In a 'Calvin and Hobbes' comic strip Calvin is taking an exam that asks him to 'explain Newton's First Law of Motion in your own words.' He writes, 'Yakka foob mog. Grug pubbawup zink wattoom gazork. Chumble spuzz.' Most of us, though, are not so quick at thinking up new roots. When a new concept needs a label, we borrow it from another language we know, or we derive it from some other word, or we use onomatopoeia, or truncation, or an acronym – all of which, as it happens, breed rootless or headless words that call for regular inflection. Brand-new roots coming out of the blue like Calvin's *could* take irregular inflection if they sounded similar enough to old irregulars; for example, Calvin's *zink* could very well be inflected as *zank* and *has zunk*. But such roots, dreamed up out of nothing, are rare; all but a handful of English roots can be traced back centuries or millennia.[25]

So the traditional definition of regularity, and the connectionist explanation based on it, have it backwards. It's not the case that a majority of English verbs are regular, which trains English speakers to use the regular suffix as the default. Rather, English speakers and their linguistic ancestors have used the regular suffix as the default for millennia, and that is *why* the majority of today's English verbs became regular. Nothing changed in the minds of speakers as regular verbs grew from a minority to the majority.

German also borrowed from Latin and French and derived many verbs from nouns, though not as much as English did. The

German-speaking lands share a long border with France but they never endured a centuries-long domination by a French-speaking elite, as Britain did after 1066. German also did not have to resort to verbing nouns as much as English did, because it has another way of adding to its verb stock: the prefixed verbs that Twain complained about. Thus today's German speakers inherited a smaller proportion of regular verbs. The difference in statistics is a sheer accident of history, a consequence of who won the Battle of Hastings. The psychology of English speakers and German speakers is the same.

The psychology of -s is also the same, though its history is very different. Old English was even more awful than German, with at least nine ways of making a plural, including the suffix -as, an ancestor of today's regular plural. When the vowels at the ends of words eroded to a bland schwa in Middle English, all the plural forms but -es and -en shriveled. The suffix with -s prevailed because it was audible, could be pronounced after both vowels and consonants, and crucially, was imported in great numbers on plural nouns borrowed from Norman French, which also happened to use a plural suffix that contained -s. Middle English speakers mentally merged the French and English -s, and hearing it on all those foreign-sounding words, reanalysed it as an all-powerful regular suffix that could apply to nonroots in general.[26]

High German (the standard dialect spoken in the southern and central parts of the country) has different statistics, because the elevation of -s to the status of a default rule came much later in its history.[27] In Old and Middle High German, -s was completely absent. It first appeared when plural nouns bearing -s were borrowed from Low German (spoken in the north and east), Dutch, English, and French, especially in the eighteenth century. By the nineteenth century, speakers started to generalize -s to all borrowed words and other 'indeclinabilia.' This offended the prescriptive grammarians of the era, who called the suffix 'strange' and 'ignorant' and urged people to stay away from it. People ignored them, and today -s is going strong as the regular default. The difference between the number of regular plurals in English and German is simply a difference in how long the -ss have been around in the two languages snatching up the new headless and rootless nouns. As with -ed and -t, their psychology is the same.

~

With this combination of psychological and historical insight we now can understand why every language – indeed every inflection in each language – has a different mixture of regular and irregular forms. Each mixture arises when unique historical events – conquest, immigration, trade, fads in speaking – are handled by an unchanging mental tool kit, which contains a frequency- and similarity-loving associative memory and a promiscuous combinatorial grammar.

A common misconception is that because Old English had more irregular verbs than Modern English, languages always evolve from irregular to regular. Languages don't consistently evolve in either direction, because different psycholinguistic processes constantly create and destroy the two kinds of words or convert one into the other. These processes have made scattered appearances throughout the book; let me collect them into two lists.

New irregular forms arise when:

- A newly coined root is similar in sound to a family of irregulars and is analogized to them, as in *spling–splang–splung*.
- An existing regular verb is similar in sound to a family of irregulars and is analogized to them, as in *kneel–knelt* and *sneak–snuck*.
- A rule is rendered opaque or obsolete by changes in pronunciation habits, and its former outputs are thereafter memorized as irregulars, as in *foot–feet* (from a rule-governed shift in the pronunciation of *oo* that had originally been triggered by a plural suffix, since deceased).
- A regular form is slurred in speech, obscuring its anatomy, as in *made* and *had*, formerly *maked* and *haved*.
- Two words merge and play musical chairs, and the inflected form of one word becomes a suppletive version of the other, as in *go–went*, the outcome of a shakeout after the merger of *go* and *wend*.
- A complex word is assembled with an irregular root as its head, as in *became, overate, chessmen, oilmice*.

New regular forms arise when:

- A newly coined root is unlike any existing word, as in *snarfed* and *moshed*.
- An irregular form slips in frequency until people can no longer recall it on demand, as in *chide–chid* and *geld–gelt*.
- A word enters the language without a root, via onomatopoeia (*pinged*), eponymy (*the Childs*), or borrowing from another language (*talismans, succumbed*).
- A word is converted from a different part of speech and lacks the right kind of root for the inflection, as in *high-sticked* and *braked*.
- A complex word is assembled without a head and hence without a percolation pipeline, such as bahuvrihi compounds, (*lowlifes, sabertooths*) and double conversions (*flied-out, grandstanded*).
- A complex word is assembled with a regular root as its head, as in *outwalked* and *carseats*.

The lists show how many kinds of brain work go into sculpting just one corner of a language.

~

German and English are sister languages, so it may not be surprising that their inflections work the same way. For all we know, the subtle grammatical phenomena we just explored may be quirks invented by one Germanic tribe rather than a design feature of the human language faculty. Do other languages show the signatures of rules as well? Let's explore a succession of tongues in an expanding circle from English.[28]

Closest to home is Dutch. Like German, Dutch is a West Germanic language that began to diverge from English in the fifth century A.D. when the Angles, Saxons, and Jutes left northern Germany for Britain. Like German and English, Dutch has strong irregular verbs and weak regular verbs. The irregulars fall into families; for example, most verbs with *ij* change the vowel to *e* and take the suffix *-en*.[29] But the irregulars are memory-bound, and just as we saw in English, the regular suffix applies when memory is sealed off by a word's structure: Verbs made from nouns, even when they sound like they belong to an irregular class, are claimed by the regular ending *-den*.

The linguist Chris Collins made up some new Dutch verbs, such as *pijlen* 'to draw arrows' (from *de pijl* 'arrow'), and *vijven* 'to throw five in dice' (from *de vijf* 'five'). Dutch speakers said their past-tense forms had to be regular *pijlden* and *vijfden*, not irregular *pelen* and *veven*.[30]

Remarkably, Dutch has *two* plurals that pass our stringent tests for regularity, *-s* and *-en*. They divide up the territory of noun roots by sound, *-s* getting the roots that end in unstressed vowels or vowel-like consonants (*l, n,* and *r*), *-en* getting the others. Within their fiefdoms each applies as the default. A few nouns are irregular, because they either take the suffix *-eren,* undergo a vowel change, or take *-s* when their sound demands *-en* or vice-versa. But all revert to the proper regular suffix when they are twisted into names or quotations. For example, irregular *rund–runderen* 'cows,' *kok–koks* 'cooks,' and *engel–engelen* 'angels' become regular *Rund–Runden* 'the Runds,' *Kok–Kokken* 'the Koks,' and *Engel–Engels* 'The Engels.'[31]

Now we step out of the Germanic family. French is a descendant of the 'Vulgar' or common Latin of the Roman Empire, and belongs to the Italic branch of the Indo-European family of languages. Italic and Germanic split apart some time between 2000 and 1000 B.C. The regular French plural suffix is *-s*, but it is now silent in most cases; in spoken French plurality is generally conveyed by the article, not the noun. But some nouns do have an audible plural: Nouns ending in *-al* or *-ail* take the irregular ending *-aux* (pronounced ō), such as *journal–journaux* 'newspapers,' *hôpital–hôpitaux* 'hospitals,' *cheval–chevaux* 'horses,' and *travail–travaux* 'works.'

Cyrus Shaoul, an MIT student, asked native French speakers to rate regular and irregular plural forms for a variety of nouns ending in *-al* and *-ail.* The Francophones liked *-aux* for the familiar irregular nouns that require it, of course, and they also liked it for unfamiliar nouns that sounded like other irregular nouns, such as the obscure *sénéchal* 'butler' and the nonsense *greval.* But they flipped their preference to the regular *-als* in an impressive number of default circumstances:[32]

Unusual-sounding nouns	*sluzjal–sluzjals* (unidentified novel objects)
Names	*Segal–Segals* (actors like Steven Segal)
Onomatopoeia	*spral–sprals* (sounds of an elephant sitting on a motorcycle)
Quotations	'*hôpital*'–'*hôpitals*' (instances of the printed word 'hospital')
Surnames	*Cheval–Chevals* (Mister Cheval and his daughter Brigitte)
Product names	*Capital–Capitals* (Kawasaki 'Capital' motorcycles)
Eponyms	*Arsenal–Arsenals* (rap artists like DJ Arsenal
Acronyms	*Original–Originals* (L'Ordre Révolutionnaire Iconoclaste Gaulois pour l'INdépendance des ALsaciens 'the Gallic Revolutionary Iconoclastic Order for the Independence of the Alsatians')

Plus ça change, plus c'est la même chose.

French falls within the Indo-European family, and a skeptic still might wonder whether systematic regularization was an invention of the ancient tribe that begot the family. Hungarian is one of the few European languages outside the family. That fact, combined with the disproportionate number of brilliant Hungarian mathematicians and scientists, led one physicist to suggest that Hungarians are an advanced race of space aliens, but that theory is no longer believed.[33] Hungarian belongs to the Uralic family, which also includes Finnish, Estonian, Lappish (now called Saami), and the Samoyedic languages spoken over a vast range of Arctic Russia. The family descended from a language spoken in the north Ural Mountains more than 7,000 years ago. Hungarian itself is a souvenir of Magyar hordes from the Eurasian steppe who invaded central Europe in the ninth century A.D.

The linguist Edith Moravcsik has made an interesting observation about irregularity in the language.[34] Several Hungarian nouns have a distinctive set of suffixes, often accompanied by a shortening of the vowel:

	Plural: *-ak +* shortened vowel	Possessed: *-a +* shortened vowel	Accusative: *-at +* shortened vowel
arany	*aranyak*	*aranya*	*aranyat*
'gold'	'gold pieces'	'his gold'	(direct object)
madár	*madarak*	*madara*	*madarat*
'bird'	'birds'	'his bird'	(direct object)
ló	*lovak*	*lova*	*lovat*
'horse'	'horses'	'his horse'	(direct object)

When they are turned into names, however, the declension
changes:

	Plural: *-ok*	Possessed: *-ja*	Accusative: *-t*
Mr Arany	*Aranyok*	*Aranyja*	*Aranyt*
	'the Aranys'	'his copy of Arany's book'	(direct object)
Mr Madár	*Madárok*	*Madárja*	*Madárt*
	'the Madárs'	'his copy of Madár's book'	(direct object)
Mr Ló	*Lók*	*Lója*	*Lót*
	'the Lós'	'his copy of Ló's book'	(direct object)

The new suffixes and the unchanged vowel match one of the
regular declensions in Hungarian, which speakers also apply to
borrowed words, as in *telefonok–telefonja–telefont*. We don't know
how pervasive these effects are, but it is remarkable to see it at all
in a language so distant from those we have examined thus far.

Most linguists think that any traces of a common ancestry
between huge families of languages such as Indo-European and
Uralic are lost in the mists of time. But a few have proposed that
Indo-European, Uralic, and Altaic (which includes Turkish,
Mongolian, and Azerbaijani) belong to a superfamily called
Eurasiatic, the legacy of a hypothetical group that peopled Eurasia
toward the end of the last Ice Age 10,000 years ago.

What lies outside of Eurasiatic? The Afro-Asiatic family,
formerly called Hamito-Semitic, originated from a language
spoken in the seventh millennium B.C. and today dominates

north Africa and the Middle East. Its two most famous languages, Arabic and Hebrew, offer not only far-flung corroborations of the regularization effect but also new evidence that a regular default is not a by-product of vocabulary statistics.

Arabic came from the language spoken by nomadic tribes in northwest and central Arabia in the centuries around the time of Jesus. The most common plural in Arabic, the 'broken plural,' imposes semi-systematic changes on the singular: *kitābun–kutubun* 'books,' *madrasatun–madārisu* 'schools.' Broken plurals apply to families of similar nouns with canonical patterns of consonants and vowels. The 'sound plural,' in contrast, is a pair of suffixes (masculine *-uun* and feminine *-aat*) that apply cleanly to a small, motley collection of nouns that don't come from standard roots. The collection includes proper names, nouns derived from verbs, diminutives (small or cute versions of things, as in *doggie* or *duckling*), unassimilated borrowings from other languages, and the names of the letters of the alphabet, which are mostly non-canonical in sound. Examples include *Othman–Othmanuun* (a man's name), *Ramadaan–Ramadaanaat* (the month of Ramadan), and *tilifunn–tilifuunaat* 'telephones.' The sound plural fits the criteria for a regular rule, and as in German, it applies as the default even though only a few examples trickle into learners' ears.[35] Just as remarkably, Arabic-speaking children often overgeneralize the sound plural, despite its scarcity.[36]

Hebrew dates from the second millennium B.C., and for nearly two millennia after the Roman destruction of the second Jewish kingdom it was preserved as the language of Jewish scripture and ritual. At the turn of the twentieth century Hebrew was revived as a living language by Jewish settlers in Palestine who wanted to shake off all trappings of the ghetto and shtetl, including the Yiddish of their parents.[37]

The psychologist Iris Berent has shown that modern Hebrew nails shut the final escape hatch for the connectionist theory that the generalizability of regular patterns comes from the statistics of regular words in a language. Several connectionist modelers have replied to our arguments about German and Arabic by saying that it may not be the *number* of regular words that is critical so much as the *scattering* of regular words in phonological space. Suppose irregulars fall into tight clusters of similar forms (*sing, ring, spring;*

grow, throw, blow; and so on), while regulars are kept out of those clusters but are sprinkled lightly and evenly throughout no-man's-land (*rhumba'd, out-Gorbachev'd, oinked,* and so on). Then one can design pattern associators that devote some of their units and connections to the no-man's-land, and they will deal properly with any subsequent strange-sounding word.[38] These models cannot be taken seriously as theories of a human child, because they have the inflections of a language innately wired in, one output node per inflection, and merely learn to select from among them. And as usual, the problem of rootless and headless words is ignored. It is interesting nonetheless to test the general idea that certain patterns of clustering among regular and irregular sounds are necessary for people to generalize the regular inflection freely.

In Hebrew most masculine nouns are pluralized with *-im,* such as *bul–bulim* 'stamps'; most feminine nouns are pluralized with *-ot,* such as *mora–morot* 'teachers.' But about two hundred nouns are irregular and take the plural of the other gender, such as masculine *kir–kirot* 'walls' and feminine *dvora–dvorim* 'bees.' By now you should be able to predict what will happen when Israelis are asked to choose the best plural for irregular-sounding nouns:[39]

In my friend's room, the <u>kirot/kirim</u> are covered with paintings.

The kir is a French drink. To prepare two <u>kirot/kirim</u>, mix two glasses of champagne and a quarter glass of Cassis liquor.

My French friends Brigitte and Jean Kir arrived for a two-week visit. The <u>Kirot/Kirim</u> will stay at my house during the first week.

Speakers know that irregular *kirot* is correct for the basic word meaning 'wall,' but switch to regular *kirim* for foreign words and names. They also apply regular *-im* to unusual-sounding made-up nouns, such as *tcharlak, krazastriyan,* and *gogof*; indeed they applied it to the strange nouns as readily as to nouns that were close in sound to existing regular nouns. Regular plurals in Hebrew are picture-perfect examples of a default rule.

Here is the punch line: Regular and irregular nouns live cheek-by-jowl in the same phonological neighborhoods. Irregular nouns do not carve out their own distinctive sounds, as in English *sing–sang, sink–sank, drink–drank* or German *singen–gesungen, sinken–gesunken, trinken–getrunken.* Most irregular nouns have sounds that are stereotypical of *regular* nouns. For example, the irregular *kir–kirot* 'walls' squats as a one-word minority in a neighborhood dominated by thirty-one regular nouns, such as *kis–kisim* 'pocket,' *min–minim* 'gender,' *pil–pilim* 'elephant,' and *shir–shirim* 'song.' Similarly, irregular *zanav–znavot* 'tails,' *valad–vladot* 'newborns,' and three other irregulars pepper a space filled by forty-three regular nouns, such as *barak–brakim* 'lightning' and *marak–mrakim* 'soups.' Irregular nouns are so well interspersed with their neighbors that no one can draw a line putting them on one side and the neighbors on the other. And that cramps the latest hope of connectionism to explain away rule-governed generalization as a by-product of the statistics of the input.

∼

Believe it or not, even Hebrew and English may belong to a discernible family with a common ancestor. A few dauntless linguists believe that Eurasiatic, Afro-Asiatic, Dravidian (the languages of south India), and South Caucasian form a super-family called Nostratic. The Nostratic speakers would have been a group of hunter-gatherers who originated in the Middle East and spread through Europe, Northern Africa, and all but the eastern part of Asia about 15,000 years ago.[40] To peer outside this superfamily, and give the words-and-rules theory one more hurdle, we can look to Chinese.

Chinese is a set of languages (we myopically call them dialects) in the Sino-Tibetan family, which also includes Tibetan and Burmese. It is famous for having no inflection whatsoever: A word keeps its sound, no matter how it is used. Some people interpret this as a refutation of any theory in which inflection is part of a universal design for language. In a message posted to an internet discussion group for child language researchers, one critic of the words-and-rules theory, alluding to the lack of inflection in Chinese, asked sarcastically, 'What the hell do Chinese speakers do with their grammar morphology genes or their dedicated neural mechanisms for regulars vs. irregulars?' The rhetorical question misdescribes the theory: It's not regular and irregular inflection per se that are thought to be biologically distinguishable, but combination and lookup more generally. Combining morphemes to form a word is morphology; combining words to form sentences is syntax. Chinese does not have much morphology (it does have some, in the form of compounding and certain derivations), but of course it does have syntax, so it does have words and rules. Better still, some of its rules work as a default, just like the ones half a world away in the Indo-European, Uralic, and Afro-Asiatic families.

In Mandarin Chinese, you can't talk about *a pen* or *some dogs*; you have to use a *classifier* or measure word, as in *yi-zhi gangbi* 'a rod of pen' or *yi-qun gou* 'a pack of dog.' English speakers sometimes have to do the same; we say *a blade of grass* (not *a grass*), *a piece of fruit, a strand of hair, a slice of bread, a stick of wood, a sheet of paper,* and *thirty head of cattle.* Chinese speakers *always* have to pick a classifier when they want to refer to a number or an amount of something.

Each classifier in Chinese tends to go with a kind of object. There are classifiers for people, animals, flat things, long flexible things, small things, one of a pair, and so on. Yet the associations are imperfect and must be memorized; they cannot be captured by rules. *Tiao* often is used for long flexible objects, such as fish, strips of paper, and pants, but it also is used for shorts and for news items, which are not long and flexible, and it cannot be used for a strand of hair, which *is* long and flexible. When Chinese speakers haven't memorized the classifier for an object, they use the classifier for a similar object. This is all familiar from the irregular verbs and suggests that people store the nouns that go with each classifier in an associative memory.

The linguist James Myers has pointed out that one classifier is different.[41] *Ge* is used in a hodgepodge of situations that have nothing in common but an inability of the speaker to draw upon memory or an analogy with something in memory. *Ge* is used with objects whose size and shape don't fit with any classifier, such as *xigua* 'watermelon.' It is used with people who don't deserve the respectful tone of the classifier for humans, such as *xiaotou* 'thief' and *pozi* 'hussy.' It is used with abstractions, such as *xiwang* 'wish' and *guojia* 'country.' It is used with nouns that have been converted from verbs, such as *zhongliao* 'completion' and *tiyan* 'learning from experience.' It is used with quotations, as in the real-life example 'You're good to me, I'm good to you, this -*ge* "good" comes to have vitality.' Combinations of *ge*- and a noun tend to be lower in relative frequency than combinations of other classifiers and their nouns. People use *ge* when they can't remember a noun or can't remember its classifier, and children overgeneralize it to inappropriate nouns.

Ge has all of the powers of a regular inflection, though it appears in a language that has no inflection. Apparently a rule assigns *ge* to anything thought of as 'noun' unless the noun already has a classifier. Myers's analysis of Chinese shows that the bewildering variation among languages can be misleading. Beneath the variation lie deep universals rooted in the nature of mental computation.

How far from English can we go and still find the fingerprints of rules? New Guinea was settled more than 40,000 years ago by an extraordinary group of early modern *Homo sapiens* who somehow crossed fifty miles of open ocean to get there. Over tens of millennia they fanned out into the isolated valleys of the highlands and splintered into tribes speaking more than 800 languages unrelated to anything spoken anywhere else on the planet. Most New Guineans had no contact with the rest of the world until the 1920s and 1930s, when prospectors, traders, and anthropologists began to explore the interior.

Around that time Margaret Mead and her second husband, the anthropologist Reo Fortune, studied a tribe called the Arapesh. She focused on gender in the sense of sex differences; he focused on gender in the sense of inflectional morphology. Mead's research has not held up well. She referred to the people as 'the gentle Arapesh,' and it turned out that the men were headhunters. Fortune's research has held up better, and the linguist Mark Aronoff recently revisited it using the tools of modern linguistic theory.[42]

Arapesh has thirteen genders. This is not as kinky as it sounds; to a linguist gender means 'kind,' as in the related words *genus, generic,* and *genre.* In Arapesh most of the genders are phono-logical. The nouns in a gender end in a particular syllable or phoneme, such as *ag*, or *r*. But one gender is different. Aronoff calls it the 'default gender,' and notes that it is used whenever the gender of a noun phrase 'cannot be determined for whatever reason.' Sound familiar?

One reason an Arapesh speaker may be unable to determine the gender of a phrase is that the phrase has no head because the noun is omitted, a bit like the English *This is nice* or *Which do you want?* A second reason is that the noun phrase is a conjunction of two nouns with different genders (somewhat like *the girl and the boy*). Conjunctions are headless, and when two nouns have different genders often it isn't clear (in any language) which noun should pass its gender up to the whole phrase. A third reason is that a noun may have an unusual sound pattern. In all of Arapesh only two nouns end in *b*. They don't match any of the other genders, and they get thrown into the default gender. A fourth reason comes from the two genders that do have something to do with sex; one has nouns referring to human females, such as *barahoku* 'granddaughter,' and one has nouns referring to human males such as *araman* 'man.' Words that designate people in a sex-neutral fashion, such as *arapeñ* 'friend,' *ašukeñ* 'elder sibling,' and *batauiñ* 'child,' don't fit either gender and get tossed into the default gender. A fifth reason is that the phrase may be headed by a sexless pronoun such as 'I' or 'you,' which don't have a gender to percolate up to the phrase as a whole. Once again we see the modus operandi of a rule.

~

In the first seven chapters we explored the tracks and traces of a rule in action, but only for two suffixes in a single language. In this chapter we have spotted them in eight other languages ranging from the closest siblings of English to the most distant strangers.

I don't mean to suggest that all languages work just like English or that they all can be explained in a simple way by the words-and-rules theory. Every construction in every language throws up a welter of complications and counterexamples and deserves a book of its own. It is striking, however, to sight rules living in the same sets of habitats – rare words, unusual words, headless constructions, converted words, children's errors – in so many historically unrelated languages. To see these deep parallels in the languages of the French and the Germans, the Arabs and the Israelis, the East and the West, people living in the Age of the Internet and people living in the Stone Age, is to catch a glimpse of the psychic unity of humankind.

9

THE BLACK BOX

Engineering students sometimes are given the problem of deducing the design of a circuit in a box from a list of its inputs and outputs. For decades that has been a pretty good definition of psychology. Though no one doubts that our thoughts and feelings are caused by the activity of the brain, that activity has been hard to study because most people don't want to hand over their brains to science until they are dead. We infer what parts of the brain must be doing by seeing what the whole person does when presented with inputs such as pictures or words or instructions. In the previous eight chapters I have argued that the brain has different subsystems for words and rules, not by peering beneath the skull but by seeking the best explanation for how people speak, understand, learn, and react to words and sentences.

Now the black box is being opened, not with a scalpel but with new technologies that allow us to see the living brain without invading it. Neuropsychologists have long studied patients with brain injuries and documented what they could no longer do. Until the advent of Computerized Axial Tomography (CT or CAT scans) they had to wait for the patient to die and be autopsied before they could learn what part of the brain had been damaged. Still newer techniques such as Positron Emission Tomography (PET) and functional Magnetic Resonance Imaging (fMRI) give pictures of the *workings* of the brain, not just its anatomy. DNA testing is beginning to pinpoint the genes responsible for inherited psychological conditions and someday will show how they affect the developing brain. This revolution has led to a new field, cognitive neuroscience, and to the declaration by

President George Bush that the 1990s were to be known as the Decade of the Brain.

If words and rules are the ingredients of language, we should be able to tell them apart in the brain. Parts of the brain that handle memory for words should be implicated in the use of irregular forms, and parts that handle rules should be implicated in the use of regular forms. This gives us another way to test the theory that rules step in when memory fails. Direct neurological damage to the memory system joins other kinds of memory failure, such as rare words, unusual words, headlessness, rootlessness, and childhood inexperience as a circumstance that should summon a rule. Moreover, because regular and irregular forms are so well matched – they have the same meaning (pastness), the same grammar (tensed), and the same complexity (one word long) – any difference in how the brain handles them can help map out the linguistic brain more finely.

In the past few years every major technique in cognitive neuroscience has been applied to the debate on regular and irregular inflection. This chapter offers a tour of the techniques and what they show about the neural seats of words and rules.

∼

It would be nice if we could pinpoint a patch of brain devoted to rules and another patch devoted to words, but that fantasy will never come true. The brain is simply not the kind of organ in which a function has to be carried out by a chunk of tissue with a recognizable shape. The kneecap has to have a certain 3-D shape to be a good kneecap, but a sense of direction or a faculty of emotional intelligence or a language instinct does not. The brain is the organ of computation, and a computational system cares about how information flows within it, not about how the system takes up space. In computers a program or file may end up in different parts of the memory or disk when loaded onto two machines or onto one machine on different occasions, and it may be fragmented across far-flung regions of the disk or memory. As long as the information is preserved and the regions are properly linked, the program can work perfectly, even though we can never draw a circle around the part of the memory or disk that contains it.

The brain is not a digital computer, of course, but that only makes the point more strongly; its circuits are not plugged into slots in a motherboard but somehow find suitable homes in the cerebral cortex as the brain develops. A mental process is a set of computations in the millions of synapses of an intricately structured neural network. A network could be contorted into all kinds of stripes, polka dots, or squiggles on the surface of the brain and still do the same thing, as long as the neurons were connected properly.

A second reason to doubt we will ever find a rule center and a word center is that neither words nor rules just dump their products into the atmosphere. They are part of a complex system and depend on connections to each other and to many other brain systems. Anyone who has recently installed a software package on a personal computer knows that even a simple program needs hundreds of files scattered all over the machine to coordinate the program with memory, input, output, and other programs. And language has even more features than the latest bloat from the software industry. The rule system in the brain must be an octopus with tentacles extending to the mouth, throat, and diaphragm (for speaking), to the ears (for understanding), to short-term memory (to hold the beginning of a sentence in mind while figuring out the end), to concepts of every kind (to plug meaningful words into the sentence), and to systems for reasoning and planning (to decide what to say and how to say it). In a man-made device each component can be in its own box and connected to the others by cables, but in the brain the system is likely to be a web of interdigitated blobs laid out along a wide swath of cortex.

Think, too, of what it would take to *see* one of the blobs. In a fantasy we might imagine that a person could enter a Zenlike trance in which he quiets his entire brain and thinks nothing but pure past-tense thoughts, making the putative rule circuit glow. In reality we have to ask the person to *do* something, like answer a question or produce a word in response to a cue. Even a task as simple as converting *walk* to *walked* forces a person to remember the instructions, read or listen to each stem, hold it in short-term memory, send the 'past tense' request to the rule system and the lexicon, activate the rule, suppress false matches with memory if there are similar irregulars, get the right suffix, join it to the stem, smooth out the sound of the junction, prepare the sequence for speech, and move the muscles, possibly while monitoring for errors at every step. A neurological patient might be unable to come out with a past-tense form if any of these abilities has been compromised. With all systems in working order, a healthy brain in a scanner might light up like a Christmas tree.

So our ability to tie the steps of language processing to circuits of the brain is still rudimentary. For now we must settle for something simpler: clues that regular and irregular words depend on different sets of brain systems (as well as some in common), and clues that irregulars depend more on the system for word memory and regulars more on the system for rules.

~

The human brain is a vast territory: billions of neurons connected by trillions of synapses in dozens of lumps, sheets, and strands, all packed into a convoluted three-dimensional shape. But we can orient ourselves, and aim at the likely habitats of words and rules, by looking at the three major canyons that carve up the cerebrum.

First, the brain has two hemispheres, and in most right-handed people, language, particularly grammar, is mostly in the left.[1]

Second, the central sulcus (fissure) subdivides each hemisphere in two. Its front bank is the motor strip, which controls movement. The motor strip is often drawn in psychology textbooks with shrunken or blown-up body parts pasted along its length, showing which patch controls which part of the body. In front of the motor strip lies the rest of the frontal lobe, which carries out the prerequisites to action: planning and organizing movements, making decisions, juggling items in short-term memory, directing attention, executing chains of reasoning, and following goals under the influence of the emotions.

The rear bank of the central sulcus is the somatosensory strip, which registers touch. The somatosensory strip also is commonly depicted with severed body parts, showing which patch of brain monitors which patch of skin. Behind the somatosensory strip, extending in a sweeping curve around to the parietal, occipital, and temporal lobes, are areas devoted to the other major senses,

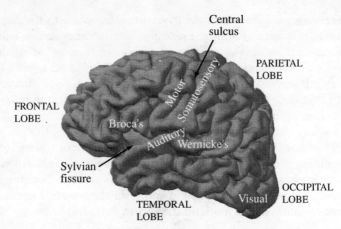

vision and hearing. Here we find not only the first stops in the cortex for the sensations, but also many areas that organize the sensations into a coherent perception of events in space and time and match components of the events to concepts of faces, people, sounds, places, tools, and living things.

The third major cleft is the Sylvian fissure, which divides the temporal lobe from the rest of the brain. The Sylvian fissure anchors the major language areas, which hang off both banks. Above the fissure toward the front lies Broca's area (actually a set of areas), thought to be involved in the planning of speech, in verbal short-term memory, and in the comprehension of complex sentences. Below the fissure toward the rear lies Wernicke's area, thought to be involved in connecting the sounds of words with their meanings. A swath of cortex from the lower part of the parietal lobe sweeping through much of the temporal lobe seems to hold words and their meanings, with the meanings of words of different categories (colors, animals, tools, and so on) concentrated in different parts. The division of language into a front part for grammar and speech planning and a back part for words and speech understanding is surely too simple. There are hints, for example, that some verb meanings have links to the frontal lobes, and that understanding sentences with subtle syntax may involve the anterior superior (front top) part of the temporal lobe. But it is a reasonable first cut.[2]

The three clefts provide compass points showing us where we

might look for the neurobiology of regular and irregular inflection. If regular forms, especially rare and new ones, are processed on the fly by rules, we might find them computed in the anterior (frontal) portions of the left Sylvian cortex. If irregular forms are stored as words, we might find them retrieved from the parietal and temporal portions of the left Sylvian cortex.

~

The outsize human brain is a vulnerable organ which can be damaged by tumors, infections, malnutrition, blocked or burst arteries, and injuries from falls, bullets, and car accidents. Many people who have suffered these tragedies participate in experiments that assess what they can and cannot do. Some do it for money, some to gain insight into what part of them has been lost and what remains, some as an altruistic contribution to science.

When a patient with a brain injury can no longer do something, it is tempting to conclude that the damaged part of the brain must be the neural center responsible for the feat the patient can no longer do. But that reasoning is unsound. Suppose a patient with a damaged X can no longer name fruits but can still name vegetables. That does not imply that X is the brain center for the names of fruit. Perhaps naming fruits, for whatever reason, is more demanding than naming vegetables, and a brain running at less than full capacity stumbles on the harder task. Decades ago the neuropsychologist Hans-Lukas Teuber pointed out that links between brain and mind ought to be based on a *double* dissociation, involving two kinds of patients and two kinds of tasks. In our example we would need to show, at a minimum, that patients with damage to area X have more trouble naming fruits than vegetables, *and* that patients with damage to area Y have more trouble naming vegetables than fruits.[3] This doesn't prove that X is for fruit and Y is for vegetables, but it does suggest that the two areas differ in the *kind* of work they do, not just in the amount of work they do, and that the difference in kind has something to do with the difference between fruits and vegetables.

One famous double dissociation in language involves regular and irregular spelling in printed words. Some patients mispronounce irregular words such as *yacht* and *aisle* (rhyming them with *matched* and *basil*) but have no trouble with nonwords such as *wug* and *dax*, whose pronunciations can be deduced from regular rules of spelling. Other patients, with damage to different parts of the brain, have the opposite problem: They can pronounce *yacht* and *aisle*, but have no idea what to do with *wug* and *dax*. The natural interpretation is that the brain contains two routes from print to sound. One uses rules, such as 'Pronounce the letter pair *ee* as the sound "ē," ' and it is needed for new and rare words, which cannot be retrieved from memory. The other memorizes entire words and their pronunciations, such as 'The string `aisle` spells the word *aisle*, which is pronounced "īl," ' and it is needed for irregularly spelled words, which defy the rules. The first kind of patient, with *surface dyslexia*, has suffered damage to the whole-word pathway; the second, with *phonological dyslexia*, has suffered damage to the rule pathway.[4]

The double dissociation gives neural reality to a distinction we might have guessed on purely computational grounds, and it challenges connectionist models, which, as with the past tense, try to capture regular and irregular forms in a single pattern associator memory. Of course spelling rules are different from rules of grammar: They are consciously taught and learned, and they show little of the abstract logic of grammar explored in chapter 6. But connectionist theorists treat them the same, so any problems they have with models for reading aloud carry over to their models for grammar.

Advertisements for pattern associators boast that regular and irregular associations are smeared across a single set of connection weights, eliminating the need for separate boxes for rules and exceptions. The problem then is how to deal with a double dissociation, such as that between patients who can no longer read novel words and patients who can no longer read words with irregular spellings. Generally modelers simulate brain damage by removing or weakening connections at random, and that leads to a *single* dissociation in which the model can no longer handle irregular words.[5] This happens because each irregular form depends on a rather small number of strong connections between particular inputs and particular outputs, making the irregular form vulnerable to damage, whereas regular forms are computed by a diffuse set of many weaker connections, offering redundancy and resistance to damage. The double dissociation suggests that the appeal to the aesthetics of a single mechanism may be misguided; the brain appears to have more than one part.

The connectionists reply that sometimes connections in a pattern associator spontaneously segregate into bundles that concentrate on regular or irregular associations. As a result, when a model is deliberately damaged at random to simulate a brain lesion, it may have more trouble with regular associations on some simulation runs and more trouble with irregular associations on other runs. But the modelers John Bullinaria and Nick Chater have shown that double dissociations occur only in artificially small toy models, where there simply aren't that many connections for the regular associations to be spread over; a small amount of damage can therefore hurt the regular associations as badly as the irregular ones. In any model with a more realistic number of connections the regular association is distributed more evenly across the connections, and a simulated lesion always hurts irregulars more. Bullinaria and Chater conclude that Teuber's logic of double dissociation is still sound.[6] In fact the logic is even sounder when the dissociation can be predicted beforehand from an understanding of what different parts of the brain do. That ensures that the double dissociation is not just a fluke, hand-picked after the fact from a mass of random data going every which way.

Michael Ullman, Greg Hickok, Marie Coppola, and I teamed up with the neuropsychologist Suzanne Corkin and the neurologists John Growdon and Walter Koroshetz, who study a variety of neurological patients.[7] We began by seeking a double dissociation in regular and irregular inflection in different kinds of aphasia. Aphasia is an impairment of language following an injury to the brain, and much of our knowledge of the organization of the language areas has come from comparisons among different types of aphasia.[8]

Agrammatism is a symptom of some forms of aphasia in which a patient has difficulty assembling words into phrases and sentences, putting the right grammatical suffixes onto their stems, and understanding complex sentences. It frequently appears after extensive damage to the anterior (front) regions of the language areas around the Sylvian fissure, including Broca's area. Agrammatic aphasics usually have trouble with single words as well, but the trouble often is less severe than their trouble with phrases and sentences. An agrammatic patient's speech often sounds like this: 'Son . . . university . . . smart . . . boy . . . good . . . good,' or 'Lower Falls . . . Maine . . . Paper. Four hundred tons a day!' Agrammatic aphasics often remember 'words' in the sense of listemes: memorized chunks that may be more than one word long, such as 'Fit as a fiddle and ready for love.'[9]

Agrammatics have great difficulty with grammatical suffixes, usually leaving them out altogether (particularly in a language like English, where bare stems are used in the infinitive and the present tense) or using the wrong one. When reading a list of words, for example, they might read *smiled* as 'smile' and *wanted* as 'wanting.' Two previous studies, one by Oscar Marin, Eleanor Safran, and Myrna Schwartz, another by William Badecker and Alfonso Caramazza, had shown that patients with impaired grammatical processing make fewer of these errors when reading *irregular* past-tense forms and plurals. Our group replicated the effect with a new sample of five agrammatic patients. The explanation is that regularly inflected words ordinarily are parsed by rules as they are read, and agrammatic patients have damage to the machinery that does the parsing. Irregular verbs are matched against memory as wholes, which the patients can still do.

A person who has suffered brain damage could have trouble with regular forms for reasons other than their regularity. To ensure that their agrammatic patients don't simply have trouble *pronouncing* an *-s* or *-ed* at the end of a word, Marin and his collaborators compared regular plurals with pluralia tantum, which have to be memorized as irregulars even though they bear the plural suffix *-s*. They compared *clues* with *news, buds* with *suds,* and *misers* with *trousers.* To ensure that the patients don't just stop reading from left to right as soon as they get to the end of a recognizable word (which would give them *smile* from *smiled*), Badecker and Caramazza gave their patients uncommon words that contained common words, such as *yearn* (which contains *year*), *dogma* (which contains *dog*), and *pierce* (which contains *pier*). To ensure that the patients don't simply have trouble with the less common or harder-to-pronounce past-tense forms, our group matched each irregular form with a regular form that had a similar ending and the same frequency in the language. For example, we matched *slid* with *tied, swept* with *slipped,* and *bought* with *stayed.* Even with all these controls, these patients had greater difficulty reading the regular forms.

Reading an inflected word aloud is different from generating it oneself, and to test patients' ability to generate past-tense forms, Ullman made up a battery of items of the form, 'Every day I *dig* a hole; Yesterday I ____ a hole.' Patients read the items or listened to them, and were asked to fill in the blanks. The verbs were regular, irregular, or nonsense words like *spuff* and *plam* (in other words, a *wug*-test). Portions of the lists of regular and irregular verbs were matched for frequency and for the sequence of consonants at the end. Rather than testing a large group of patients given a diagnostic label such as 'Broca's aphasia' (which often lumps together patients with huge messy lesions and a hodgepodge of symptoms), we did a case study of a patient whose lesion was confined to anterior regions of the brain and the basal ganglia. His symptoms of agrammatism were unmistakable, but his ability to name things, though worse than control subjects, was reasonably good. That suggests that his mental grammar was more impaired than his mental dictionary, and as we predicted, he had far more trouble inflecting regular verbs than irregular verbs, was almost incapable of inflecting novel words like *plam*, and never overgeneralized the rule to irregular verbs, which would have resulted in errors like *digged.*

The other half of the double dissociation comes from patients with *anomia,* a difficulty in retrieving and recognizing words despite fluent and generally grammatical speech. Anomic patients often have their words stuck on the tip of the tongue, and they resort to circumlocutions, pronouns, and generic words such as *something* and *stuff.* Here is a transcript of one anomic patient trying to name some objects:

> [A clock:] Of course, I know that. It's the thing you use, for counting, for telling the time, you know, one of those, it's a . . . [But doesn't it have a name?] Why, of course it does. I just can't think of it. Let me look in my notebook.

[His elbow:] That's the part of my body where, my hands and shoulders, no, that's not it. No, doctor, I just can't get it, isn't that terrible?

[A wallet:] This is a kind of bag you use to hold something; you may hold materials in it and keep it in your pocket.[10]

Anomia is often associated with extensive damage to *posterior* parts of the brain, especially the junction of the parietal and temporal lobes, and often with damage to large parts of the temporal lobe as well.[11] Sometimes patients with posterior lesions have *jargon aphasia* in which they speak in their own neologisms, such as *nose cone* for *phone call* or words that no one recognizes at all. Interestingly, they often stick regular suffixes onto their jargon, a self-administered *wug*-test. One patient, struggling to name a box of matches, said, 'Waitresses. Waitrixies. A backland and another bank. For bandicks er bandicks I think they are, I believe they're zandicks, I'm sorry, but they're called flitters landocks.' He does it with verbs as well as nouns: 'She wikses a zen from me,' 'He mivs in a love-beautiful home.'[12] This suggests that regular inflection may be computed in a part of the brain that is distinct from the parts in which words are handled.

We tested six aphasic patients with anomia, but focused on one whose lesion was confined to the posterior parts of the brain. This picture shows the approximate size and shape of the lesion of the anomic patient and, for comparison, of the lesion of the agrammatic patient discussed earlier.

Area damaged in patient with anomia

Area damaged in patient with agrammatism

Just as one would expect if anomic patients suffered greater damage to their mental dictionary than to their mental grammar, they had more trouble inflecting irregular verbs than regular ones, were relatively good at inflecting novel verbs like *plam* (as much as 80 percent of the time), and interestingly, made overgeneralization errors like *digged*, just as children do. For example, the patient with the circumscribed lesion made the error 25 percent of the time. In these three symptoms, the anomic patient is the mirror image of the agrammatic patient.

The psychologists William Marslen-Wilson and Lorraine Tyler had doubly dissociated words from rules in the brain in a different way. Recall from chapter 5 that when intact people hear a word, they are primed to recognize related words. After hearing *swan*, for example, people recognize *goose* more quickly, presumably because the mental dictionary entries overlap or are linked. According to both the pattern associator theory and the words-and-rules theory, a pair of irregular words such as *find* and *found* should be associated in a similar way, and sure enough, in experiments where the words are spoken, *found* primes *find* just as *swan* primes *goose*. Regular *walked* primes its stem too, but according to the words-and-rules theory it is for a different reason: The brain unconsciously analyses *walked* into *walk* and *-ed*, and the stem *walk* primes itself. We know the priming is caused by grammatical relatedness, not mere overlap in sound, because overlapping but unrelated words such as *gravy* and *grave* do not prime each other.

If regular and irregular priming work in different ways in the brain, different neurological patients might show priming by regular forms but not irregular forms, and vice-versa. The technique does not require the patients to speak aloud (they just press a button if the item is a word), so it bypasses any remaining worry that irregular forms are more easily pronounced than regular ones.

Marslen-Wilson and Tyler discovered two agrammatic patients in whom *walked* did not prime *walk* (regular inflection), though *found* did prime *find* (irregular inflection), and *swan* primed *goose* (semantically related words). Presumably the patients' circuitry for grammatical analysis was impaired, so *walked* and *walk* struck them as no more related than *gravy* and *grave*. But the associations in their mental dictionaries, such as *swan* to *goose*, were not as impaired, and by the same token neither were the associations between *found* and *find*. In a third patient the dissociation went the other way: *walked* primed *walk* but *swan* failed to prime *goose*, and as expected, *found* failed to prime *find*.

The patterns of damage in the patients' brains were diffuse and hard to delineate, so we cannot use the double dissociation to identify the brain areas normally responsible for regular and irregular priming. The first two patients, who lost priming of regular verbs, had massive damage to the left hemisphere but no damage to the right hemisphere. Presumably they lost the areas in the left hemisphere responsible for grammatical processing, but retained a partial knowledge of words and their relationships in the right hemisphere. (In healthy people words presented to the right hemisphere can often prime other words in the same category, suggesting that words and their meanings are stored in the right hemisphere in addition to the left.[13]) The third patient, who lost priming of irregular words and semantically related words, had extensive damage to the right hemisphere and patchy damage to the left. Perhaps both copies of his mental dictionary were damaged, but just enough grammatical machinery survived in the left hemisphere to analyse the regular forms. Whether or not this anatomy is correct, it is clear that regular and irregular verbs depend on different sets of areas of the brain.

∼

Not all brain damage comes in the form of a lesion from a stroke. Neurodegenerative diseases, the result of genes, aging, viruses, autoimmune attacks, environmental toxins, and unknown causes, can affect some parts of the brain more than others. The most common neurodegenerative disorder is Alzheimer's disease, which strikes about a tenth of people over sixty-five and half of those over eighty-five. In Alzheimer's disease, deposits called plaques accumulate around neurons, and tangled filaments accumulate within them. Neurons die, neurotransmitters are depleted, and brain tissue is chronically inflamed. Sufferers slowly lose their memory, judgment, and knowledge of who and where they are.[14]

The course of Alzheimer's disease varies from patient to patient, but one frequent pattern interested us. Memory loss is an early and noticeable symptom of the disease, and it includes memory for words. Patients have difficulty in retrieving uncommon words, in naming objects, and in supplying the word that goes with a definition. Yet many patients speak fluently and grammatically, understand sentences with relatively complex syntax, and even convert ungrammatical sentences into grammatical ones.[15] The greater impairment in word retrieval than in grammatical processing may be caused by the distribution of the neurofibrillary tangles in the cortex. Typically the tangles are more numerous in the temporal lobes and adjacent parts of the parietal lobes than they are in the frontal lobes, as shown in this diagram, where darker shades indicate more tangles:[16]

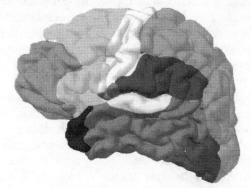

We predicted that patients with Alzheimer's disease who have particular difficulty with word retrieval should look like anomic patients when producing past-tense forms, and indeed they did. The Alzheimer's patients had more difficulty inflecting irregular verbs than regular verbs, were surprisingly good at the *wug*-test (84 percent correct), and often overgeneralized the regular past tense to irregular verbs in childlike errors such as *swimmed* (27 percent of the time).[17] The psychologists David Balota and Richard Ferraro showed that the same thing happens in reading aloud: Alzheimer's patients often regularize irregular spellings, for example, pronouncing *pint* as if it rhymed with *mint*.[18]

Is there a contrasting neurodegenerative disease that might

dissociate words from rules in the other direction? Ullman thought of a possibility. Many neuroscientists believe that the brain has two major memory systems, one for facts – 'knowing that' – and one for skills – 'knowing how.'[19] The fact system, also called declarative memory, needs the hippocampus (a seahorse-shaped organ embedded in the inner surface of the temporal lobe) and adjacent structures to form memories; once formed, the memories are permanently stored in the cortex, largely in the temporal and parietal lobes. There are the parts of the brain that are hit earliest and hardest by Alzheimer's disease. The skill system, also called procedural memory, underlies motor skills such as reaching and walking, but also cognitive and perceptual skills such as scanning, sorting, ordering, predicting, and generating associations. The skill system needs the basal ganglia, a set of organs buried in the cerebrum that receive input from all over the cortex and send their output primarily to the frontal lobes (via the thalamus, the relay station in the center of the brain). Most areas in the frontal lobe have corresponding areas in the basal ganglia, and these two parts of the brain work together as parts of a single system.[20]

Many neurons in the basal ganglia transmit signals to one another by releasing the neurotransmitter chemical called dopamine. In Parkinson's disease the cells that manufacture dopamine degenerate and the basal ganglia malfunction. (The most famous young sufferers of Parkinson's disease are the boxer Muhammad Ali and the actor Michael J. Fox, who was diagnosed with the disease in 1998 when he was thirty-seven.) People with Parkinson's disease have tremors and difficulty initiating movement, and when they do budge, their movements are often slow and rigid. They also may be impaired in the kinds of tests that tap frontal lobe functions, such as planning, sequencing, and paying close attention. Interestingly, their speech is often grammatically simplified, with more nouns and verbs and fewer grammatical morphemes such as prepositions. They have difficulty understanding sentences via their syntax, such as *It was the boy that the girl tickled* and *The eagle that the hawk chased was fast.*[21] Yet their vocabularies often are less impaired and sometimes are not impaired at all.[22] In some ways Parkinson's disease is a mirror image of Alzheimer's. The degeneration affects the skill system rather than the fact system, it has a bigger effect in the frontal lobes than in the temporal and parietal lobes, and it compromises grammatical processing more than word lookup.

As with all neurological diseases, Parkinson's patients differ from one another, and we focused on a sample of patients with slowness in moving the right side of their bodies. The right side is controlled by the left hemisphere of the brain, and Ullman figured that these patients were likely to have more of a malfunction in their left basal ganglia, which in turn should compromise the language processing areas of the left frontal lobe. As predicted, these patients were slightly better at inflecting irregular verbs than regular verbs (even when the verbs were equated for ease of pronunciation, as in *passed* and *lost*), were poorer still at inflecting novel verbs like *plam*, and never overgeneralized the rule to irregulars in errors such as *swimmed*.[23] All three outcomes differ from those of the Alzheimer's patients, completing a double dissociation that mirrors the one between agrammatic and anomic patients.

In all of these demonstrations damage to a part of the brain causes a difficulty in retrieving words or applying rules, suggesting that those parts of the brain may be necessary for those parts of language. But to be sure that a part of the brain really is linked to some part of the mind, neuroscientists like to show the opposite as well: that activity in a part of the brain causes a particular experience or behavior. In the case of a role for the skill system in carrying out rules of language, a very different neurological disorder offered the possibility of linking brain activity to overt behavior.

Huntington's disease is an inherited neurodegenerative disorder made famous by the folk singer Woody Guthrie, who died of the disease in 1967 (his last days were portrayed in the movie *Alice's Restaurant*, written by his son Arlo). People who carry the gene begin to notice symptoms in their forties, when neurons in the basal ganglia start to die. Unlike the degeneration in Parkinson's disease, the dying neurons are in circuits that ordinarily *suppress* movement, keeping the body under control. As a result, Huntington's patients may suffer from involuntary and unsuppressable movements – hence the old name for the disease, Huntington's Chorea, from the Greek word for 'dance' that is also seen in *choreograph* and *terpsichorean*.[24]

Ullman tested several patients with Huntington's disease and made a remarkable discovery: They seemed to *overapply* the past-tense rule, as if the disease led to excess activity in the circuitry that executes mental rules as well as in the circuitry that executes movement. The patients often applied the rule to irregular verbs, resulting in errors like *digged*. Unlike the similar errors of patients with anomia and Alzheimer's disease, though, the errors of the Huntington's patients could not be attributed to a difficulty in retrieving *dug*; they had little difficulty retrieving words in general. The errors come from a failure to suppress the rule, not from a failure to retrieve the irregular form. Moreover, the patients often overapplied the suffix to *regular* verbs, or applied it too strenuously, resulting in errors such as *lookeded* and *lookid* that were rare in all the other patient groups. These errors weren't simply stutters or other exaggerations of the physical movements of the tongue and mouth, because in an error like *digged* or *dugged* the correct form does not call for *-ed*. Also, the patients never added an extra suffix to irregular forms that happen to end in *t* or *d*, like *kept*; that is, they virtually never made errors like *kepted* or *kepid*. This suggests that we were observing a compulsion to add the suffix *-ed*.

All this adds up to suggest that irregular and regular inflection, and words and rules more generally, depend on different systems in the brain. Moreover, if Ullman is right, these two systems may be subdivisions of the brain's two major systems for remembering information: Words may be a part of the 'knowing that' system; rules may be a part of the 'knowing how' system.[25]

～

If the 1990s will be remembered as the Decade of the Brain and the dawn of cognitive neuroscience, the first decade of the 2000s may be remembered as the Decade of the Gene and the dawn of cognitive genetics. New techniques for analysing the human genome are beginning to identify the genes that shape the brain to learn and feel in particular ways. Two recent discoveries of genes tied to language and thought will probably be the first of many, even if they never lead to the scenario in the cartoon on the following page.

ROBOTMAN® by Jim Meddick

In the early 1990s Noam Chomsky's hypothesis that language has a genetic basis was thrown into the spotlight when newspapers reported the discovery of a large English family, the KEs, in which half the members had a congenital difficulty with speech and language.[26] The syndrome is called Specific Language Impairment (SLI), and like most labels slapped on children with behavioral problems, it means little, only that the problems with language are not side effects of hearing impairment, autism, retardation, or some other identifiable condition. Specific Language Impairment is a family of ailments that strike about 3 percent to 5 percent of children. They speak late, articulate poorly, and have trouble learning to read.[27] Everything improves as they get older, but they sometimes struggle with language all their lives in the way normal people struggle with a foreign language. They often make errors in speaking, particularly with grammatical morphemes, as in *Carol is cry in the church.* Acquaintances that strike you as chronically tongue-tied and inarticulate may have a grown-up form of the impairment. Canadian Prime Minister Jean Chrétien, beloved from coast to coast for mangling the two national languages with equal proficiency, is a prime suspect.[28]

Chrétien's son and brother also have language problems, and every study that has looked at the relatives of people with language impairment has found that the impairment runs in families.[29] In the KE family the inheritance pattern was striking. Among the thirty-one members spanning three generations, half were impaired, and their distribution in the family tree would make any geneticist predict that the syndrome is controlled by a single dominant gene, or a string of genes lying next to each other on a chromosome. That prediction was stunningly confirmed in 1998 when a team of geneticists took blood samples from twenty-seven members of the family and found a small stretch on the long arm of chromosome 7 that correlated perfectly with having the impairment.[30] They called it *SPCH1*, the first genetic region specifically linked to a speech and language disorder. The region contains several genes whose products are active in the brain, including a protein that may play a role in the growth and differentiation of neural pathways, a molecule that makes neurons stick to other cells, a molecule used as a signal in tissue development, and a kinase, one of a large family of enzymes that change the function of a protein. (Many kinases are thought to have a role in neural development and plasticity.) The geneticists don't yet know whether it is a mutation in one of these genes, or a deletion of several of them, that causes the disorder.

A single gene rarely targets a trait exactly, and *SPCH1* is no exception. Its effect is more like a sloppy brain lesion than a surgical excision of a single organ. The brains of the impaired family members are abnormal in several areas, particularly the frontal lobe and basal ganglia. On top of their deficits with language, the impaired children have difficulty carrying out sequences of mouth or face movements, and they score lower, on average, than their unimpaired relatives on nonverbal intelligence tests. Yet many of the impaired family members have intelligence scores in the normal range, and some test higher than some of their unimpaired relatives. That suggests that the language impairment is not simply a consequence of an overall dulling of the brain; instead it appears to be one of several abilities compromised by the genetic defect. Nor is the language deficit reducible to the articulatory problems that the impaired members of the family had as children; they make errors in writing, comprehension, and judging the grammaticality of sentences, not just in speaking.[31]

The impaired members of the KE family often omit or misuse inflections (a common problem among language-impaired people), but their ability to name objects was not as severely impaired. They should, then, find regular nouns and verbs harder to inflect than irregular ones, and should have trouble inflecting novel words in a *wug*-test. Ullman and the linguist Myrna Gopnik tested the prediction, as did the psychologist Faraneh Vargha-Khadem and her colleagues. Novel words indeed were vexing; some of the impaired members were at a complete loss as to what to do with them, and most of the others inflected them less than 10 percent of the time. Contrary to our prediction, however, regular verbs were no harder on average than irregular verbs; both were fairly hard. Ullman and Gopnik discovered one reason why regular words were no harder then irregular words: Some of the family members consciously applied a rule they had been taught in school. One muttered 'add an *s*' to herself, another proudly announced that he remembered to use the rule drilled into him by his teacher.

The other members of the family, Ullman and Gopnik conjectured, may have done passably well with regular verbs for a different reason. Unlike stroke victims, people with SLI grow up with their impairment and have opportunities to compensate by using other strategies. They may memorize regular past-tense forms as if they were irregular, and recall them from memory when they need them. That would explain why they were baffled at the *wug*-test but did better with real regular verbs. This conjecture led to a successful prediction: The impaired family members should be highly sensitive to the frequency of a regular past-tense form in the language, doing well only with the common past-tense forms, unlike unimpaired people, who inflect rare and common regular verbs with equal ease.[32]

Ullman replicated this test with another group of language-impaired children who offered cleaner scientific tests of theories about language. The KE family first came to the attention of researchers because of their striking pattern of inheritance, not because of the details of their impairment. The psychologist Heather van der Lely has screened many language-impaired children and selected a few whose impairments are strictly confined to language, indeed to the grammatical computations underlying language. Fewer than a fifth of children given the label of Specific Language Impairment meet those criteria. The children in her group are average or above average in nonverbal intelligence, and they speak clearly and accurately. Van der Lely found that 78 percent have first degree relatives with a history of language impairment. Half of their siblings were affected, brothers and sisters equally, and though often one parent was affected, in no case were both parents affected. The pattern suggests that 'Grammatical SLI,' as van der Lely calls it, may be caused by a single dominant gene.[33]

One boy, AZ, showcases the specificity of Grammatical SLI. His nonverbal IQ ranges from 119 to 131, putting him in the top 10 percent of the population. Yet when he was first tested at the age of ten, his ability to complete sentences, and to understand sentences whose meanings depended on their syntax (such as *The boy is tickled by the girl*) was at the level of a five-year-old. When speaking, he left out inflections 75 percent of the time, as in *My Dad go to work*. He often left out entire phrases, as in *The dog was poking in*, meaning poking his head in a jar. And he avoided recursive sentence structures common in the speech of four-year-olds, such as *Can you ask Mum if I can have an ice cream?*

AZ's problems with language were concentrated in grammar. His vocabulary was below average but not as dramatically so as his grammatical abilities. And he had no trouble whatsoever reasoning with words or using language in socially appropriate ways. He was fine in tests of deductive reasoning; for example, when told that 'Mary has never flown,' he correctly inferred that Mary has never been in a helicopter. He completed verbal analogies such as 'Kipper is to fish as cheddar is to ____.' And he never made the egocentric error typical of younger children: opening a conversation with *he* or *she* without stopping to think that the listeners have no idea who *he* or *she* refers to.

Ullman and van der Lely gave the group of grammatical SLI children, who were nine to twelve years old, a list of verbs to put in the past tense, and compared their performance with the performance of control groups of unimpaired children matched in tests of sentence comprehension (around five or six years old) and matched in vocabulary (around seven or eight years old). Obviously the impaired children would have been trounced by a control group matched in age; these younger control groups, matched instead in language abilities, were intended to highlight qualitative differences in the impaired children's language, as opposed to mere delays in their timetable that would have made them like younger children.[34]

The grammatical SLI children were desperately bad in the *wug-*

tests, inflecting only about 7 percent of the verbs. They were almost as bad at inflecting low-frequency regular verbs such as *to flap*, succeeding 11 percent of the time. The control groups, which were much younger, did three to seven times better. The impaired children did better with higher-frequency regular verbs like *rob* than with lower-frequency regular verbs like *flap*, unlike the control children, who were no better with the common regular verbs than with the rare ones.[35] The impaired children were no better with low-frequency regular verbs than with low-frequency irregular verbs; the control groups were up to twice as good.

Evidently the SLI children were memorizing their regular forms. In an ingenious follow-up van der Lely found a way to corroborate this conclusion. Recall that both adults and unimpaired children say that a monster who eats mice is a *mice-eater*, but that a monster who eats rats is a *rat-eater*, not a *rats-eater*. That is because *mice* is a stored root, just like any other simple word, and is available for insertion into a compound, whereas *rats* is formed by a rule that creates a complex word later in the processing stream. Yet the SLI children, unlike the control groups, were happy to say *rats-eater*; they said it almost as often as they said *mice-eater*. That suggests that their mental representations of regular and irregular forms work the same way.[36]

All this suggests that the loss of certain genes can interfere with the development of normal grammatical circuitry in the brain, including the ability to inflect new and uncommon regular verbs. Children lacking these genes can learn to compensate by relying more on memory.

～

Are there genetic disorders that go the other way, with preserved language and impaired intelligence? That double dissociation would be good evidence that the human genome codes for a brain in which language is a distinct system. If language were simply another accomplishment of a general-purpose intelligence, then any impairment of intelligence would have to impair language as well.

In a recent paper the psychologist Ursula Bellugi and her colleagues discuss a girl they have worked with for several years:

In describing her future aspirations, Crystal, a 16-year-old adolescent, states: 'You're looking at a professional book writer. My books will be filled with drama, action, and excitement. And everyone will want to read them. I'm going to write books, page after page, stack after stack . . . I'll start on Monday.' Crystal describes a meal as 'a scrumptious buffet,' an older friend as 'quite elegant,' and her boyfriend as 'my sweet petunia'; when asked if someone could borrow her watch, she replies, 'My watch is always available for service.' Crystal can spontaneously create original stories – she weaves a tale of a chocolate princess who changes the sun color to save the chocolate world from melting; she recounts with detail a dream in which an alien from a different planet emerges from a television. Her creativity extends to music; she has composed the lyrics to a love song.

In view of her facility with language, proclivity for flowery, descriptive terms, and professed focus on drama and action, her aspiration may seem plausible; but in fact Crystal has an IQ of 49, with an IQ equivalent of 8 years. At the age of 16, she fails all Piagetian seriation and conservation tasks (milestones normally attained in the age range of 7 to 9 years); has reading, writing and math skills comparable to those of a first or second grader; demonstrates visuospatial abilities of a 5-year-old; and requires a babysitter for supervision.[37]

Crystal has Williams syndrome, a rare form of retardation accompanied by heart and circulatory defects, an elfin or pixielike face, and abnormal calcium metabolism. Together with their excellent language skills, people with Williams syndrome have other islands of preserved ability: They are friendly to strangers, good at recognizing faces, and competent at inferring what other people are thinking.[38]

Recently the genetic defect behind Williams syndrome was identified: a deletion of about ten adjacent genes on the long arm of chromosome 7 (the same chromosome as *SPCH1*, though in a different place).[39] Different parts of the syndrome can be traced to different missing genes. The absence of a gene for a protein called elastin causes blood vessel defects, and astonishingly, the absence of a kinase gene, *LIM-kinase1*, is responsible for their terrible spatial abilities. People who have lost only the elastin and *LIM-*

kinase1 genes, but not the other genes, have circulatory problems and terrible trouble in spatial reasoning, such as arranging blocks, assembling toys, or copying simple shapes. But they are not retarded; in fact they are unimpaired in every other way. *LIM-kinase1* is active in fetal and adult brains, and helps to regulate the tiny filaments found in the fingerlike projections of growing neurons.[40] Presumably *LIM-kinase1* plays an important role in the development of the neural networks used in spatial reasoning, possibly in the parietal lobes. The other missing genes, perhaps, are necessary for the development of other parts and processes of the brain, though not for language or face perception. Neuroimaging studies have shown that the brains of people with Williams syndrome are smaller overall, and are different in many subtle ways.

Children with Williams syndrome are slow in beginning to talk, but they take off in late childhood and adolescence. Their speech is grammatically complex and largely without error, and they understand sentences whose meanings depend on their grammatical structure, such as *The truck is pushed by the car* and *Is every dancer pinching her?* In all of the grammatical tests on which children with Specific Language Impairment do poorly, children with Williams syndrome do well.[41]

The vocabularies of the children are good for their mental age, and they can generate lists of words (say, animals) as quickly as normal adolescents. Yet something about their word use is not quite normal. Listeners are struck by their recherché and slightly off-target word choices, such as *toucan* for a parrot, *evacuate the glass* for emptying it, and *concierge* for an usher. When asked to list members of a category they come up with unusual examples, such as *shrike* and *spearhawk* for birds and *teriyaki* and *chop suey* for foods. They easily think of the secondary meanings of ambiguous words, such as 'fastener' for *nut* or 'weapon' for *club*. It's not that their mental dictionaries are thoroughly disordered; when they see the word *hen*, they are quicker to recognize *farm*, just like other people. But the fine points that govern word choice in the rest of us are not quite in place, and the children have shown other anomalies in how they learn and react to words.[42]

When I first heard Bellugi talk about Williams syndrome, I shot up in my seat when she casually mentioned that their only obvious grammatical errors consist of overgeneralizations like *catched* and *sleeped*. It makes perfect sense: Their grammar is running smoothly, but their word fetcher doesn't have the usual bias to fetch frequent and appropriate words quickly. Irregular verbs survive on that bias, so occasionally an irregular form doesn't pop into mind quickly enough, and the rule is ready and waiting to step in. Bellugi and her collaborator and husband, the linguist Ed Klima, sent me their data on Crystal's verb use. Indeed she overgeneralized *-ed* to irregular verbs 16 percent of the time, more than three times the average rate of unimpaired preschool children. That's just one sample from one child, but the finding has now been replicated and extended in two new groups of Williams syndrome children. Hilary Bromberg, working with Ullman, Marcus, and Kara Kelly and Karen Levine of the Children's Hospital in Boston, and Harald Clahsen, working with Mayella Almazan in England, found that people with Williams syndrome inflected regular nouns and verbs extremely well, did beautifully in *wug*-tests, and frequently overgeneralized *-s* and *-ed* to irregular nouns and verbs.[43]

What about evidence that the memory system for irregular forms is out of order? The psychologist Annette Karmiloff-Smith showed that French-speaking Williams syndrome children have trouble guessing the gender of nonsense French nouns from their sounds — for example, guessing that *bicron* is masculine and *faldine* is feminine. In French the gender of a new word is not dictated by rule; speakers analogize it to the closest-sounding words in memory, just as everyone does with irregular words. The failure of children with Williams syndrome to make good guesses about gender is another hint that their patterns of word associations in memory are unusual.[44]

The explanation is not perfect. The anomalies in the mental lexicons of people with Williams syndrome are still poorly understood, and there is no direct evidence that they are sluggish in retrieving common appropriate words, which is what the explanation requires. But overall, the genetic double dissociation is striking, suggesting that language is both a specialization of the brain and that it depends on generative rules that are visible in the ability to compute regular forms. The genes of one group of children impair their grammar while sparing their intelligence;

the genes of another group of children impair their intelligence while sparing their grammar. The first group of children rarely generalize the regular pattern; the second group of children generalize it freely.

~

Neuroscientists often depend on brain lesions and genetic knockouts to understand what different parts of the brain are for. They are more confident, though, when they can also record from a part of the brain and actually see it springing into activity when doing its purported job. Can we spy on parts of a healthy human brain and see them using words or rules?

Two techniques are widely used in cognitive neuroscience, differing in their ability to measure the brain in time and in space. Electroencephalography is like a radio broadcast: You can follow the action moment by moment, but are never sure where anything is happening. Functional neuroimaging is like Victorian photography: The pictures are filled with detail, but the subjects have to stay motionless while the picture is taken or they will be blurred beyond recognition. Both techniques recently have been applied to regular and irregular inflection.

The electroencephalogram, or EEG, is familiar to many people from hospitals. Electrodes are pasted all over the scalp and weak electrical signals coming from the brain are sent along a Medusa-like bundle of wires to an amplifier and, in the old days, to a set of pens wiggling madly over a running sheet of graph paper. Nowadays the signals are digitized and stored in a computer. The electrical signals come from neighborhoods of neurons that are active at the same time; they generate electrical currents, which are conducted by the tissues of the brain, skull, and scalp. Those tissues are pretty good conductors, so a signal measured at any one part of the scalp is a cacophony of billions of neurons screaming with different rhythms from all over the brain. But if you present a word to a person hundreds of times, begin measuring the signal from the moment the word is presented, and average the signals, then all the screaming not elicited by the word cancels out and you have a picture of the brain's electrical response to the word itself. This response is called an Evoked

Potential or an Event-Related Potential, ERP for short. Generally one can't tell where in the brain an ERP comes from, but activity in different parts of the brain will show up as stronger or weaker signals at the different electrodes, so activity in one part can sometimes be distinguished from activity in another part.

The ERP signal is a train of peaks and troughs of voltage that come from different way stations in the brain. The early ones are echoes of the processing of the raw sights and sounds, but the later ones reflect the recognition and analysis of the word, and they can vary up or down when the person is paying attention to the word or is surprised by it. Many of these blips have been identified, named, and linked with particular stages of cognitive processing. One example is the N400, a negative blip in the signal about 400 milliseconds after a word is presented. A word evokes an N400 when it makes no sense in context. For example, as you read the sentence *He spread his warm bread with socks,* your brain gives off an N400 four tenths of a second after your eyes alight on *socks.* A similar response can be elicited by a nonword such as *fep* or *blicket.*

A different kind of blip, called a Left Anterior Negativity or LAN, builds up more gradually, often peaks later, and is picked up most strongly by electrodes at the front of the head on the left side. A word evokes an LAN when it makes a sentence ungrammatical. When you read the sentence *The teacher is being fallen,* your brain gives off an LAN between three and seven tenths of a second after your eyes hit *fallen.*[45]

Harald Clahsen and Thomas Münte, one of the discoverers of the LAN, reasoned that these electrical signatures can tell us whether the brain thinks it is dealing with a misselected word or a violation of grammar. They and their colleagues showed German-speaking subjects a set of words with correct and incorrect plural suffixes. The incorrect ones were expected to get a rise out of the brain, and they were either irregular nouns with a regular suffix such as *Bauer-s* 'farmers' (it should be *Bauer-n*), or regular nouns with an irregular suffix such as *Auto-n* 'cars' (it should be *Auto-s*). The illicit regular suffix elicited an LAN, as if the brain was recoiling from an incorrectly applied rule of grammar. But the illicit irregular suffix elicited an N400, as if the brain was recoiling from a weird word. This is exactly what you would expect if regular suffixes are applied by rule and irregular suffixes are stored on words. The team got similar results in two replications, one showing German participles to German speakers and one showing Italian participles to Italian speakers. Ullman, working with Aaron Newman and Helen Neville (another discoverer of the LAN), got similar results by showing English speakers a list of verbs missing their past-tense markers. Together these studies show that the difference between words and rules can be read from the electrical startles of the healthy brain.[46]

~

As ethereal as our thoughts may feel to us as we think them, they are incarnated in living flesh that must be bathed in blood to get its energy and oxygen. When brain tissue is working harder, it calls more oxygenated blood its way. That is the basis for two amazing technologies of functional neuroimaging, Positron Emission Tomography (PET) and functional Magnetic Resonance Imaging (fMRI).[47]

In Positron Emission Tomography, a person engaged in a particular task, such as reading words or looking at pictures, is injected with a glob of mildly radioactive water, which soon circulates to the brain. When an oxygen atom in the water molecule decays, it emits a positron (the positively charged antimatter version of the electron), which soon collides with a nearby electron, annihilating them both and sending gamma rays shooting out in opposite directions. A ring of gamma-ray detectors surrounds the head, and two of them pick up the simultaneous arrival of the gamma rays, revealing the spot between them at which the annihilation occurred. The various spots are accumulated for about forty seconds, painting a picture of the blood flow in a cross-sectional slice of the brain aligned with the detectors. The picture is shown in color, with the active areas in yellows and reds, the quiescent ones in greens and blues.

Unfortunately the picture displays *all* the brain areas that were active in those forty seconds, and thus picks up everything the person may have been thinking and feeling in the interval: itches, daydreams, curiosity about the point of the experiment, claustrophobia, impure thoughts about the attractive technician, and so on. The blobs showing the brain areas for reading or understanding cannot be distinguished from blobs showing the brain areas for everything else. One solution is to scan people's brains twice, once while they are *not* doing the task, once when they are doing it, and subtract the first image from the second. Better still, an image of the brain doing a simple task can be subtracted from an image of the brain doing a slightly more complicated task, revealing the sites of the extra mental processes required by the more complicated task. For example, if you subtract an image of a person reading nonsense words such as *bluck* from an image of the person reading real words such as *black*, the difference in blood flow should pick out the parts of the brain that handle the meaning of a word as opposed to its look and sound. Naturally, this logic is only as good as the psychologist's theory of which tasks engage which mental processes. If the two tasks in fact are equally complicated and evoke overlapping sets of brain areas, rather than one task evoking a subset of the brain areas evoked by the other task, an image of the difference between them will be uninterpretable.

In Magnetic Resonance Imaging a person slides his head into the bore of a strong magnet, which pulls many of the atoms in the brain into alignment with its magnetic field. Radio waves are then sent through the brain, which makes the atoms tilt; when the waves are turned off, they wobble back into alignment, giving off a weak radio signal. The molecules in the brain become tiny radio transmitters, each kind of molecule having a characteristic frequency, and the radio signals are picked up by receivers surrounding the head. By playing with the shape of the magnetic field and the frequencies of the radio pulses, engineers can arrange for the molecules to announce *where* they are, not just what they are, and a computer can generate a crisp black-and-white photograph of a cross-section of the brain. By then comparing the radio signature of oxygenated hemoglobin (the molecule in the blood that carries oxygen to tissues) with the signature of deoxygenated (spent) hemoglobin, the computer can color in the parts of the brain receiving more oxygenated blood. This provides the 'f' in fMRI: a picture of the functioning of the brain, not just its anatomy. Functional magnetic resonance imaging is slowly taking over from PET in cognitive neuroimaging because it uses no radioactivity, gives sharper pictures, and does not need as much time to build up the image.

The next step is obvious: scan people's brains while they are generating regular and irregular past-tense forms, and see whether different areas light up, as predicted by the words-and-rules theory. I had planned such an experiment with one of the major PET research centers, but other people had the idea too, and we were scooped by four different labs. The good news is that all four found that regular and irregular forms are computed in different parts of the brain. The bad news is that they disagree on which parts handle the regulars and which parts handle the irregulars.[48]

Each study produced a different pizza of active and inactive blobs for the regular and irregular tasks. And the simplest pattern one might have hoped for – more activity in left frontal areas with regular verbs, more activity in left parietal and temporal areas for irregular verbs – did not leap out of the combined data. There are intriguing hints of it in some of the studies, as well as hints of basal ganglia involvement in regular inflection,[49] but no pattern was consistent across all the studies. I can imagine many reasons for the discrepancies: The experiments used different neuro-imaging techniques, different languages, different tasks, different subtractions, and different designs, each with strengths and weaknesses. I can also imagine a more interesting reason: Language processing embraces many more steps and areas than the simple front-back distinction allows for, and the scans are picking up the various areas and blurring them into a single image. For example, one area of the frontal lobe seems to be engaged when people try to think of words that meet some criterion, such as an action that goes with some object.[50] Perhaps it is also engaged in directing a search for the irregular past-tense form that goes with some stem, and this task requirement confounds the equating of frontal areas with regular inflection. Often, when a cognitive process is first put under the beam of a scanner, the early studies contradict each other. But the kinks eventually get worked out, and I suspect this will happen with regular and irregular inflection. At the very least, the worst nightmare for the words-and-rules theory did not come true: regular and irregular verbs lighting up the same brain areas.

One other technique has cognitive neuroscientists excited, and it may help to reconcile the neuroimaging results with all the others. Magnetoencephalography, or MEG, has the promise of combining the moment-by-moment precision of ERP with the localization in space of PET and fMRI. In theory it can provide a movie of which part of the cerebral cortex is most active at every step between stimulus and response.

MEG feeds off the same neural events that create the ERP signal: an electrical current flowing down the dendrites of a swatch of neurons that are simultaneously active. You may remember from high school science that an electrical current gives rise to a magnetic field wrapped around it, like the fingers of your right hand curling around the axis of your outstretched thumb. The magnetic field thrown off by neural activity, unlike the electrical field, is not badly distorted as it passes through the tissues of the brain, skull, and scalp, and if it could only be recorded, the source of the field could be reconstructed by computer, a bit like guessing the position of a magnet under a piece of cardboard from the curving lines of iron filings above it. The technical problem is that the brain's magnetic field is unimaginably weak, and it is swamped by other fields, such as that of the Earth; measuring it has been compared to listening for the footsteps of an ant during a rock concert. When wires are cooled to within a few degrees of absolute zero, however, they become superconductors that can be traversed by infinitesimal currents, and with a lot of wizardry they can be fashioned into detectors of these weak magnetic fields. The detectors line a head-shaped plastic cavity bathed in liquid helium, and when a person inserts his head, the magnetic activity of his brain can be recorded.[51]

MEG would seem to be perfect for watching language unfold in time, and in our first experiments with the technique, Jaemin Rhee, Ullman, and I have caught a glimpse of an interesting trajectory. About a quarter of a second after people see a word and begin to generate its past-tense form, their brains are active in the left temporal-parietal regions, where presumably the word stems are recognized and memory is searched for any irregular forms. About a tenth of a second later, with regular verbs and only regular verbs, the activity shifts to the left frontal lobe, where we suspect that the suffixing operation is carried out.[52] This is exactly the trajectory we had predicted from the earlier experiments with neurological patients and ERPs, and the results might also help make sense of the neuroimaging studies, which blurred together different blips of brain activity that we can now see are nicely separated in time.

Whatever the outcome of the past-tense treasure hunt, I hope it will be emblematic of a trend in intellectual life in the coming millennium that the biologist E. O. Wilson has called consilience: the unification of the arts and sciences by an understanding of mind, brain, and human nature.[53] Regular and irregular inflection has long been mulled over by novelists and poets, dictionary writers and editors, philologists and linguists. Now this topic straight out of the humanities is being probed with the cutting-edge tools of molecular genetics and imaging of the brain. Some people fear this kind of development as a crass 'reductionism' that will marginalize the humanities and plough under the richness of their subject matter, but it is far from that. Without an understanding of the contents of the mind from psychology, linguistics, and all the other disciplines they touch, neuroscientists would not know where to begin in studying the human brain, and their technologies would be expensive toys. Ultimately all knowledge is connected, and insight into a phenomenon can come from any direction, from the outcome of the Battle of Hastings to the sequence of a kinase gene.

A DIGITAL MIND IN
AN ANALOG WORLD

The ingredients of language are words and rules. Words in the sense of memorized links between sound and meaning; rules in the sense of operations that assemble the words into combinations whose meaning can be computed from the meanings of the words and the way they are arranged. I have tried to convince you of this simple idea, and to illuminate some of the wonders of language, by exploring the ins and outs of a single curiosity in which the two ingredients may be contrasted.

Regular and irregular forms in English are the same size – one word long – and express the same ideas – past tense or plural. Yet the human mind treats them differently. Irregular forms fail to show up, and the regular pattern makes itself available, in a variety of cases that have nothing in common but a failure of access to information in memory. We have seen the regular form surface when a word is new, rare, unusual, without a standard root, or without a way for information in the root to apply to the whole word. We have seen it surface when the memories of words are freshly formed in children and when they have decayed from disease in adults.

This is an odd assortment of circumstances, some rather exotic. Surely the mind is not equipped with features designed to give rise to each of them. The simplest explanation is that regular inflection is computed by a mental operation that does not *need* access to the contents of memory: a symbol-processing rule, which attaches a suffix to any word that bears the mental symbol 'verb' or 'noun.'

We also have seen that the power of a rule to serve as a default, stepping in when memory and analogy fail, can be observed in languages from all over the world. It does not depend on a word's being frequent in the language, nor on their having distinctive sounds. Children have a sense of the kinds of words to which a rule may apply and the kinds of words to which a rule may not apply, even when they have never encountered those kinds of words before. All this suggests that a rule does not gain its power from having been pounded into the child's mind. Instead it may gain its power from the very nature of the child's mind.

I believe that regular and irregular forms show us the mental mechanisms that lie behind the two principles of language. A memory system stores and retrieves words, implementing Ferdinand de Saussure's principle of the arbitrary sign. A system of symbolic computation generates grammatical combinations of words, implementing Wilhelm von Humboldt's principle of the infinite use of finite media. Together they explain the vast expressive power of language, the ability to convey an unlimited number of new ideas.

I want to leave you with a remarkable parallel between regular and irregular inflection and something completely different. The parallel cannot be a coincidence, and it hints that the distinction between regular and irregular forms may expose even deeper principles about the nature of the mind and how it reflects the world.[1]

~

People think in categories, like 'furniture,' 'vegetable,' 'grand-mother,' and 'turtle.' The categories underlie much of our vocabulary – such as the words *turtle* and *furniture* – and they underlie much of our reasoning. We are not dumbfounded by every new turtle we see; we categorize it as a 'turtle' and expect it to have certain traits, like being slower than a hare and with-drawing into its shell when frightened. This means that before-hand we did not mindlessly record every turtle we had seen, like a video camera; we must have abstracted what turtles have in common. To understand mental categories is to understand much of human reasoning.

Concepts in the mind pick out categories in the world, and the simplest explanation of concepts is that they are conditions for membership in a category, a bit like definitions in a dictionary. An 'odd number' is an integer that when divided by two leaves a remainder. A 'bachelor' is an unmarried adult male. A 'grand-mother' is the mother of a parent. 'A 'turtle' is a reptile with a broad flattened body enclosed in a shell formed of a dorsal carapace and a ventral plastron, united at the sides.

The power of a definition is that it transcends the particulars of experience. People can recognize a new turtle when they see one, as long as it conforms to the definition. Psychologists call these categories 'classical' or 'Aristotelian' categories, after the Greek philosopher who emphasized logic and definitions as the basis of knowledge. For decades psychologists studied concept learning in humans and animals by presenting them with drawings of colored shapes, indicating which ones belonged to a category such as 'large red square,' and measuring how long it took the subjects to infer the category.[2]

All this was challenged by the Austrian philosopher Ludwig Wittgenstein in a famous passage from his *Philosophical Investiga-tions*, a collection of ruminations published after his death in 1951:[3]

66. Consider for example the proceedings we call 'games.' I mean board-games, card-games, ball-games, Olympic games, and so on. What is common to them all? – Don't say: 'There *must* be something common, or they would not be called "games"' – but *look and see* whether there is anything common to all. – For if you look at them you will not see something that is common to *all*, but similarities, relationships, and a whole series of them at that. To repeat: don't think, but look! – Look for example at board-games, with their multifarious relationships. Now pass to card-games; here you will find many correspondences with the first group, but many common features drop out, and others appear. When we pass next to ball-games, much that is common is retained, but much is lost. – Are they all 'amusing'? Compare chess with noughts and crosses [tic-tac-toe]. Or is there always winning and losing, or competition between players? Think of patience [solitaire]. In ball games there is winning and losing; but when a child throws his ball at the wall and catches it again, this feature has disappeared. Look at the parts played by skill and

luck; and at the difference between skill in chess and skill in tennis. Think now of games like ring-a-ring-a-roses; here is the element of amusement, but how many other characteristic features have disappeared! And we can go through the many, many other groups of games in the same way; can see how similarities crop up and disappear.

And the result of this examination is: we see a complicated network of similarities overlapping and criss-crossing: sometimes overall similarities, sometimes similarities of detail.

67. I can think of no better expression to characterize these similarities than 'family resemblances'; for the various resemblances between members of a family: build, features, color of eyes, gait, temperament, etc. etc. overlap and criss-cross in the same way. – And I shall say: 'games' form a family.

And Wittgenstein did not live to see Doom, professional wrestling, or Six Degrees of Kevin Bacon.[4] As he noted, a category can be extended to embrace new cases 'as in spinning a thread we twist fibre on fibre. And the strength of the thread does not reside in the fact that some one fibre runs through its whole length, but in the overlapping of many fibres.'

In the 1970s the psychologist Eleanor Rosch brought Wittgenstein's ideas into psychology by showing that many human concepts picked out family resemblance categories rather than classical categories.[5]

First, with most categories it is almost impossible to find a set of membership conditions. If the definition of a 'turtle' includes having a shell, what do we do with leatherbacks and other soft-bodied turtles? If a 'bachelor' is an 'unmarried man,' does that mean the Pope is a bachelor? A 'chair' needn't have legs or a seat or a back; think of that staple of the 1970s bachelor pad, the beanbag chair. Nor must it be capable of supporting a seated human – think of the Hollywood prop that disintegrates into smithereens when the bad guy smashes it over the head of the good guy. The general point is illustrated here by Opus the Penguin:

BLOOM COUNTY by Berkeley Breathed. *Reprinted with the artist's permission.*

Second, the members of a category are not created equal, which is what one would expect if they were admitted into the category by meeting the definition. Everyone agrees that a blue jay is somehow a better example of a bird than a chicken or a penguin, and that an armchair is a better example of furniture than a grandfather clock. The best member of all is called the prototype, such as the sparrow for 'bird' and a wrench for 'tool,' and it sums up the category in people's minds. Dictionaries often show a prototype next to the definition of a category. Next to the entry for *bird* you are likely to see a picture of a sparrow or a robin, not a picture of a turkey or a kiwi.

Third, the categories of the mind have fuzzy borders. People aren't quite sure whether garlic, parsley, seaweed, or edible flowers should count as vegetables, and the Reagan administration created a ruckus when it justified cut-backs in funding for school lunches by reclassifying ketchup as a vegetable. If a clamp is a tool, why not a ball of string? Is a scorpion a bug? Is a sport utility vehicle a car or a truck? Is synchronized swimming a sport?

Fourth, most of our everyday categories, and not just games, show Wittgenstein's family resemblance and crisscrossing features. Many vegetables are green, but carrots aren't; many are crunchy when raw, but spinach isn't. As for chairs, this cartoon from *The New Yorker* says it all:

"*Attention, everyone! I'd like to introduce the newest member of our family.*"

Fifth, categories have stereotyped features: traits that everyone associates with the category, even if they have nothing to do with the criteria for membership. When people think of a grandmother, they think of gray hair and chicken soup, not of a node in a genealogical tree.

Many experiments have confirmed that everyday concepts act like family resemblance categories.[6] People are comfortable with the very idea that categories have better and worse members: They have no trouble rating the 'goodness' of the members of a category on a scale of 1 (best) to 7 (worst). For example, they give a robin an average rating of 1.1 on the bird scale, and a chicken a rating of 3.8. Football was judged a fine example of a sport, earning a rating of 1.2; wrestling was a so-so example, eking out a 4.7. A carrot is a vegetable par excellence (1.1), but parsley is a more dubious instance (3.8). Murder is an excellent crime at 1.0, but vagrancy is not so good at 5.3. The ratings of different people agree closely.

When people are shown pictures of objects and asked to press a button if the object belongs to a named category, they press the 'fruit' button more quickly to a picture of an apple than to a picture of a watermelon. That suggests that the category 'fruit' is more easily evoked in people's minds by the apple. Rosch asked people to make up sentences with category words such as 'bird.' Typical responses were 'I heard a bird twittering outside my window' and 'Three birds sat on the branch of the tree.' Then she replaced the word 'bird' by various species: sparrow, penguin, eagle, ostrich. The absurdity of 'I heard a penguin twittering outside my window' and 'Three ostriches sat on the branch of the tree' shows that it must have been prototypical birds that had popped into the subjects' minds. Children have similar intuitions: When they first learn a word, they use it with prototypical members of a category: *bird* is used with sparrows, *vegetable* with carrots or celery.

It's also easy to show in the lab that people are fuzzy about borderline cases. The psychologists Michael McCloskey and Sam Glucksberg asked subjects to give true-or-false verdicts on category membership. Everyone agreed that cancer is a disease, that apples are fruit, and that flies are insects. But when it came to deciding whether stroke is a disease, a pumpkin is a fruit, or a leech is an insect, half the subjects went one way and half went the other – and when they were asked again a month later, many changed their minds.[7]

Does this mean that people's heads are stuffed with fuzz and that classical categories are fictions? Surely not. People can learn categories with clean definitions, crisp edges, and no family resemblance, such as 'odd number.' They can learn that a dolphin is not a fish, though it has a strong family resemblance to the fishes, and that a seahorse is a fish, though it looks more like a little horse. They can understand that Tina Turner is a grandmother, though she lacks all the usual traits, and that my childless great-aunt Bella was not a grandmother, though she had gray hair and made a mean chicken soup. Though people refer to women in their third trimester as 'very pregnant,' they also understand what it means when parents say to their daughters, 'You can't be just a little bit pregnant.'[8]

The psychologists Sharon Armstrong and Henry and Lila Gleitman replicated Rosch's experiments using the most classical, Aristotelian categories they could find, 'odd number' and 'woman.' The subjects rated '7' as an excellent example of an odd number, and '447' as not such a good example; they thought that a 'housewife' was an excellent example of a woman, and a 'policewoman' not such a great example. The same gradations emerged in their real-time mental processes: They pushed an 'odd number' button more quickly when '3' flashed on the screen than when '2,643' did. Surely those students would not have made it into the prestigious University of Pennsylvania if they really thought that numbers could be more or less odd, and indeed in a questionnaire they averred that a number was either even or odd, with no in-between cases. So they must have been capable of turning their fuzziness on and off. Family resemblance categories are real, but so are classical categories; they live side by side in people's minds, as two ways of construing the world.[9]

~

What does this have to do with regular and irregular verbs? The psychologist Dan Slobin and the linguist Joan Bybee were the first to point out that classes of irregular verbs with similar past-tense forms, such as *sing–sang, ring–rang, drink–drank* and *bind–bound, find–found, grind–ground,* are just like Wittgenstein's family resemblance categories.[10] All five of their distinguishing traits can be found in the irregulars.

First, despite the contortions of centuries of language scholars, no one has been able to craft a set of rules that properly pick out the different kinds of irregular verbs. As Mark Twain said of a German grammar book, there are more exceptions to a rule than there are instances of it. For example in English the largest family of irregular verbs are no-changers like *rid–rid, cut–cut,* and *set–set.* They all end in *t* or *d,* but there is no hope of lassoing the family with a rule stipulating that verbs ending in *t* or *d* belong to it. Next to *hit–hit, slit–slit, split–split,* and *quit–quit* we find regular *flit–flitted, twit–twitted,* and *pit–pitted.* Near *let* and *set* we find regular *fret, sweat,* and *whet.* Beside *cut* and *shut* we find *butt, jut,* and *strut.* Adjacent to *hurt* we find *blurt* and *spurt*; near *burst* we find regular *bust.* All the other classes of irregulars have rule defeaters too, as we saw in chapter 5.

Second, in every irregular family some members are more equal than others. *Hit* and *split* are full-fledged no-changers, but in the minds of Americans *spit–spit* and *forbid–forbid* are so-so. Ditto for the other families:

Good Examples	Poor Examples
bleed–bled, feed–fed	*plead–pled, speed–sped*
burn–burnt, bend–bent	*learn–learnt, lend–lent, rend–rent*
deal–dealt, feel–felt, mean–meant	*kneel–knelt, dream–dreamt*
freeze–froze, speak–spoke	*weave–wove, heave–hove*
get–got, forget–forgot	*beget–begot, tread–trod*
write–wrote, drive–drove, ride–rode	*smite–smote, strive–strove, stride–strode*

Many of the classes have a prototype or best kind of member. For the *ing–ung* family it is verbs that fit the pattern *s-consonant-consonant-ing*, such as *string*. Bybee discovered that people are most tempted to grant an irregular form to a made-up verb when the verb matches the prototype, as in *spling–splung* and *skring–skrung*.

Third, in the halo around the poor relations in an irregular family there are verbs so poor that no one knows whether they belong in the family at all:

He has *stridden* around the park three times.
They seem to have *striven* to baffle their readers.
I don't know how she *bore* the guy.
I *forwent* the pleasure of grading papers last night.
The mice *throve* in the compost.

Fourth, the members of irregular families resemble each other in crisscrossing ways, rather than by sharing any trait. Take the second-biggest family of irregular verbs, the *ring–rang–rung* family, which change *ĭ* to *ă* or *ŭ*. Most of the members end with the consonant *ng*, which is velar (pronounced at the velum or soft palate) and nasal (pronounced through the nose): *shrink, sink, stink, cling, fling, sling, slink, sting, string, swing,* and *wring*. That screams for a rule that states, 'Change *ing* to *ung*.' But the rule runs afoul of the crisscrossing resemblances, as we saw in chapter

4. Some family members end in a consonant that is velar but not nasal: *stick, dig, sneak,* and *strike.* Others end in a consonant that is nasal but not velar: *win, spin, swim,* and *begin.* The rule would miss them all.

The other families of irregulars criss and cross as well. *Blow–blew, grow–grew,* and *throw–threw* begin with a cluster of consonants and end in ō. *Draw, fly,* and *slay* have the consonants but not the vowel, and *know* has the vowel but not the consonants. Incidentally, the spelling of *know,* which once reflected its pronunciation, shows that the word *used* to have a consonant cluster, like its relatives, before English speakers stopped pronouncing the *k* in words like *knee, knife, knob,* and *knuckle.* The *ow–ew* class started out neat and became ragged, a fact to which we will return.

Fifth, irregular families have stereotyped features that run in the family but play no role in defining the past-tense form. Take the verbs that change *d* to *t,* such as *bend–bent.* In principle any verb ending in *d* could see it replaced by *t:* Our language could have given us *sled–slet, fold–folt,* and so on. In reality almost all of these verbs end in *-end: lend–lent, send–sent, spend–spent, bend–bent.* Similarly, one can imagine a language in which any ā could become ŏo (as in *foot*), but in English the ās that do give way to ŏo are preceded by a tongue-tip consonant and followed by *k: take–took, shake–shook, forsake–forsook.*

Clusters of irregular verbs pass all five tests of Wittgenstein's family resemblance categories. In his book *Women, Fire, and Dangerous Things* (a family resemblance category in an Australian aboriginal language), the linguist George Lakoff called attention to the fuzziness that lies at the heart of that traditional bastion of rules, grammar. He cited irregular verbs as the ultimate proof of the bankruptcy of the two-thousand-year-old Aristotelian tradition in Western thought that seeks precise definitions for everything in sight.[11]

But Lakoff did not notice that right next door to the irregulars are the *regular* verbs, and they pass all the tests of *classical* categories. Other than verbs with an irregular form in memory, all verbs are members of the regular family in equal standing, simply by meeting the criterion 'is a verb.' As we have seen, regular verbs can have any sound: sounds that are strange in English, as in *ploamphed, oinked,* and *out-Gorbachev'd;* sounds that are already associated with irregular verbs, as in *high-sticked* and *flied out;* and sounds that have rarely or never been heard before, as in *Borked* and *anastomosed.* People find *ploamphed* to be as good a past tense of *ploamph* as *plipped* is of *plip,* and they produce and approve the past-tense forms of rare verbs like *balk* as readily as they do with common verbs like *walk.* The regular verbs do not fall into clusters, have no stereotypes, no family resemblance, and aside from occasional interference from irregular verbs, no fuzzy examples.

Why on earth should irregular verbs act like games and furniture and vegetables, and regular verbs act like grandmothers and odd numbers? Are we seeing the outward signs of some deep common cause, or is it all a coincidence, worthy of attention only from conspiracy buffs? I believe there is something beneath the similarities, and that the facts of regularity and irregularity offer glimmers of insight into the nature of our conceptual categories.[12] These facts shed light on the mental machinery that computes our conceptual categories and on the things in the world that our conceptual categories are good at picking out.

~

Regular and irregular forms coexist but require different computational mechanisms: symbol combination for regular forms, associative memory for irregular forms. The same may be true for classical and family resemblance categories.

Before Rumelhart and McClelland built their pattern associator for the past tense, they built one for conceptual categories. It learned concepts like 'dog,' 'cat,' and 'bagel' by picking up associations among the perceptual features (furry, four-legged, and so on) that tend to co-occur in them.[13] For example, the concept 'cat' was implicit in a pattern of strong connections among units that stand for the typical traits of cats, such as whiskers, meowing, and pointy ears. The pattern associator reproduced most of the signatures of family resemblance categories that Rosch had demonstrated in human beings, such as responding to prototypical cats more strongly than to atypical cats. Many subsequent models have had similar success.[14] That is because a family resemblance category is held together by crisscrossing traits, and a pattern associator is a gadget for learning how traits crisscross.

But just as pattern associators for the past tense are good at some things and not so good at others, so too are pattern associators for concepts. A model that is good at picking up stereotypes is apt to project the stereotype onto atypical objects. One model, for example, when taught that a plate had broken, ignored the teacher and concluded that the object was either a window or a vase, because all the broken objects in its training set were windows or vases.[15] Another, when told that an office had drapes, concluded that it wasn't an office, because all the offices in its training lacked drapes.[16] Gary Marcus has shown that standard pattern associators cannot generalize from 'a skunk has skunk babies,' 'a cat has cat babies,' and 'a bear has bear babies' to 'a greeble has greeble babies' (where a greeble is a newly encountered animal), because they lack a variable, 'X,' that would allow them to learn that 'an X has X babies.'[17] These failures are reminiscent of the past tense models' habit of turning out strange blends, or nothing at all, when fed rare or unusual words.

The facts about verbs and the facts about concepts converge to suggest that the human mind is a hybrid system, learning fuzzy associations and crisp rules in different subsystems. Most of the recent models of human categorization in cognitive psychology (which are designed to capture people's speed and accuracy when learning artificial categories in the lab) are built out of two parts: a pattern associator for categories based on families of similar exemplars, and a rule selector for categories based on rules. The psychologists were forced to these hybrid models because with some categories subjects quickly figure out a rule (such as 'rectangles that are taller than they are wide'), whereas with other categories subjects go by their gut feelings, memorizing some of the examples and classifying the new ones according to how similar they are to the memorized ones. No model that uses a single mechanism to capture people's behavior with every kind of category does as well as the hybrid models.[18] Some modelers even link the rule system to the frontal cortex and the exemplar-based system to the temporal and posterior cortex, much as we did for rules and words in the preceding chapter.

~

Why do we have these two ways of knowing? It is unlikely that natural selection equipped us with mental machinery that is completely out of synch with the world in which we live. Might the difference between classical and family resemblance categories reflect a difference between two kinds of things in the world, or at least two ways of reasoning about things in the world? In the case of the past-tense system, we know the ancestry and logic of the verbs in considerable detail. Perhaps they have something to teach us about the different kinds of conceptual categories.

Irregular forms are relics of history. They fall into families because originally they were generated in matched sets by rules, but the rules died long ago and the families have been disintegrating ever since. Vowels drift, consonants get swallowed, words lose their popularity, dialects break apart or coalesce. After centuries or millennia irregular forms are no longer the orderly outputs of a rule, nor are they a list of unrelated sounds; they are a family resemblance category. A clear example is the verb *know–knew*, which used to have a consonant cluster like its siblings *grow–grew, blow–blew,* and *throw–threw,* but then lost its first consonant, messing up the class.

Children are born into a linguistic world that throws the members of a family resemblance category at them, and they cope quite well. With their pattern-loving memories they reproduce most of their parents' irregular verbs. Occasionally they lose an irregular like *chide–chid* or *seem–sempt,* but occasionally they add one like *kneel–knelt* or *sneak–snuck.* The adopted word shares some traits with its new family, because it was that similarity that attracted it in the first place. But each adoptee brings some unique traits in with it, so the class remains ragged. The next generation also finds itself with a family resemblance category to commit to memory.

Regular past-tense forms, in contrast, have no history. In fact they barely have an existence. Only the past-tense *rule* exists. Children don't have to cope with learning the quirks of regular forms because they don't have to learn regular forms at all. The rule creates them when they are needed, and then they can be thrown away, because the rule is always around to create them again the next time. Now, that is an exaggeration – children have to remember a few regular forms to learn the rule to begin with, and adults certainly do remember many regular forms alongside the rule. Yet once the rule is acquired, the forms don't *need* to exist for speakers to use and understand them. The category of regular forms is not a real category but a virtual category: the list of forms that *would* be created if the rule were allowed to work its way through all the verbs in a person's vocabulary. Children never see the category; what they learn is not a class of regular forms left behind by previous speakers but a rule that matches the rule in other people's heads.

The past-tense rule itself would hardly be worth the trouble were it not part of the magnificent *system* of rules we call grammar. Children are wired to learn that system, which allows us all to convey an infinite number of brand-new thoughts. The category of regular forms is a by-product of the rule system.

The two kinds of conceptual categories, I think, harmonize with two kinds of things in the world in the same way that regular and irregular verbs harmonize with two kinds of things in the minds of other speakers.

To see this we have to begin at the beginning. Why does the mind even *have* categories like 'birds' and 'games'? No two inhabitants of the world are identical, and one can imagine a mind that treated every object as a unique individual, just as we treat our friends as unique individuals. In fact we don't have to imagine such a mind; Jorge Luis Borges has imagined him for us, in his story 'Funes the Memorious':

> We, at one glance, can perceive three glasses on a table; Funes, all the leaves and tendrils and fruit that make up a grape vine. He knew by heart the forms of the southern clouds at dawn on the 30th of April, 1882, and could compare them in his memory with the mottled streaks on a book in Spanish binding he had only seen once and with the outlines of the foam raised by an oar in the Río Negro the night before the Quebracho uprising. . . . A circle drawn on a black-board, a right triangle, a lozenge – all these are forms we can fully and intuitively grasp; Ireneo could do the same with the stormy mane of a pony, with a herd of

cattle on a hill, with the changing fire and its innumerable ashes, with the many faces of a dead man throughout a long wake. . . .

Not only was it difficult for him to comprehend that the generic symbol *dog* embraces so many unlike individuals of diverse size and form; it bothered him that the dog at three fourteen (seen from the side) should have the same name as the dog at three fifteen (seen from the front). His own face in the mirror, his own hands, surprised him every time he saw them. Swift relates that the emperor of Lilliput could discern the movement of the minute hand; Funes could continuously discern the tranquil advances of corruption, of decay, of fatigue. He could note the progress of death, of dampness.[19]

Why are we not like Funes the Memorious? Are we just an anal retentive species that likes to put things into pigeonholes for the sheer orderliness of it all? And if we are, how do we decide on the pigeonholes? There are a frightful number of ways to sort objects into categories – alphabetically, in pairs, according to height, and so on. Why 'birds'?

The answer is that people form categories that give them an advantage in reasoning about the world by allowing them to make good predictions about aspects of an object they have not directly seen. We cannot bring every object home and put it under a microscope or send tissue samples out for lab testing. We have to observe a few traits that the object wears on its sleeve and infer the traits that we cannot see directly. Good categories let us do that. If Tweety has feathers and a beak, Tweety is a bird; if Tweety is a bird, Tweety is warm-blooded, can fly, and has hollow bones. Bad categories do not: If we knew only that Tweety's name begins with a 'T,' nothing of interest would follow.[20]

These inferences work only if the world is properly structured. If a capricious god had assembled every object with a unique, random combination of traits, like the numbers on a lottery ticket, inference would be impossible. The blood of a feathered friend would be cold as often as hot, its bones solid as often as hollow. Luckily for us, we don't live in that world. We live in a lawful world in which traits tend to hang together in the same way in many objects.

Our mental categories are useful because they reflect the lawfulness of the world. In theory, laws could be apprehended in different ways. At one extreme one could extract the underlying laws directly and use them in chains of deduction. An example is using the laws of solid geometry and thermodynamics to predict that small animals lose heat faster than large ones, because heat is lost at surfaces and small things have a greater ratio of surface area to volume than large things. At the other extreme we can assemble an enormous database by measuring every trait of every object we can find, and when faced with a new object, find the closest old object and predict it is similar. If we learn that sparrows lose heat quickly, we guess that starlings do too.

The first method seems powerful and insightful, the second one mindless and drudgelike. But often mortal knowers have no choice but to use the second. As the poet John Ciardi wrote,

Who could believe an ant in theory?
A giraffe in blueprint?
Ten thousand doctors of what's possible
Could reason half the jungle out of being.

Many things we find around us could not be deduced by any body of laws, because they are shaped by myriad events of history no longer visible to us.

Take birds. In the course of evolution a species begins in a population of inter-breeding organisms adapting to an ecological niche. Natural selection 'engineers' the organisms to compete well in that environment, and sexual reproduction homogenizes them. If we could go back and look at the last common ancestors of the birds, they would be as similar as a single species is today. They would be genetically similar because they descended from a common set of ancestors, bred with one another, and underwent natural selection for traits such as wings and a streamlined shape. But that uniformity did not last for long. Their descendants went off in different directions, some becoming nocturnal like the kiwi, some taking to the sea like penguins, some growing large like ostriches. The aftermath of this radiation is a family resemblance category. All birds have common traits such as beaks and wings and feathers because they inherited them from their common

ancestor. But other traits crisscross, rather than running throughout the class, because each species has a unique history in which
some traits were lost and others acquired.

And this brings us to the parallel between irregular verbs and
family resemblance categories. Irregular families were once generated by rules but then accumulated idiosyncrasies, and now they
must be memorized individually. Classes of animal species were
once adapted to a single niche but then dispersed and accumulated idiosyncrasies, and now each species must be learned
about through observation. In each case the surviving similarities
in the family members are too useful to ignore, and the
memory system extracts the patterns rather than filing each
item in a separate slot in memory. The patterns determine the
better and worse members, and they allow a knower to guess that
newly encountered similar items belong to the family.

A comparison of the history of words and the history of species
may strike you as far-fetched, but it has a distinguished back-
ground. Darwin himself illustrated his key idea – that the
similarities and differences among organisms could be explained
by their family history – by analogy to how words change in
languages:

> The formation of different languages and of distinct species, and
> the proofs that both have been developed through a gradual
> process, are curiously parallel. . . . We find in distinct languages
> striking homologies due to community of descent, and analogies
> due to a similar process of formation. . . . The frequent presence
> of rudiments, both in languages and in species, is still more
> remarkable. The letter *m* in the word *am*, means I; so that in the
> expression *I am*, a superfluous and useless rudiment has been
> retained. In the spelling also of words, letters often remain as the
> rudiments of ancient forms of pronunciation. . . . We see
> variability in every tongue, and new words are continually
> cropping up; but as there is a limit to the powers of the memory,
> single words, like whole languages, gradually become extinct. As
> Max Muller has well remarked: 'A struggle for life is constantly
> going on amongst the words and grammatical forms in each
> language. The better, the shorter, the easier forms are constantly
> gaining the upper hand, and they owe their success to their own

inherent virtue.' To these more important causes of the survival of certain words, mere novelty and fashion may be added; for there is in the mind of man a strong love for slight changes in all things. The survival or preservation of certain favoured words in the struggle for existence is natural selection.[21]

The analogy lives today in modern biology and linguistics. Similar statistical techniques are used to find the best groupings of organisms and to find the best groupings of languages, based on the co-occurrences of their traits. When biologists are unsure of which species to lump together in a genus or family, they sometimes take hundreds of measurements of animals' parts and feed them into an algorithm that finds the best categories in which to lump them.[22] Similarly, when linguists are unsure of which languages to lump together in a family, they sometimes feed hundreds of sets of cognate words into an algorithm that finds the best families in which to lump them. These algorithms are not literally pattern associator memories, but they rely on the same principle: Entities that share many traits probably come from the same category and should be treated alike.[23]

Not all family resemblance categories start off in lockstep and then diversify, but probably all of them are governed by hidden laws that make them similar and historical contingencies that make them different. Today's chairs did not descend from some ancestral ur-chair, so what makes them similar? It is that they must hold up a human bottom, and that forces most of them to have a stable, accessible, elevated, weight-bearing platform. At the same time chairs differ because of local variations in styles, tastes, materials, and expertise. Games are similar because they are meant to amuse, and they differ because of countless historical and local circumstances – the invention of playing cards, the invention of the computer, the availability of ice, grass, or water, the locals' taste for spectacle, violence, or brain work.

If we evolved a taste for family resemblance categories because they really do exist in the world as a product of history, why did we also evolve a taste for classical categories? I think it is because classical categories are by-products of rules in the mind that allow us to exploit laws in the world. The rules thereby allow us to deduce predictions about how things in the world work. Classical categories are not free-floating definitions, useful only for pigeon-holing things. They always are part of a *system* of interlocking rules that churn out handy deductions or computations. Just as regular verbs are products of a rule system (grammar), classical categories are products of their own rule systems. Odd numbers belong to arithmetic, triangles to geometry, grandmothers to kinship, dolphins to biological taxonomy, pregnancy to physiology, presidents to law. Each system allows a person to deduce unobserved traits from observable ones, not by remembering that they co-occurred but by cranking through a chain of implications. Using the rules of arithmetic, one can deduce that a set of forty-three objects cannot be divided into two equal parts. Using the rules of kinship, one can deduce that one's grandmother is the daughter of one's great-grandparents. Using the laws of physiology, one can deduce that a pregnant woman will become a parent, unless she has a miscarriage, abortion, or stillbirth. Using the laws of zoology, one can deduce that dolphins suckle their young and periodically surface to breathe. Using the laws of the land, one can deduce that the President of the United States was born in the United States more than thirty-five years ago.

Since these rule systems are, like grammar, combinatorial and recursive, they allow us to reason about an unlimited range of cases, often far from our experience. The laws of kinship allow us to say something about our family tree not just a hundred years ago but a hundred thousand or a million years ago. We can predict that if the United States still exists in 2804, there will be a presidential election that year.

When we use a system of rules, we have to turn off the family resemblance system, just as we seal off our memory for similar verbs when applying a rule to a verb that has to be regular. Within our systems of reasoning about kinship and law, a grandmother doesn't have to be grandmotherly nor a president presidential. It doesn't matter that a dolphin looks like a fish, or that the sides of a real-world triangle are not infinitesimally thin or perfectly straight. The human mind can think in *idealizations*, reducing an object to an austere description of the variables manipulated by the rule system, such as generation and gender in the case of kinship or the outcome of the electoral process in the case of law.

Science in particular depends on the mind's ability to think in idealizations, such as point masses, frictionless planes, perfect vacuums, and randomly interbreeding populations of organisms. The laws of science can be categorical statements uncluttered by the grubby details of the objects they refer to, and that allows them to be chained together in long inferences that lead to counterintuitive but correct conclusions – for example, that heat consists of moving molecules and that people and fish are cousins. That would never happen if the only form of human reasoning were the habit of generalizing similar traits to similar objects.

Of course not all people know formal science, but everyone knows a folk science (often blended with religion), in which the world and its parts are explained by elaborate interactions of hidden forces, traits, and essences. *Homo sapiens* has been said to occupy the 'cognitive niche' in nature:[24] We use knowledge of cause and effect to think up novel, complex sequences of behavior that defeat the defenses of plants and animals. People in all cultures, including hunter-gatherers whose lifestyle resembles that of our evolutionary ancestors, transcend their experience of concrete events, dig beneath appearances to ferret out laws, and combine these laws in their mind's eye to manipulate the world to their advantage. They assemble complicated traps, snares, and weapons. They recognize a few scratches on the ground as the tracks of an animal of a certain size, species, and condition, and predict its destination so they can ambush it. They remember a flower in the spring and return to it in the fall to dig up the underground tuber that has invisibly grown in the interim. They extract juices and powders from plants and animals and turn them into medicines and poisons.[25] None of these acts of creation would be possible if the mind simply remembered objects and expected similar ones to be behave similarly. They depend on abstract, combinatorial reasoning, of the kind made possible by rules and variables.

Some rule systems help us deal with the material world, but many help us deal with one another. The problem with fuzzy boundaries is that people can claim to see the edges of the boundary in different places. A child doesn't go to bed one evening and wake up as an adult. At some point the child may deem himself mature enough to drink or drive, whereas others may not want to take the chance of letting him. Love grows and deepens with time, but on a given day one lover may see the relationship as having ripened to a lifelong exclusive commitment, while the other lover – and interested third parties – may have a different opinion. Several people may be wise and powerful enough to merit the leadership of a group, but when a decision has to be made for the whole group, only one voice can prevail. People stave off border disputes around socially touchy categories by implementing rules that artificially sharpen the borders. They make up conditions for adulthood, marriage, and rank, complete with rites of passage that make entry into the categories instantaneous.

We have seen that much of the richness of language comes from the tension between words and rules. In the same way, much of the richness of the public sphere of life comes from tensions between family resemblance categories built from experience and the classical categories defined by science, law, or custom. Family resemblance categories resonate with common sense, but leave us groping when faced with something that is neither fish nor fowl. Classical categories offer neat divisions, but are bound to seem legalistic, pedantic, or abstruse. Is a fertilized ovum with a full complement of human DNA a person? What about a cell scraped from a cheek that has a full complement of human DNA and which can be cloned into a person? In surrogate births, who is the real mother: the woman who donated the egg or the woman who bore the child? Is the perpetrator of a crime innocent if he is freed on a technicality? Should a difficult court case be resolved by appealing to the most similar precedent or by appealing to constitutional principle? In 1999 President Bill Clinton was impeached for perjury after he denied having sex with his intern, Monica Lewinsky, despite their having engaged in fellatio. Clinton had treated 'sex' as a classical category – a list of anatomical configurations stipulated by the law – and his adversaries treated it as a family resemblance category.

We have digital minds in an analog world. More accurately, a part of our minds is digital. We remember familiar entities and their graded, crisscrossing traits, but we also generate novel mental products by reckoning with rules. It is surely no coincidence that the species that invented numbers, ranks, kinship terms, life stages, legal and illegal acts, and scientific theories also invented grammatical sentences and regular past-tense forms. Words and rules give rise to the vast expressive power of language, allowing us to share the fruits of the vast creative power of thought.

GLOSSARY

ablaut. The process of inflecting a verb by changing its vowel: *sing–sang–sung*.

adjective. The part-of-speech category comprising words that typically refer to a property or state: *the BIG BAD wolf; too HOT*.

adverb. The part-of-speech category comprising words that typically refer to the manner or time of an action: *tread SOFTLY; BOLDLY go; He will leave SOON*.

affix. A prefix or suffix.

agrammatism. A symptom of aphasia in which the patient has trouble producing well-formed words and grammatical sentences, and trouble understanding sentences whose meanings depend on their syntax, such as *The dog was tickled by the cat*.

agreement. The process in which a verb is altered to match the number, person, and gender of its subject or object: *He SMELLS* (not *SMELL*) versus *They SMELL* (not *SMELLS*).

anomia. A symptom of aphasia in which the patient has difficulty retrieving or recognizing words.

aphasia. A family of syndromes in which a person suffers a loss or impairment of language abilities following damage to the brain.

Aristotelian category. See **classical** category.

article. The part-of-speech category comprising words that modify a noun phrase, such as *a, the,* and *some*. Often subsumed in the **determiner** category.

associationism. The theory that intelligence consists in associating ideas that have been experienced in close succession or that resemble one another. The theory is usually linked to the British empiricist philosophers John Locke, David Hume, David Hartley, and John Stuart Mill, and it underlies behaviorism and much of connectionism.

auxiliary. A special kind of verb used to express concepts related to the truth of the sentence, such as tense, negation, question/statement, necessary/possible: *He MIGHT complain; He HAS complained; He IS complaining; He DOESN'T complain; DOES he complain?*

back-formation. The process of extracting a simple word from a complex word that was not originally derived from the simple word: *to bartend* (from *bartender*), *to burgle* (from *burglar*).

bahuvrihi. A headless compound that refers to someone by what he has or does rather than by what he is: *flatfoot, four-eyes, cutthroat.*

behaviorism. A school of psychology, influential from the 1920s to the 1960s, that rejected the study of the mind as unscientific, and sought to explain the behavior of organisms (including humans) with laws of stimulus-response conditioning. Usually associated with the psychologist B. F. Skinner.

blocking. The principle that forbids a rule to apply to a word if the word already has a corresponding irregular form. For example, the existence of *came* blocks a rule from adding *-ed* to *come,* thereby preempting *comed.*

Broca's aphasia. An aphasia characterized by difficulty in articulation, fluency, grammar, and the comprehension of complex sentences.

Broca's area. A region in the lower part of the left frontal lobe that has been associated with speech production, the analysis of complex sentences, and verbal short-term memory.

canonical root. A root that has a standard sound pattern for simple words in the language, a part-of-speech category, and a meaning arbitrarily related to its sound.

case. A distinction among noun forms corresponding approximately to the distinction among subjects, objects, indirect objects, and the objects of prepositions. In English it is the difference between *I* and *me, he* and *him,* and so on.

CAT scan. Computerized Axial Tomography. The construction of a cross-sectional picture of the brain or body from a set of X-ray data.

central sulcus. The groove in the brain that separates the frontal lobe from the parietal lobe, also called the Central fissure and the Rolandic fissure.

ChilDES. The Child Language Data Exchange System. A computer database of transcripts of children's speech (http://childes.psy.cmu.edu), developed by the psycholinguists Brian MacWhinney and Catherine Snow.

classical category. A category with well-specified conditions of membership, such as 'odd number' or 'President of the United States.'

coda. The consonants at the end of a syllable: *task, pomp*.

cognate. A word that resembles a word in another language because the two words descended from a single word in an ancestral language, or because one language originally borrowed the word from the other.

cognitive neuroscience. The study of how cognitive processes (language, memory, perception, reasoning, action) are carried out by the brain.

collocation. A string of words commonly used together: *excruciating pain; in the line of fire*.

compound. A word formed by joining two words together: *blackbird; babysitter*.

conjugation. The process of inflecting a verb, or the set of the inflected forms of a verb: *quack, quacks, quacked, quacking*.

connectionism. A school of cognitive psychology that models cognitive processes with simple neural networks subjected to extensive training. Much, but not all, of contemporary connectionism is a form of associationism.

consonant. A phoneme produced with a blockage or constriction of the vocal tract.

conversion. The process of deriving a new word by changing the part-of-speech category of an old word: *an impact* (noun) → *to impact* (verb); *to read* (verb) → *a good read* (noun).

cortex. The surface of the cerebral hemispheres of the brain, visible as gray matter, containing the bodies of neurons and their synapses with other neurons; the main site of neural computation underlying the higher cognitive, perceptual, and motor processes.

declension. The process of inflecting a noun, or the set of the inflected forms of a noun: *duck, ducks*.

default. The action taken in a circumstance that has no other action specified for it. For example, if you don't dial an area code before a telephone number, the local area code will be used as the default.

derivation. The process of creating new words out of old ones, either by affixation (*break + able* → *breakable; sing + -er* → *singer*), or by compounding (*super + woman* → *superwoman*).

determiner. The part-of-speech category comprising articles and similar words: *a, the, some, more, much, many*.

diphthong. A vowel consisting of two vowels pronounced in quick succession: *bite; loud; mAke.*

Early Modern English. The English of Shakespeare and the King James Bible, spoken from around 1450 to 1700.

empiricism. The approach to studying the mind that emphasizes learning and environmental influence over innate structure. A second sense, not used in this book, is the approach to science that emphasizes experimentation and observation over theory.

eponym. A noun derived from a name: *a SCROOGE; a SHYLOCK.*

ERP. Event-related potential. An electrical signal given off by the brain in response to a stimulus such as a word or picture, measured by electrodes pasted to the scalp.

family resemblance category. A category whose members have no single trait in common, but in which subsets of members share traits, as in a family. Examples include tools, furniture, and games.

fMRI. Functional Magnetic Resonance Imaging. A form of MRI that depicts the metabolic activity in different parts of the brain, not just the brain's anatomy.

generative linguistics. The school of linguistics associated with Noam Chomsky that attempts to discover the rules and principles that govern the form and meaning of words and sentences in a particular language and in human languages in general.

generative phonology. The branch of generative grammar that studies the sound pattern of languages.

gerund. A noun formed out of a verb by adding *-ing: His incessant WHINING.*

grammar. A database, algorithm, protocol, or set of rules that governs the form and meaning of words and sentences in a language. Not to be confused with the guidelines for how one 'ought' to speak that are taught in school and explained in style manuals.

grammatical morphemes. Morphemes, typically short and frequent, that express inflectional categories such as person, number, and tense, or that help to define the grammatical structure of a sentence. Examples include prefixes, suffixes, auxiliaries, prepositions, articles, and conjunctions.

head. The special word in a phrase, or special morpheme in a word, that determines the meaning and properties of the whole: *The MAN in the gray flannel suit; red-winged blackBIRD.*

headless. A phrase or word lacking a head: *The few, the proud* (noun phrases lacking head nouns); *a lowlife, a ne'er-do-well* (compounds lacking a noun that refers to the person).

homophones. Words that are identical in sound.

idiom. A phrase whose meaning cannot be predicted from the literal meaning of its parts: *go bananas; keep tabs on; take a leak.*

imperative. The form of a verb used in making a command: *LEAVE now!*

Indo-European. The group of language families that includes most of the languages of Europe, southwest Asia, and northern India; thought to be descended from a language, Proto-Indo-European, spoken by a prehistoric people.

infinitive. The form of a verb that lacks a tense and that stands for the verb as a whole: *to EAT, We can EAT.*

inflection. The process of altering a word to express its current use or grammatical role in a sentence: *dogs* (plural inflection); *walked* (past-tense inflection); *walking* (progressive inflection); *walks* (third-person present-tense inflection).

intransitive. A verb that may appear without an object: *We DINED; She THOUGHT that he was smart;* as opposed to a **transitive** verb, which may appear with one, as in *He DEVOURED the steak; I TOLD him to leave.*

irregular. A word with an idiosyncratic inflected form instead of the one usually created by a rule of grammar; *brought* (not *bringed*); *mice* (not *mouses*); as opposed to **regular** words, which simply obey the rule (*walked, rats*).

lexicon. A set of words or a dictionary. The mental lexicon is a person's knowledge of the words of his or her language.

linguist. A scholar or scientist who studies how languages work. Does not refer here to a person who speaks many languages.

listeme. An uncommon but useful term corresponding to one of the senses of *word.* It refers to an element of language that must be memorized because its sound or meaning does not conform to some general rule. All morphemes, word roots, irregular forms, collocations, and idioms are listemes.

long vowel. A vowel that takes about twice as long to pronounce as the other vowels; in English, the tense vowels in *bait, beet, bite, boat,* and *boot.*

mass noun. A noun that refers to an unmeasured quantity of stuff, rather than a single thing, and which ordinarily cannot take a plural: *mud; milk; anguish; evidence.*

MEG. Magnetoencephalography: The measurement of the magnetic signals given off by the brain.

Middle English. The language spoken in England from shortly after the Norman invasion in 1066 to around the time of the Great Vowel Shift in the 1400s.

Modern English. The variety of English spoken since the eighteenth century. See also **Early Modern English.**

mood. Whether a sentence is a statement, an imperative, or a subjunctive.

morphemes. The smallest meaningful pieces into which words can be cut: *un micro wave ability.*

morphology. The component of grammar that builds words out of pieces (morphemes). Morphology is often divided into inflection and derivation.

MRI. Magnetic Resonance Imaging. A technique that constructs pictures of cross-sections of the brain or body. See also **fMRI.**

neural network. A kind of computer model, loosely inspired by the brain, consisting of interconnected units that send signals to one another and turn on or off depending on the sum of their incoming signals. The connections have strengths that increase or decrease during a training process.

neurons. The information-processing cells of the nervous system, including brain cells and the cells whose axons (output fibers) make up the nerves and spinal cord.

neurotransmitter. A chemical that is released by a neuron at a synapse and that excites or inhibits the other neuron at the synapse.

noun. The part-of-speech category comprising words that typically refer to a thing or person: *dog, cabbage, John, country.*

nucleus. The vowel or vowels at the heart of a syllable: tr*AI*n; t*A*p.

number. The distinction between singular and plural: *chipmunk* versus *chipmunks.*

Old English. The language spoken in England from around 450 to 1100. Also called Anglo-Saxon, after the tribes speaking the language that invaded Britain around 450.

onset. The consonants at the beginning of a syllable: *STR*ing; *PL*ay.

participle. A form of the verb that cannot stand by itself, but needs to appear with an auxiliary or other verb: *He has* EATEN (perfect participle); *He was* EATEN (passive participle); *He is* EATING (progressive participle).

part of speech. The syntactic category of a word: noun, verb, adjective, preposition, adverb, conjunction.

passive. A construction in which the usual object appears as the subject, and the usual subject is the object of the preposition *by* or absent altogether: *I was robbed; He was nibbled to death by ducks.*

pattern associator memory. A common kind of neural network or connectionist model consisting of a set of input units, a set of output units, and connections between every input unit and every output unit, sometimes via one or more hidden layers of units. Pattern associator memories are designed to memorize the outputs for each of a set of inputs, and to generalize from similar inputs to similar outputs.

perfect. A verb form used for an action that has already been completed at the time the sentence is spoken: *John HAS EATEN.* See also **pluperfect.**

person. The distinction between *I* (first person), *you* (second person), and *he/she/it* (third person).

PET. Positron Emission Tomography. A technique for constructing pictures of cross-sections of the brain or body in which areas with different kinds or amounts of metabolic activity are shown in different colors.

phoneme. A vowel or consonant; one of the units of sound corresponding roughly to the letters of the alphabet that are strung together to form a morpheme: *b a t; b ea t; s t ou t.*

phonetics. How the sounds of language are articulated and perceived.

phonology. The component of grammar that determines the sound pattern of a language, including its inventory of phonemes, how they may be combined to form legitimate words, how the phonemes must be adjusted depending on their neighbors, and patterns of intonation, timing, and stress.

phrase. A group of words that behaves as a unit in a sentence and that typically has some coherent meaning: *in the dark; the man in the gray suit; dancing in the dark; afraid of the wolf.*

pluperfect. A construction used for an action that had already been completed at some time in the past: *When I arrived, John had EATEN.* See also **perfect.**

pluralia tantum. Nouns that are always plural, such as *jeans, suds,* and *the blues.* The singular is *plurale tantum.*

predicate. A state, event, or relationship, usually involving one or more participants, often identified with the verb phrase of a sentence: *The gerbil* ATE THE PEANUT.

preposition. A part-of-speech category comprising words that typically refer to a spatial or temporal relationship: *in, on, near, by, for, under, before.*

preterite. The simple past-tense form of a verb: *He walked; We sang.* It is usually contrasted with a verb form that indicates a past event using a participle, such as *He has walked* or *We have sung.*

productivity. The ability to speak and understand new word forms or sentences, ones not previously heard or used.

progressive. A verb form that indicates an ongoing event: *He is* WAVING *his hands.*

psycholinguist. A scientist, usually a psychologist by training, who studies how people understand, produce, or learn language.

recursion. A procedure that invokes an instance of itself, and thus can be applied, ad infinitum, to create or analyse entities of any size: 'A *verb phrase* can consist of a verb followed by a noun phrase followed by a *verb phrase.*'

regular. See **irregular.**

rime. The part of a syllable consisting of the vowel and any following consonants; the part that rhymes: MOON; JUNE.

root. The most basic morpheme in a word or family of related words, consisting of an irreducible, arbitrary pairing between a sound and a meaning: ELECTRICITY, ELECTRICAL, ELECTRIC, ELECTRIFY, ELECTRON.

rootless. A word that has no root but gets its sound in some other way, such as by onomatopoeia, quotation, truncation, eponymy, being an acronym, or conversion from another part-of-speech category.

schwa. The neutral vowels in ARRIVE, MOTHER, and ACCIDENT.

semantics. The components of a rule or lexical entry that define the meaning of a morpheme, word, phrase, or sentence. Does not refer here to haggling over exact definitions.

stem. The main portion of a word, the one that prefixes and suffixes are stuck on to: WALKS, BREAKABLE, ENSLAVE.

stress. Emphasis on a syllable in pronunciation, making it louder, longer, higher in pitch, more distinctly articulated, or some combination: *América, Cánada, Massachúsetts.*

strong verbs. The irregular verbs in the Germanic languages (including

English) that undergo a vowel change in the past tense and do not end in a *t* or a *d*: *sing–sang, wear–wore*.

subjunctive. A verb form that indicates a hypothetical or counterfactual state of affairs: *It is important that he* GO; *Let it* BE; *if I* WERE *a carpenter*.

suppletion. An inflected form that is phonologically unrelated to its root and instead comes from some other word: *go–went, be–was, good–better, person–people*.

Sylvian fissure. The huge horizontal cleft that separates the temporal lobe from the rest of the brain.

synapse. A connection between neurons; the site at which activity from one neuron affects the activity of another. Changes in the strengths of synapses are thought to be the neural basis of learning and memory.

syncretism. Distinct inflections that have the same form: He WALKED (past tense), *He has* WALKED (perfect participle), *He is being* WALKED (passive participle); *the* CATS (plural), *the* CAT'S *pyjamas* (possessive), *the* CATS' *mother* (plural possessive).

syntax. The component of grammar that arranges words into phrases and sentences.

tense. Relative time of occurrence of the event described by the sentence, the moment at which the speaker utters the sentence, and often, some third reference point: present (*He eats*), past (*He ate*), future (*He will eat*).

tense vowel. A vowel pronounced with the muscle at the root of the tongue advanced toward the front of the mouth. In English, the long vowels are all tense.

transitive. See **intransitive.**

umlaut. The process of shifting the pronunciation of a vowel toward the front of the mouth. In German, vowels that undergo umlaut (or that underwent it in earlier historical periods) are indicated by two dots: *ä, ö, ü*.

unvoiced. See **voicing.**

verb. The part-of-speech category comprising words that typically refer to an action or state: *hit, break, run, know, seem*.

voiced. See **voicing.**

voicing. Vibration of the vocal folds in the larynx simultaneously with the articulation of a consonant; the difference between *b, d, g, z, v* (voiced) and *p, t, k, s, f* (unvoiced).

weak verbs. In the Germanic languages, the verbs that form the past tense or participle by adding *t* or *d*. They include weak irregular verbs such as *sleep–slept, hit–hit,* and *bend–bent,* and all the regular verbs.

wug-test. A test of linguistic productivity in which a person is given a novel word and encouraged to use it in some inflected form: 'Here is a wug; Now there are two of them; there are two . . .'

NOTES

Chapter 1. The Infinite Library

1. Saussure, 1916/1959.
2. Witnessed by a friend, Margit Maus.
3. Pinker, 1994, 149–151; Miller, 1991.
4. Marslen-Wilson, 1987.
5. Rayner & Pollatsek, 1994.
6. Levelt et al., 1998.
7. Miller, 1967, 82.
8. Miller, 1991.
9. Borges, 1964a.
10. Humboldt, 1836/1972; Chomsky, 1966.
11. Eco, 1995.
12. Eco, 1995; Borges, 1965.
13. Languages differ in their division of labor between simple words and grammatical combinations, and some, such as Native American languages, have fewer words and more rules. But even in these languages speakers cannot deduce the meanings of most words from their sound, and must commit whole words to memory.
14. Empson, 1959.
15. Chomsky, 1959; Lenneberg, 1964.
16. Marcus, Pinker, Ullman, Hollander, Rosen, & Xu, 1992.
17. *Buyed* from Lenneberg, 1964; *holded* from Cazden, 1972; *stealed* from Lila Gleitman, personal communication; *heared* and *goed* from Gary Marcus's analysis of transcripts of children's speech in the Child Language Data Exchange System (ChilDES) (MacWhinney & Snow, 1985, 1990, http://childes.psy.cmu.edu/childes; see Marcus et al., 1992.)
18. Lederer, 1990.
19. Pinker & Prince, 1988.
20. Hogg, 1988; Murray, 1998.

21. An additional two children produced *glanged*.
22. Marcus et al., 1992; Aronoff, 1976; Kiparsky, 1982.
23. Allen, 1980.
24. See also Prasada & Pinker, 1993.
25. Pinker & Prince, 1988.

Chapter 2. Dissection by Linguistics

1. See also Dronkers, Pinker, & Damasio, 1999.
2. Aronoff, 1976; di Sciullo & Williams, 1987; Pinker, 1994, chap. 5.
3. Aronoff, 1994.
4. Lieber, 1992.
5. *Atlantic Monthly*, November 1997.
6. Bernstein, 1977.
7. *Boston Globe*, October 4, 1998, C5.
8. Seuss, 1987.
9. Williams, 1981; Selkirk, 1982.
10. *Boston Globe*, summer 1998.
11. Pinker, 1994, chap. 5.
12. Pinker & Prince, 1988.
13. Williams, 1981; Selkirk, 1982; Pinker & Prince, 1988.
14. Carstairs, 1987; Carstairs-McCarthy, 1994; Aronoff, 1994.
15. Carstairs, 1987; Carstairs-McCarthy, 1994; Bybee, 1985; Spencer, 1991.
16. Zwicky, 1975; Pinker & Prince, 1988.
17. Pinker & Prince, 1988.
18. Dell, 1995.
19. Fromkin, 1971; Garrett, 1980; Stemberger, 1985.
20. McGuinness, 1997.
21. Cassidy, 1985.
22. Melissa Bowerman, personal communication, July 30, 1998.
23. Pinker & Prince, 1988.
24. Pinker, 1984/1996, chap. 5.
25. Pinker & Prince, 1988; Kim et al., 1991, 1994.

Chapter 3. Broken Telephone

1. Hymes, 1964.
2. Discussions of the history of the English language in this chapter

are based on the following sources: *Oxford English Dictionary*; Cassidy & Ringler, 1891/1971; Curme, 1935/1983; Görlach, 1991; Jespersen, 1938/1982; Johnson, 1986; Levin, 1964; Pyles & Algeo, 1982; Williams, 1975; Chomsky & Halle, 1968/1991.

3. Francis & Kučera, 1982.

4. Quirk et al., 1985.

5. Quirk et al., 1985, 307.

6. *Newsweek*, November 22, 1993, 34.

7. Raymond, 1991.

8. Lieber, 1980.

9. Lederer, 1990.

10. Beard & McKie, 1982.

11. *Tantrum, tantra.* Originally published in *The Enigma*, the official publication of the National Puzzlers' League. For more information, send a SASE to Francis Heaney, 509 E. 5th Street, New York NY 10009, or visit http://www.puzzlers.org. My colleague, the linguist and poet Samuel Jay Keyser, has discovered an alternative solution, a false advocate: *fit, prophet.* Keyser's solution has the advantage of scanning properly in the poem, though the spellings bend the rules.

12. Lists of irregular verbs and their variants in contemporary English may be found in Curme, 1935/1983, Quirk et al., 1985, and Pinker & Prince, 1988. Nonstandard dialectal forms also are provided in Mencken, 1936, 1919/1986. Data on variation and uncertainty in the choice of past forms may be found in Ullman, 1993; Haber, 1975, 1976; Quirk, 1970; and Murray, 1998.

13. Jackendoff, 1983; Lakoff, 1987; Pinker, 1989.

14. Bybee & Slobin, 1982; Rumelhart & McClelland, 1986; Pinker & Prince, 1988; Stemberger, 1983; MacWhinney, 1978.

15. Stemberger, 1983.

16. Stemberger, 1983; Stemberger & MacWhinney, 1986b, 1988; Bybee & Slobin, 1982.

17. Bybee & Slobin, 1982; Kuczaj, 1977, 1978; Marcus et al., 1992.

18. *Tread* from P. Rozin and A. Fallon in *Psychological Review*, 94, 1987.

19. Peter Hotton in the *Boston Globe*, February 24, 1991.

20. Bernstein, 1977.

21. Chomsky & Halle, 1968/1991, 253.

22. Myers, 1987.

23. Stevens & Keyser, 1989.

24. Baldi, 1983; Williams, 1975; Crystal, 1997; Comrie, Matthews, & Polinsky, 1996.

25. Renfrew, 1987.

26. Levin, 1964; Johnson, 1986; Pyles & Algeo, 1982; Baldi, 1983; Watkins, 1992.

27. The use of the excellent word *muzzy* as a technical term for uncertain past-tense forms was suggested by Lyn R. Haber, 1975, 1976.

28. The *Spring has sprung* verse is quoted often and with many variations, though the original is obscure. In *Verse and Worse: A Private Collection* (1952), Arnold Silcock reproduces an anonymous poem called 'The Budding Bronx': Der spring is sprung / Der grass is riz / I wonder where dem boidies is. / Der little boids is on der wing, / Ain't dat absoid? / Der little wings is on der boid!

29. Ann Taylor, 'Deep Things,' in Jane and Ann Taylor, *Original Poems for Infant Minds*, 1804.

30. Jane Holtz Kay, architecture critic of *The Nation*, in the *Boston Globe*, April 1990.

31. *Boston Globe,* January 7, 1992.

32. *Boston Globe,* February 20, 1997.

33. Camille Paglia in *Salon*, March 22, 1999.

34. *Boston Globe*, May 27, 1995.

35. Bybee & Slobin, 1982; Marcus et al., 1992.

36. Bernstein, 1977, 452.

37. Cullen Murphy, 'The Lay of the Language,' *Atlantic Monthly*, May 1995.

38. From Jan Freeman's column 'The Word,' *Boston Globe*, February 21, 1999.

39. Curme, 1935/1983, 282.

40. Pyles & Algeo, 1982, 201.

41. Reported by M. H. Greenblatt, in *Espy*, 1975. Possibly apocryphal, since Dizzyisms were eagerly recorded, traded, and no doubt embroidered and fabricated. As Yogi Berra, another oft-quoted ballplayer, may or may not have remarked, 'I never said most of the things I said.' In a radio broadcast that was carefully transcribed and reprinted in Staten, 1992, Dean consistently said *swung*, not *swang*.

42. Murray, 1998.

43. *New York Times Magazine,* July 16, 1995.

44. Hogg, 1988.

45. Thanks to Stephanie Shattuck-Hufnagel for *skun* and to Carol Miller for *tug*. Special thanks to Walter Kent, an earwitness, for confirming Dizzy Dean's pronunciation of *slud*.
46. *New York Times Magazine*, July 16, 1995.
47. *Boston Globe*, May 3, 1998.
48. From Ann Landers's syndicated column, February 1999.
49. By C. Sigman and P. DeRose.
50. Pyles & Algeo, 1982.
51. Heine, Claudi, & Hunnemeyer, 1991.
52. Westcott, 1970; Jakobson, 1971; Kuryłowicz, 1964; Swadesh, 1964.
53. Cooper & Ross, 1975; Tanz, 1971; Pinker & Birdsong, 1979; Woodworth, 1991.
54. Westcott, 1970.

Chapter 4. In Single Combat

1. Curme, 1935/1983; Jespersen, 1938/1982; Pyles & Algeo, 1982; Williams, 1975.
2. Curme, 1935/1983; Mencken, 1936, 1919/1986.
3. Bybee & Moder, 1983.
4. Rosten and Ross, 1989.
5. Robert Winder, *The Independent*, April 1994, 16.
6. Lederer, 1990.
7. MacWhinney & Snow, 1985, 1990.
8. Xu & Pinker, 1995.
9. Michael Gazzaniga's daughter, Francesca, December 9, 1991.
10. Chomsky & Halle, 1968/1991.
11. Rumelhart & McClelland, 1986.
12. Hobbes, 1651/1957.
13. Eco, 1995, 281.
14. See Eco, 1995, chap. 14, for citations of the numerous essays in which Leibniz developed and refined these schemes.
15. Newell, 1990; Anderson, 1993.
16. Hume, 1748/1955.
17. Chomsky & Halle, 1968/1991; Halle & Mohanan, 1985. The assumption that irregular patterns are the product of rules has been retained in descendants of the theory, such as the Distributed Morphology framework of Morris Halle and Alec Marantz (Halle & Marantz, 1993).

18. From Halle & Mohanan, 1985, 104.
19. Lahiri & Marslen-Wilson, 1991.
20. Chomsky & Halle, 1968/1991, 49.
21. Chomsky & Halle, 1968/1991, 54.
22. From 'English' by T. S. Watt, originally published in the *Manchester Guardian*, excerpted here from Lederer, 1990.
23. Bybee & Slobin, 1982; Pinker & Prince, 1988.
24. Bybee & Slobin, 1982; Wittgenstein, 1953; Rosch, 1978; see also chap. 10.
25. Xu & Pinker, 1995.
26. Bybee & Moder, 1983.
27. Rumelhart & McClelland, 1986.
28. Sampson, 1987.
29. Rumelhart, McClelland, & the PDP Research Group, 1986; McClelland, Rumelhart, & the PDP Research Group, 1986; Quinlan, 1991; Smolensky, 1988.
30. Schneider, 1987.
31. Hartley, 1775/1973.
32. Pinker & Prince, 1988.
33. Pinker & Mehler, 1988; Prasada & Pinker, 1993; Pinker, 1997, 112–131; Marcus et al., 1995; Marcus, 1997, 1998, in press a, b, c; Fodor & Pylyshyn, 1988; Fodor & McClaughlin, 1990; Minksy & Papert, 1988; Lachter & Bever, 1988; Anderson, 1993; Newell, 1990; Ling & Marinov, 1993; Hadley, 1994a, b.
34. Dewdney, 1997.
35. Kim et al., 1991, 1994; Marcus et al., 1995.
36. Marcus et al., 1992.
37. An example of this approach can be found in Westermann & Goebel, 1995. Ironically, Rumelhart and McClelland, 1986, and the designers of a follow-up model, MacWhinney and Leinbach, 1991, discussed the desirability of building a separate network for the lexicon, which would have brought the models closer to the words-and-rules theory. They did not implement the suggestion, however, and it has been ignored in subsequent debates.
38. Wickelgren, 1969. Wickelish representations are also used by Patterson, Seidenberg, & McClelland, 1989, and by Mozer, 1991.
39. Prince & Pinker, 1989.
40. Prasada & Pinker, 1993; Sproat, 1992; Marcus, in press c.
41. Smolensky, 1990, 1995; Shastri & Ajjanagadde, 1993; Shastri, 1999;

Pollack, 1990; Hadley & Hayward, 1994; Hummel & Holyoak, 1997; Hummel & Biederman, 1992; Halford et al., 1997.

42. Daugherty & Hare, 1993; Daugherty & Seidenberg, 1992; Daugherty, MacDonald, Petersen, & Seidenberg, 1993; Goebel & Indefrey, 1994; Hare & Elman, 1992; Hare, Elman, and Daugherty, 1995; Hoeffner, 1992; MacWhinney & Leinbach, 1991; Plunkett & Marchman, 1991, 1993; Forrester & Plunkett, 1994; Sproat, 1992; Westermann and Goebel, 1995; Plunkett & Nakisa, 1997; Nakisa & Hahn, 1996. For critiques, see Marcus et al., 1992, 1995; Prasada & Pinker, 1993; Kim et al., 1994; Marcus, 1995a, 1997, 1998, in press a, b, c; Ling & Marinov, 1993.

43. Egedi & Sproat, 1991; Sproat, 1992; see chap. 5. Also Marcus & Halberda, unpublished research discussed in chap. 7 of Marcus, in press c.

44. Aronoff, 1976; Bresnan, 1978; Jackendoff, 1975, 1997; Lieber, 1980; Spencer, 1991.

Chapter 5. Word Nerds

1. Baayen & Renouf, 1996.
2. Sproat, 1992; Allen, Hunnicutt, & Klatt, 1987, chap. 3; Ritchie, et al., 1991; Karttunen, Koskenniemi, & Kaplan, 1987; Karttunen, Kaplan, & Zaenen, 1992.
3. Francis & Kučera, 1982.
4. Bybee, 1985.
5. Bybee, 1985.
6. Example from Jane Grimshaw; see Pinker & Prince, 1988.
7. Baayen, 1994; Baayen & Lieber, 1991; Baayen & Renouf, 1996; Lieber, 1992.
8. The reason there are 62 hapax legomena among irregular past-tense forms, but earlier I said there are only 17 irregular verbs with a frequency of one, is that the first number embraces verbs that occur only once in the past tense, regardless of how many times they occur in other tenses, whereas the latter embraces verbs that occur only once in any form whatsoever.
9. Ullman, 1999.
10. Prasada, Pinker, & Snyder, 1990.
11. Seidenberg & Bruck, 1990; Beck, 1997; Shenkman, 1994; Ullman, 1993. In all the experiments people took longer to produce low-

frequency irregular forms than high-frequency irregular forms, and this effect did not occur with regular forms. In some experiments low- and high-frequency regulars took equal time; in others the low-frequency regular forms were also slower, but not to the same degree as low-frequency irregulars; in still others low-frequency forms were *faster* than high-frequency forms, especially when the list of items contained many irregulars. See Beck, 1997, and the discussion on page 152 and note 22 later in the chapter.

12. Prasada, Pinker, & Snyder, 1990; also unpublished experiments in my lab conducted by Michael Ullman and Marie Coppola.

13. Stemberger, 1983.

14. Ullman, 1993, 1999; Ullman & Pinker, 1990; see also Stemberger, 1989.

15. Seidenberg & Bruck, 1990.

16. Marslen-Wilson, 1989; Feldman, 1995.

17. Stanners, Neiser, Hernon, & Hall, 1979; Kempley & Morton, 1982; Napps, 1989; Münte et al., 1998; Marslen-Wilson, 1998, also reported in Marslen-Wilson, Hare, & Older, 1995, and described in Orsolini & Marslen-Wilson, 1997. A fifth study, by Fowler, Napps, & Feldman, 1985, sort of replicated the effect: Regular forms consistently primed their stems more strongly than irregulars did, but the differences were not statistically significant. One reason is that some of the derived words they classified as regular probably were not.

18. Described in Marslen-Wilson, 1998; Marslen-Wilson, Hare, & Older, 1995; and Orsolini & Marslen-Wilson, 1997.

19. Marslen-Wilson & Tyler, 1998.

20. Seidenberg, 1992; Daugherty & Seidenberg, 1992.

21. Marcus et al., 1992.

22. As mentioned in note 11, sometimes high-frequency regular verbs are, paradoxically, *slower* to produce than low-frequency regular verbs. One explanation is that stored forms always inhibit the rule, even if they are identical to the form the rule is trying to create. Just as *broke* blocks the creation of *breaked*, an entry for *walked* that is stored in memory may block the creation of *walked* by rule, slowing down the rule process (compared to, say, *stalked*, whose memory entry is too weak to slow down the rule). Evidence for this explanation is the fact that the harmful effects of high frequency tend to occur when the word list has a high percentage of irregular

forms, encouraging subjects to go to their mental lexicons on every trial; see Beck, 1997. For another demonstration of a reverse word frequency effect, and a similar explanation, see Balota, Law, & Zevin, 1999.

23. Ullman, 1993, 1999; Ullman & Pinker, 1990; Pinker & Prince, 1994.
24. Burani, Salmaso, & Caramazza, 1984; Katz, Rexer, & Lukatela, 1991; Sereno & Jongman, 1997; Baayen, Dijkstra, & Schreuder, 1997.
25. Schreuder & Baayen, 1995; Baayen, Dijkstra, & Schreuder, 1997; Baayen, 1994. For similar models see Caramazza, Laudanna, & Romani, 1988; Burani & Caramazza, 1987; Laudanna & Burani, 1995; Taft, 1994.
26. Beck, 1997; see also note 22.
27. Baayen, Dijkstra, & Schreuder, 1997.
28. Baayen, Dijkstra, & Schreuder, 1997; Schreuder & Baayen, 1995; Baayen & Schreuder, in press.
29. Hinton, McClelland, & Rumelhart, 1986, 82.
30. Pinker & Prince, 1988.
31. Pinker & Prince, 1988; Kim et al., 1991.
32. Pinker & Prince, 1988.
33. Ullman, 1999.
34. Ullman, 1993, 1999.
35. Prasada & Pinker, 1993.
36. Borowsky, 1989.
37. Nonetheless, Wickelgraphs – units composed of three letters, like the Wickelphones composed of three phonemes – continued to be used in several connectionist models that were developed after the Rumelhart-McClelland model, such as in Patterson, Seidenberg, & McClelland, 1989, and Mozer, 1991.
38. Egedi & Sproat, 1991; Sproat, 1992.
39. MacWhinney & Leinbach, 1991.
40. Hare, Elman, & Daugherty, 1995.
41. Hare & Elman, 1992; Nakisa & Hahn, 1996.
42. James, 1890/1950, 270.

Chapter 6. Of Mice and Men

1. *Boston Globe Magazine*, January 10, 1993.
2. Bernstein, 1977, 189.
3. *Boston Globe*, March 29, 1998.

4. Lakoff, 1987, cited in Kim et al., 1991; MacWhinney & Leinbach, 1991; Harris, 1992; Shirai, 1997; Daugherty et al., 1993.

5. Kim et al., 1991, 1994; Kim, 1999.

6. *San Francisco Chronicle*, November 1991.

7. Harris, 1992; Daugherty et al., 1993; Shirai, 1997.

8. Kim, 1999; Kim et al., 1994; Marcus et al., 1995.

9. Kiparsky, 1982; Williams, 1981; di Sciullo & Williams, 1987; Selkirk, 1982; Lieber, 1980, 1992; Pinker & Prince, 1988; Kim et al., 1991, 1994; Marcus et al., 1995. See also Spencer, 1991.

10. From the Simon and Garfunkel song, 'A Simple Desultory Philipic.'

11. McCarthy & Prince, 1990, in press.

12. Saussure, 1916/1959.

13. Curme, 1935/1983.

14. Pinker, 1989, 118–122; Gropen, Pinker, Hollander, Goldberg, & Wilson, 1989.

15. Recorded by Annie Senghas from a statistics lecture by Harvard professor Donald Rubin.

16. *Boston Globe* column by Mike Barnicle, May 16, 1995.

17. From the food section of the *Boston Globe*, June 26, 1991.

18. From John Gillespie Magee, Jr.'s 'High Flight.'

19. Spencer, 1991.

20. Bernstein, 1977, 335.

21. Richard Russo, *Nobody's Fool*. New York: Vintage, 1994, 40–41.

22. J. R. R. Tolkien, *The Fellowship of the Ring*, Part 1 of *The Lord of the Rings*. New York: Ballantine Books, 1954.

23. From 'How I Came West, and How I Stayed,' in *The Atlantic*, October 1991, 89–96. Quote from p. 90.

24. *Newsweek*, August 7, 1989.

25. Adams, 1988.

26. *Top Shelfs* overheard by Gary Marcus. *Supermans* overheard by John J. Kim. *Sea Wolfs* used in the *Boston Globe* by columnist Alex Beam, November 1992. *Gold Maple Leafs* from an ad in the financial pages of the *Boston Globe*.

27. *Popular Photography*, October 1996, 28.

28. Heard by Beth Levin.

29. Heard by Roslyn Pinker.

30. *Sports Illustrated*, December 6, 1989.

31. Mencken, 1919/1986, 532; see also Curme, 1935/1983, 274.

32. *New York Times*, May 3, 1998.

33. Noticed by Roslyn Pinker.

34. Bernstein, 1977, 81.

35. Mencken, 1936, 439, n. 2. See also Curme, 1935/1983, 274.

36. Fowler, 1965, 206.

37. Francis & Kučera, 1982.

38. *Boston Globe*, October 1989.

39. From David Calhoun in *Scientific American*, April 1994.

40. Kim et al., 1991.

41. April 7, 1992.

42. *New York Times Magazine*, January 30, 1994.

43. Kim et al., 1991.

44. Kim et al., 1991. The *kleed* experiment replicated an earlier study performed by the linguists Greg Carlson, Jay Keyser, and Tom Roeper, 1977; also described in Kim et al., 1991.

45. *Oxford English Dictionary*, New ed.

46. Mithun, 1988; Bybee, 1985.

47. Based on the book by Anita Loos.

48. Tiersma, 1982; Bybee, 1985.

49. Marchand, 1969; Quirk et al., 1985; Kiparsky, 1982.

50. Kiparsky, 1982; Gordon, 1985.

51. *Boston Globe*, July 13, 1998.

52. *Boston Globe*, March 23, 1995.

53. Reported by Gregory Murphy.

54. Spencer, 1991.

55. Lederer, 1990, 22.

56. Kiparsky, 1982; Gordon, 1985.

57. Kiparsky, 1982; Gordon, 1985; Spencer, 1991.

58. Selkirk, 1982.

59. Senghas, Kim, Pinker, & Collins, 1991; Senghas, Kim, & Pinker, 1999.

60. Churma, 1983.

61. Alegre & Gordon, 1996, 1997.

62. Lieber, 1992.

63. Alegre & Gordon, 1996.

64. Senghas, Kim, Pinker & Collins, 1991; Senghas, Kim, & Pinker, 1999; Alegre & Gordon, 1996.

65. di Sciullo & Williams, 1987.

Chapter 7. Kids Say the Darnedest Things

1. Originally published in *The Nation*; also in S. Kinsolving, *Dailies and Rushes: A Collection of Poems.* Boston: Grove/Atlantic, 1999. Reprinted with the permission of the poet.
2. Barbara Vine, *A Dark-Adapted Eye.* New York: Penguin, 1993, 186.
3. The history of the study of past tense overgeneralization errors, and most of the original data discussed in this chapter, are presented in detail in Marcus, Pinker, Ullman, Hollander, Rosen, & Xu, 1992.
4. Brown, 1973; Pinker, 1984/1996, 1994, chap. 9.
5. Examples discovered by Emily Wallis and Jennifer Ganger.
6. Brown, 1973.
7. Berko, 1958.
8. *Ated* from the MacWhinney corpus in the Child Language Data Exchange System (ChiLDES), MacWhinney & Snow, 1985, 1990, http://childes.psy.cmu.edu/childes; *poonked* from the Kuczaj corpus: Kuczaj, 1976. *Lightninged* and *spidered* from data collected in collaboration with Jennifer Ganger, 1998. For other examples see p. 17 in chapter 1. Thanks to Jeremy Wolfe for *presseded,* and to Jim Hillenbrand for *pukeded.*
9. Cazden, 1968; Marcus, 1995b.
10. Examples from the child called Sarah studied by Brown, 1973; taken from ChiLDES.
11. *Specialer* and *powerfullest* from my nephew, Eric Boodman, at age three and a half.
12. Thanks to Madeleine V. MacDonald for *oneth, cirtangle,* and *theirselves.*
13. Pinker & Prince, 1988.
14. Thanks to Katya Rice for *to verse.*
15. Cazden, 1968; Marcus et al., 1992.
16. Strauss, 1982.
17. Brown, 1973; Marcus et al., 1992.
18. Pinker, 1984/1996.
19. Aronoff, 1976; Kiparsky, 1982; Pinker, 1984/1996; Marcus et al., 1992.
20. Pinker, 1984/1996; Marcus et al., 1992.
21. From transcripts in ChiLDES originally collected by Kuczaj, 1976, 1977; reported in Marcus et al., 1992.
22. Cazden, 1972.
23. Zwicky, 1970.

24. Morgan & Travis, 1989; Morgan, Bonamo, & Travis, 1995; see also Marcus, 1993.
25. Stromswold, 1994.
26. Marcus et al., 1992.
27. Marcus et al., 1992.
28. From the data of Kuczaj, 1976, 1977.
29. Lachter & Bever, 1988.
30. Kuczaj, 1978; see also Marcus et al., 1992.
31. Error data from Joseph Stemberger; calculation of rate from Marcus et al., 1992.
32. Ullman, 1993, 1999; Ullman & Pinker, 1990.
33. Bybee, 1985.
34. Marcus et al., 1992.
35. Marcus, Pinker, & Larkey, 1995.
36. Tramo et al., 1995; Bouchard, 1994.
37. Ganger, 1998; Ganger, Pinker, Baker, & Chawla, 1999.
38. Rumelhart & McClelland, 1986.
39. For example, Plunkett & Marchman, 1991, 1993; see Marcus, 1995a.
40. Marcus, 1995b.
41. Marcus et al., 1992.
42. Aronoff, 1994.
43. Kim et al., 1994.
44. Gordon, 1985.
45. Kim et al., 1994.
46. Marcus et al., 1992.
47. Darwin, 1874, 101–102.

Chapter 8. *The Horrors of the German Language*

1. On universals and variation see Greenberg, Ferguson & Moravcsik, 1978; Comrie, 1981; Hawkins, 1988; Shopen, 1985; Bybee, 1985; Aronoff, 1994; Whaley, 1997.
2. Cited by Halle, 1973.
3. Orsolini & Marslen-Wilson, 1997; Say, 1998; Caramazza, Laudanna, & Romani, 1988. An English analogue may be found in such nouns as *wolf–wolves* and *scarf–scarves* where the plural stem is irregular *wolv-* or *scarv-* but is inflected with the regular -s plural. What's the evidence that the -s in these hybrid plurals indeed is the regular -s, as opposed to an irregular soundalike? Senghas, Kim, and I

(Senghas et al., 1991; Senghas, Kim, & Pinker, 1999) found that people refuse to put these plurals into compounds, a sign that the plural as a whole is thought of as regular, despite its containing an irregular stem. Compounds containing hybrid plurals such as *wolves project* and *scarves-wearer* sounded as bad as compounds containing pure regular plurals, such as *graves permit* and *cloves-cutter.*

4. *National Review*, December 8, 1997.
5. Rumelhart & McClelland, 1986, 230–231; Bybee, 1991, 86–87; Plunkett & Marchman, 1991, 67; 1993, 55; see Marcus et al., 1995, for quotations.
6. Emphasis added. Twain, 1880/1979.
7. See Marcus et al., 1995, for a full explanation of the sources and calculations behind the vocabulary statistics mentioned in this chapter.
8. Marcus et al., 1995; Wiese, 1996; Wunderlich, 1992; Köpcke, 1988.
9. MacWhinney & Leinbach, 1991, 137; Bybee, 1991, 86–87. See Marcus et al., 1995, for quotations.
10. Wiese, 1996; Wunderlich, 1992; Clahsen & Rothweiler, 1992; Clahsen et al., 1992; Marcus et al., 1995.
11. Clahsen & Rothweiler, 1992.
12. Marcus et al., 1995.
13. Twain, 1880/1979.
14. Bybee, 1995; Marcus et al., 1995.
15. One exception is suffixed words. For example, most words with feminine suffixes such as *-e, -schaft, -keit,* and *-ung* take *-en.* The explanation is that the suffixes themselves like words with their own irregular plural forms, and serve as the heads of the complex words that contain them.
16. Janda, 1991; Marcus et al., 1995.
17. Janda, 1991; Wiese, 1996; Wunderlich, 1992; Bornschein & Butt, 1987.
18. Van Dam, 1940.
19. Curme, 1935/1983, 114.
20. Clahsen et al., 1992, 1996; see also Marcus et al., 1995.
21. Clahsen et al., 1992, 1996.
22. Marcus et al., 1995.
23. Pyles & Algeo, 1982.
24. Sellar & Yeatman, 1930/1970.

25. Bauer, 1983.
26. Keyser & O'Neil, 1985.
27. Bornschein & Butt, 1987; Janda, 1991.
28. On the world's language families and their origins see Crystal, 1997; Comrie, 1990; Bright, 1992; Comrie, Matthews, & Polinsky, 1996.
29. Donaldson, 1987; Booij, 1977.
30. Collins, 1991.
31. Collins, 1991.
32. Shaoul, 1993.
33. Attributed to Fritz Houtermans by Otto Frisch in his memoir *What Little I Remember*. New York: Cambridge University Press, 1979.
34. Edith Moravcsik, personal communication, June 1994; see also Kiefer, 1985; Dressler, 1985; Moravcsik, 1975.
35. McCarthy & Prince, 1990.
36. Omar, 1973.
37. W. Chomsky, 1957.
38. Hare, Elman, & Daugherty, 1995; Nakisa & Hahn, 1996; Forrester & Plunkett, 1994; Plunkett & Nakisa, 1997.
39. Berent, Pinker, & Shimron, 2000.
40. Comrie, Matthews, & Polinsky, 1996; see also Pinker, 1994, chap. 7.
41. Myers, 1998.
42. Fortune, 1942; Aronoff, 1994.

Chapter 9. The Black Box

1. For reviews of language and the brain see Dronkers, Pinker, & Damasio, 1999; Pinker, 1994, chap. 10; Gazzaniga, Ivry, & Mangun, 1998; Damasio & Damasio, 1989; Caplan, 1987; Goodglass, 1993; and the chapters on language in Gazzaniga, 1995.
2. Dronkers, Pinker, & Damasio, 1999; Damasio & Damasio, 1989; Damasio et al., 1996; Caramazza & Shelton, 1998; Gazzaniga, Ivry, & Mangun, 1998; Goodglass, 1993.
3. Teuber, 1955.
4. Coltheart, 1985; Coltheart et al., 1993.
5. One exception is a simulation of the past tense by the psychologist Virginia Marchman (1993), which seemed to go the other way. But 60 percent of Marchman's 'irregular' items were no-change verbs such as *hit–hit*, which use a highly predictable and uniform mapping shared with the regular verbs. This artificial word list, and

the fact that the model didn't do well with the regular verbs even *before* it was lesioned, explain the anomalous result.

6. Bullinaria and Chater, 1995. For attempts to model a double dissociation in a pattern associator model of reading aloud see Patterson, Seidenberg, & McClelland, 1989; Plaut, McClelland, Seidenberg, & Patterson, 1996; for critiques see Besner et al., 1990; Coltheart et al., 1993; Spieler & Balota, 1997; Balota & Spieler, 1997.

7. Ullman et al., 1993, 1997.

8. Dronkers, Pinker, & Damasio, 1999; Gazzaniga, Ivry, & Mangun, 1998; Damasio & Damasio, 1989; Caplan, 1987; Goodglass, 1993.

9. Gazzaniga, Ivry, & Mangun, 1998.

10. Gardner, 1974, 75–76.

11. Damasio et al., 1996; Goodglass, 1993; Alexander, 1997.

12. Butterworth, 1983.

13. Hagoort, Brown, & Swaab, 1996; Chiarello, 1991.

14. Goldman & Côté, 1991.

15. Murdoch et al., 1987; Illes, 1989.

16. Arnold et al., 1991; Kemper, 1994; see also the chapters on Alzheimer's disease in Feinberg & Farah, 1996.

17. Ullman et al., 1997.

18. Balota & Ferraro, 1993.

19. Squire, Knowlton & Musen, 1993; Mishkin, Malamut, & Bachevalier, 1984; Gazzaniga, Ivry, & Mangun, 1998.

20. Middleton & Strick, 1994; Wise, Murray, & Gerfen, 1996; Côté & Crutcher, 1991.

21. Illes, 1989; Lieberman et al., 1992; Grossman et al., 1992; Kemmerer, 1999.

22. Growdon & Corkin, 1986.

23. Ullman et al., 1997.

24. Reiner et al., 1988; Young & Penney, 1993; Côté & Crutcher, 1991.

25. A similar suggestion was made by the neurologists Antonio and Hannah Damasio, 1992.

26. Gopnik & Crago, 1991; Vargha-Khadem et al., 1995.

27. Leonard, 1998.

28. 'PM May Have Excuse for Language Gaffes,' Andrew Duffy, *Montreal Gazette,* October 18, 1997.

29. Bishop, North, & Donlan, 1995; Leonard, 1998; van der Lely & Stollwerck, 1996; Stromswold, 1998.

30. Fisher et al., 1998.

31. Vargha-Khadem et al., 1995; Gopnik & Crago, 1991; Ullman & Gopnik, 2000.
32. Ullman & Gopnik, 2000. See also Oetting & Rice, 1993, and Oetting & Horohov, 1997, for other demonstrations that the frequency of regularly inflected forms has a greater effect on SLI children than on controls.
33. Van der Lely, Rosen, & McClelland, 1998; van der Lely & Stollwerck, 1996; van der Lely & Ullman, 1996, 1999; van der Lely, 1997.
34. Van der Lely & Ullman, 1996, 1999.
35. With the youngest group there was a small frequency effect, but it was entirely due to differences in the familiarity of the stems, not of the past-tense forms; see van der Lely & Ullman, 1996, 1999.
36. Van der Lely & Christian, 1998.
37. Rossen et al., 1996.
38. Karmiloff-Smith et al., 1995.
39. Ewart et al., 1993; Frangiskakis et al., 1996.
40. Rosenblatt & Mitchison, 1998.
41. Rossen et al., 1996; Clahsen & Almazan, 1998.
42. Rossen et al., 1996; Tyler et al., 1997; Stevens & Karmiloff-Smith, 1997.
43. Bromberg et al., 1994; Clahsen & Almazan, 1998.
44. Karmiloff-Smith, Grant, & Berthoud, 1993.
45. Gazzaniga, Ivry, & Mangun, 1998; Garrett, 1995.
46. Weyerts et al., 1997; Penke et al., 1997; Gross et al., 1998; Newman, Neville, & Ullman, 1998, 1999.
47. Gazzaniga, Ivry, & Mangun, 1998; Martin, Brust, & Hilal, 1991; Posner & Raichle, 1994.
48. Jaeger et al., 1996; Ullman, Bergida, & O'Craven, 1997; Kemmerer, Dapretto, & Bookheimer, 1999; Indefrey et al., 1997.
49. See Kemmerer et al., 1999.
50. Buckner & Petersen, 1996.
51. Papanicolaou, Simos, & Basile, 1998; Roberts, Poeppel, & Rowley, 1998; Levelt et al., 1998.
52. Rhee, Pinker, & Ullman, 1999.
53. Wilson, 1998.

Chapter 10. A Digital Mind in an Analog World

1. For a more complete exploration of the ideas in this chapter see Pinker & Prince, 1996.
2. Smith & Medin, 1981.
3. Wittgenstein, 1953.
4. The object of Six Degrees of Kevin Bacon is to connect the actor Kevin Bacon to some other actor or actress with the shortest chain of films in which they co-appeared. Take, for example, Marlon Brando. He appeared in *The Godfather* with Al Pacino, who co-starred in *Sea of Love* with Ellen Barkin, who appeared with Bacon in *Diner* – a chain of three links.
5. Rosch, 1978, 1988; Smith & Medin, 1981.
6. Rosch, 1978, 1988; Smith & Medin, 1981.
7. McCloskey & Glucksberg, 1978.
8. Armstrong, Gleitman, & Gleitman, 1983; Rey, 1983; Pinker, 1997, chaps. 2 and 5; Marcus, in press a, b, c; Smith, Langston, & Nisbett, 1992; Smith, Medin, & Rips, 1984; Sloman, 1996; Goel, 1995.
9. Armstrong, Gleitman, & Gleitman, 1983.
10. Bybee & Slobin, 1982; Pinker & Prince, 1988.
11. Lakoff, 1987.
12. Pinker & Prince, 1996.
13. McClelland & Rumelhart, 1985.
14. Gluck & Bower, 1988; Kruschke, 1992.
15. Pinker, 1989, pp. 354–356.
16. Rumelhart, Smolensky, McClelland, and Hinton, 1986.
17. Marcus, in press b, c.
18. Ashby et al., 1998; Erikson & Kruschke, 1998; Nosofsky, Palmeri, & McKinley, 1994; Sloman, 1996; Goel, 1995.
19. Borges, 1964b, 63–64, 65. For a description of a real-life Funes the Memorious see Luria, 1968.
20. Quine, 1969; Rosch, 1978; Bobick, 1987; Anderson, 1990; Shepard, 1987; Tenenbaum, 1999.
21. Darwin, 1874, 106; see also Kelly, 1992.
22. Ridley, 1986.
23. Weng & Sokal, 1995; Chen, Sokal, & Ruhlen, 1995; Warnow, 1997; Warnow, Ringe, & Taylor, 1996.
24. Tooby & DeVore, 1987.
25. Pinker, 1997, chaps. 5 and 8; Brown, 1991.

REFERENCES

Adams, C. 1988. *More of The Straight Dope.* New York: Ballantine Books.

Alegre, M. A., & Gordon, P. 1996. Red rats eater exposes recursion in children's word formation. *Cognition, 60,* 65–82.

Alegre, M. A., & Gordon, P. 1997. Why compounds researchers aren't rats eaters: Semantic constraints on plurals inside compounds. Unpublished manuscript, Department of Psychology, University of Pittsburgh.

Alexander, M. P. 1997. Aphasia: Clinical and anatomic aspects. In T. E. Feinberg & M. Farah (Eds.), *Behavioral neurology and neuropsychology.* New York: McGraw-Hill.

Allen, J., Hunnicutt, M. S., & Klatt, D. H. 1987. *From text to speech: the MITalk System.* New York: Cambridge University Press.

Allen, W. 1980. The Kugelmass episode. In W. Allen, *Side effects.* New York: Random House.

Anderson, J. R. 1990. *The adaptive character of thought.* Mahwah, NJ: Erlbaum.

Anderson, J. R. 1993. *Rules of the mind.* Mahwah, NJ: Erlbaum.

Armstrong, S. L., Gleitman, L. R., & Gleitman, H. 1983. What some concepts might not be. *Cognition, 13,* 263–308.

Arnold, S., Hyman, B., Flory, J., Damasio, A., & Hoesen, G. V. 1991. The topographical and neuroanatomic distribution of neurofibrillary tangles and neuritic plaques in the cerebral cortex of patients with Alzheimer's disease. *Cerebral Cortex, 1,* 103–116.

Aronoff, M. 1976. *Word formation in generative grammar.* Cambridge, MA: MIT Press.

Aronoff, M. 1994. *Morphology by itself: Stems and inflectional classes.* Cambridge, MA: MIT Press.

Ashby, F. G., Alfonso-Reese, L. A., Turken, A. U., & Waldron, E. M.

1998. A neuropsychological theory of multiple systems in category learning. *Psychological Review, 105,* 442–481.

Baayen, R. H. 1994. Productivity in language production. *Language and Cognitive Processes, 9,* 447–469.

Baayen, R. H., Dijkstra, T., & Schreuder, R. 1997. Singulars and plurals in Dutch: Evidence for a parallel dual-route model. *Journal of Memory and Language, 37,* 94–117.

Baayen, R. H., & Lieber, R. 1991. Productivity and English derivation: A corpus-based study. *Linguistics, 29,* 801–843.

Baayen, R. H., & Renouf, A. 1996. Chronicling the *Times*: Productive innovations in an English newspaper. *Language, 72,* 69–96.

Baayen, R. H., & Schreuder, R. in press. The balance of storage and computation in the mental lexicon: The case of morphological processing in language comprehension. In S. Nootebohm, F. Weerman, & F. Wijnen (Eds.), *Storage and computation in the language faculty.* Dordrecht: Kluwer.

Badecker, W., & Caramazza, A. 1991. Morphological composition in the lexical output system. *Cognitive Neuropsychology, 8,* 335–367.

Baldi, P. 1983. *An introduction to the Indo-European languages.* Carbondale, IL: Southern Illinois University Press.

Balota, D. A., & Ferraro, R. 1993. A dissociation of frequency and regularity effects in pronunciation performance across young adults, older adults, and individuals with senile dementia of the Alzheimer types. *Journal of Memory and Language, 32,* 573–592.

Balota, D. A., Law, M. B., & Zevin, J. D. 1999. The attentional control of lexical processing pathways: Reversing the word-frequency effect. Unpublished manuscript, Department of Psychology, Washington University, St. Louis.

Balota, D. A., & Spieler, D. H. 1997. The utility of item-level analyses in model evaluation: A reply to Seidenberg and Plaut. *Psychological Science, 9,* 238–240.

Bauer, L. 1983. *English word formation.* New York: Cambridge University Press.

Baynes, K., & Iven, C. 1991. Access to the phonological lexicon in an aphasic patient. Paper presented at the Twenty-ninth Annual meeting of the Academy of Aphasia, Rome, October 13–15.

Beard, H., & McKie, R. 1982. *A gardener's dictionary.* New York: Workman.

Beck, M.-L. 1997. Regular verbs, past tense and frequency: tracking down a potential source of NS/NNS competence differences. *Second Language Research, 13,* 93–115.

Berent, I., Pinker, S., & Shimron, J. 1999. Default nominal inflection in Hebrew: Evidence for mental variables. *Cognition, 72,* 1–44.

Berko, J. 1958. The child's learning of English morphology. *Word,* 14, 150–177.

Bernstein, T. 1977. *The careful writer: A modern guide to English usage.* New York: Atheneum.

Besner, D., Twilley, L., McCann, R. S., & Seergobin, K. 1990. On the connection between connectionism and data: Are a few words necessary? *Psychological Review, 97,* 432–446.

Bishop, D. V. M., North, T., & Donlan, C. 1995. Genetic basis of Specific Language Impairment. *Developmental Medicine and Child Neurology, 37,* 56–71.

Bobick, A. 1987. *Natural object categorization.* MIT Artificial Intelligence Laboratory Technical Report 1001.

Booij, G. 1977. *Dutch morphology.* Dordrecht: Foris.

Borges, J. L. 1964a. The Library of Babel. In J. L. Borges, *Labyrinths: Selected stories and other writings.* New York: New Directions.

Borges, J. L. 1964b. Funes the memorious. In J. L. Borges, *Labyrinths: Selected stories and other writings.* New York: New Directions.

Borges, J. L. 1965. The analytical language of John Wilkins. In J. L. Borges, *Other inquisitions 1937–1952.* Austin: University of Texas Press.

Bornschein, M., & Butt, M. 1987. Zum Status des *s*-Plurals im gegenwärtigen Deutsch. In W. Abraham & N. Arhammar (Eds.), *Linguistik in Deutschland.* Akten des 21. Linguistischen Kolloquiums (135–154). Tübingen: Narr.

Borowsky, T. 1989. Structure preservation and the syllable coda in English. *Natural Language and Linguistic Theory, 7,* 145–166.

Bouchard, T. J., Jr. 1994. Genes, environment, and personality. *Science, 264,* 1700–1701.

Bresnan, J. 1978. A realistic transformational grammar. In M. Halle, J. Bresnan, & G. Miller (Eds.), *Linguistic theory and psychological reality.* Cambridge, MA: MIT Press.

Bright, W. (Ed.). 1992. *International encyclopedia of linguistics.* 4 vols. New York: Oxford University Press.

Bromberg, H. S., Ullman, M., Marcus, G., Kelly, K. B., & Levine, K. 1994. The dissociation between lexical memory and grammar in Williams syndrome: Evidence from inflectional morphology. Paper presented at the Williams Syndrome Professional Conference, San Diego, July.

Brown, D. E. 1991. *Human universals.* New York: McGraw-Hill.

Brown, R. 1973. *A first language: The early stages.* Cambridge, MA: Harvard University Press.

Buckner, R. L., & Petersen, S. E. 1996. What does neuroimaging tell us about the role of prefrontal cortex in memory retrieval? *Seminars in the Neurosciences, 8,* 47–55.

Bullinaria, J. A., & Chater, N. 1995. Connectionist modeling: Implications for cognitive neuropsychology. *Language and Cognitive Processes, 10,* 227–264.

Burani, C., & Caramazza, A. 1987. Representation and processing of derived words. *Language and Cognitive Processes, 2,* 217–227.

Burani, C., Salmaso, C., & Caramazza, A. 1984. Morphological structure and lexical access. *Visible Language, 18,* 342–352.

Butterworth, B. 1983. Lexical representation. In B. Butterworth (Ed.), *Language production,* Vol. 2. New York: Academic Press.

Bybee, J. L. 1985. *Morphology: A study of the relation between meaning and form.* Philadelphia: Benjamins.

Bybee, J. L. 1991. Natural morphology: The organization of paradigms and language acquisition. In T. Huebner and C. Ferguson (Eds.), *Crosscurrents in second language acquisition and linguistic theories.* Amsterdam: Benjamins.

Bybee, J. L. 1995. Regular morphology and the lexicon. *Language and Cognitive Processes, 10,* 425–455.

Bybee, J. L., & Moder, C. L. 1983. Morphological classes as natural categories. *Language, 59,* 251–270.

Bybee, J. L., & Slobin, D. I. 1982. Rules and schemes in the development and use of the English past tense. *Language, 58,* 265–289.

Caplan, D. 1987. *Neurolinguistics and linguistic aphasiology: An introduction.* Cambridge: Cambridge University Press.

Caramazza, A., Laudanna, A., & Romani, C. 1988. Lexical access and inflectional morphology. *Cognition, 28,* 297–332.

Caramazza, A., & Shelton, J. A. 1998. Domain-specific knowledge systems in the brain: the animate-inanimate distinction. *Journal of Cognitive Neuroscience, 10,* 1–34.

Carlson, G., Keyser, S. J., & Roeper, T. 1977. Dring, drang, drung. Unpublished manuscript, Department of Linguistics, University of Rochester.

Carstairs, A. 1987. *Allomorphy in inflection.* New York: Croom Helm.

Carstairs-McCarthy, A. 1994. Inflection classes, gender, and the principle of contrast. *Language, 70,* 737–788.

Cassidy, F. G. (Ed.). 1985. *Dictionary of American Regional English.* Cambridge, MA: Harvard University Press.

Cassidy, F. G., & Ringler, R. N. (Eds.). 1891/1971. *Bright's Old English Grammar & Reader.* 3d ed. New York: Holt, Rinehart and Winston.

Cazden, C. B. 1968. The acquisition of noun and verb inflections. *Child Development, 39,* 433–448.

Cazden, C. B. 1972. *Child language and education.* New York: Holt, Rinehart and Winston

Chen, J., Sokal, R. R., & Ruhlen, M. 1995. Worldwide analysis of genetic and linguistic relationships of human populations. *Human Biology, 67,* 595–612.

Chiarello, C. 1991. Interpretation of word meanings by the cerebral hemispheres: One is not enough. In P. J. Schwanenflugel (Ed.), *The psychology of word meanings.* Mahwah, NJ: Erlbaum.

Chomsky, N. 1959. A Review of B. F. Skinner's 'Verbal Behavior.' *Language, 3,* 26–58.

Chomsky, N. 1966. *Cartesian linguistics: A chapter in the history of rationalist thought.* New York: Harper & Row.

Chomsky, N., & Halle, M. 1968/1991. *The sound pattern of English.* Cambridge, MA: MIT Press.

Chomsky, W. 1957. *Hebrew: The eternal language.* Philadelphia: Jewish Publication Society.

Churma, D. G. 1983. Jets fans, Raider Rooters, and the interaction of morphosyntactic processes. In J. F. Richardson, M. Marks, & A. Chukerman (Eds.), *Papers from the parasession on the interplay of phonology, morphology, and syntax.* Chicago: Chicago Linguistics Society, University of Chicago Press.

Clahsen, H., & Almazan, M. 1998. Syntax and morphology in Williams syndrome. *Cognition, 68,* 167–198.

Clahsen, H., Marcus, G. F., Bartke, S., & Wiese, R. 1996. Compounding and inflection in German child language. In G. Booij and J. van Marle (Eds.), *Yearbook of morphology 1995.* Dordrecht: Kluwer.

Clahsen, H., & Rothweiler, M. 1992. Inflectional rules in children's grammars: evidence from the development of participles in German. *Morphology Yearbook,* 1–34.

Clahsen, H., Rothweiler, M., Woest, A., & Marcus, G. F. 1992. Regular and irregular inflection in the acquisition of German noun plurals. *Cognition, 45*, 225–255.

Collins, C. 1991. Notes on Dutch morphology. Unpublished manuscript, Department of Brain & Cognitive Sciences, MIT.

Coltheart, M. 1985. Cognitive neuropsychology and the study of reading. In A. Young (Ed.), *Functions of the right cerebral hemisphere.* London: Academic Press.

Coltheart, M., Curtis, B., Atlins, P., & Haller, M. 1993. Models of reading aloud: Dual-route and parallel distributed processing approaches. *Psychological Review, 100*, 589–608.

Comrie, B. 1981. *Language universals and linguistic typology.* Chicago: University of Chicago Press.

Comrie, B. 1990. *The world's major languages.* New York: Oxford University Press.

Comrie, B., Matthews, S., & Polinsky, M. 1996. *The atlas of languages.* New York: Facts on File.

Cooper, W. E., & Ross, J. R. 1975. World order. In R. E. Grossman, L. J. San, & T. J. Vance (Eds.), *Papers from the parasession on functionalism.* Chicago: Chicago Linguistics Society, University of Chicago Press.

Côté, L., & Crutcher, M. D. 1991. The basal ganglia. In E. R. Kandel, J. H. Schwartz, & T. M. Jessel (Eds.), *Principles of neural science.* 3d ed. Norwalk, CT: Appleton & Lange.

Crystal, D. 1997. *The Cambridge Encyclopedia of Language.* 2d ed. New York: Cambridge University Press.

Curme, G. O. 1935/1983. *A grammar of the English language. Vol. 1. Parts of speech.* Essex, CT: Verbatim.

Damasio, A. R., & Damasio, H. 1992. Brain and language. *Scientific American, 267*, September.

Damasio, H., & Damasio, A. R. 1989. *Lesion analysis in neuropsychology.* New York: Oxford University Press.

Damasio, H., Grabowski, T. J., Tranel, D., Hichwa, R. D., & Damasio, A. R. 1996. A neural basis for lexical retrieval. *Nature, 380*, 499–505.

Darwin, C. R. 1874. *The descent of man and selection in relation to sex.* 2d ed. New York: Hurst & Company.

Daugherty, K. G., & Hare, M. 1993. What's in a rule? The past tense by some other name might be called a connectionist net. In M. C. Mozer, P. Smolensky, D. S. Touretzky, J. L. Elman, and A. S.

Weigend (Eds.), *Proceedings of the 1993 Connectionist Models Summer School.* Mahwah, NJ: Erlbaum.

Daugherty, K. G., MacDonald, M. C., Petersen, A. S., & Seidenberg, M. S. 1993. Why no mere mortal has ever flown out to center field but people often say they do. *Proceedings of the Fifteenth Annual Conference of the Cognitive Science Society.* Mahwah, NJ: Erlbaum.

Daugherty, K., & Seidenberg, M. 1992. Rules or connections? The past tense revisited. *Proceedings of the Fourteenth Annual Conference of the Cognitive Science Society.* Mahwah, NJ: Erlbaum.

Dell, G. S. 1995. Speaking and misspeaking. In L. R. Gleitman & M. Liberman (Eds.), *An invitation to cognitive science.* 2d ed. Vol. 1. *Language.* Cambridge, MA: MIT Press.

Dewdney, A. K. 1997. *Yes, we have no neutrons: An eye-opening tour through the twists and turns of bad science.* New York: Wiley.

di Sciullo, A. M., & Williams, E. 1987. *On the definition of word.* Cambridge, MA: MIT Press.

Donaldson, B. 1987. *Dutch reference grammar.* Leiden: Martinus Nijhoff.

Dressler, W. 1985. Typological aspects of natural morphology. *Wiener Linguistische Gazette, 35–36,* 3–26.

Dronkers, N., Pinker, S., & Damasio, A. R. 1999. Language and the aphasias. In E. R. Kandel, J. H. Schwartz, & T. M. Jessell (Eds.), *Principles of neural science,* 4th ed. Norwalk, CT: Appleton & Lange.

Eco, U. 1995. *The search for the perfect language.* Cambridge, MA: Blackwell.

Egedi, D. M., & Sproat, R. W. 1991. Connectionist networks and natural language morphology. Unpublished manuscript, Linguistics Research Department, Lucent Technologies, Murray Hill, NJ.

Empson, W. 1959. Invitation to Juno. In W. Empson, *Collected Poems.* New York: Harcourt Brace.

Erikson, M. A., & Kruschke, J. K. 1998. Rules and exemplars in category learning. *Journal of Experimental Psychology: General, 127,* 107–140.

Espy, W. R. 1975. *An almanac of words at play.* New York: Clarkson Potter.

Ewart, A. K., Morris, C. A., Atkinson, D., Jin, W., Sternes, K., Spallone, P., Dean Stock, A., Leppert, M., & Keating, M. T. 1993. Hemizygosity at the elastin locus in a developmental disorder, Williams syndrome. *Nature Genetics, 5,* 11–16.

Feinberg, T. E., & Farah, M. (Eds.). 1996. *Behavioral neurology and neuropsychology*. New York: McGraw-Hill.

Feldman, L. B. 1995. (Ed.). *Morphological aspects of language processing*. Mahwah, NJ: Erlbaum.

Fisher, S. E., Vargha-Khadem, F., Watkins, K. E., Monaco, A. P., & Pembrey, M. E. 1998. Localisation of a gene implicated in a severe speech and language disorder. *Nature Genetics, 18,* 168–170.

Fodor, J. A., & McClaughlin, B. 1990. Connectionism and the problem of systematicity: Why Smolensky's solution doesn't work. *Cognition, 35,* 183–204.

Fodor, J. A., & Pylyshyn, Z. 1988. Connectionism and cognitive architecture: A critical analysis. *Cognition, 28,* 3–71. Reprinted in S. Pinker & J. Mehler (Eds.), *Connections and symbols*. Cambridge, MA: MIT Press.

Forrester, N., & Plunkett, K. 1994. Learning the Arabic plural: The case for minority default mappings in connectionist networks. In A. Ram and K. Eiselt (Eds.), *Proceedings of the Sixteenth Annual Conference of the Cognitive Science Society*. Mahwah, NJ: Erlbaum.

Fortune, R. F. 1942. *Arapesh*. Publications of the American Ethnological Society, no. 19. New York: J. Augustin.

Fowler, C. A., Napps, S. E., & Feldman, L. 1985. Relations among regular and irregular morphologically related words in the lexicon as related by repetition priming. *Memory and Cognition, 13,* 241–255.

Fowler, H. W. 1965. *A dictionary of modern English usage*. 2d ed., revised by Sir Ernest Gowers. New York: Oxford University Press.

Francis, N., & Kučera, H. 1982. *Frequency analysis of English usage: Lexicon and grammar*. Boston: Houghton Mifflin.

Frangiskakis, J. M., Ewart, A. K., Morris, A. C., Mervis, C. B., Bertrand, J., Robinson, B. F., Klein, B. P., Ensing, G. J., Everett, L. A., Green, E. D., Proschel, C., Gutowski, N. J., Noble, M., Atkinson, D. L., Odelberg, S. J., & Keating, M. T. 1996. LIM-Kinase1 hemizygosity implicated in impaired visuospatial constructive cognition. *Cell, 86,* 59–69.

Fromkin, V. 1971. The non-anomalous nature of anomalous utterances. *Language, 47,* 27–52.

Ganger, J. 1998. Genes and environment in language acquisition: A study of early vocabulary and syntactic development in twins. Doctoral dissertation, Department of Brain & Cognitive Sciences, MIT.

Ganger, J., Pinker, S., Baker, A., & Chawla, S. 1999. A twin study of early vocabulary and grammatical development. Paper presented at the Biennial Meeting of the Society for Research in Child Development, Albuquerque, NM.

Gardner, H. 1974. *The shattered mind.* New York: Vintage.

Garrett, M. F. 1980. Levels of processing in sentence production. In B. Butterworth (Ed.), *Language production.* Vol. 1, *Speech and talk.* New York: Academic Press.

Garrett, M. F. 1995. The structure of language processing: Neuropsychological evidence. In M. Gazzaniga (Ed.), *The cognitive neurosciences.* Cambridge, MA: MIT Press.

Gazzaniga, M. S. (Ed.). 1995. *The cognitive neurosciences.* Cambridge, MA: MIT Press.

Gazzaniga, M. S., Ivry, R. B., & Mangun, G. R. 1998. *Cognitive neuroscience: The biology of the mind.* New York: Norton.

Gluck, M. A., & Bower, G. H. 1988. From conditioning to category learning: An adaptive network model. *Journal of Experimental Psychology: General, 117,* 37–50.

Goebel, R., & Indefrey, P. 1994. The performance of a recurrent network with short term memory capacity learning the German -s plural. Paper presented at the Workshop on Cognitive Models of Language Acquisition, Tilburg, The Netherlands, April 21–23.

Goel, V. 1995. *Sketches of thought.* Cambridge, MA: MIT Press.

Goldman, J., & Côté, L. 1991. Aging of the brain: Dementia of the Alzheimer's type. In E. R. Kandel, J. H. Schwartz, & T. M. Jessel (Eds.), *Principles of neural science.* 3d. ed. Norwalk, CT: Appleton & Lange.

Goodglass, H. 1993. *Understanding aphasia.* San Diego: Academic Press.

Gopnik, M., & Crago, M. 1991. Familial aggregation of a developmental language disorder. *Cognition, 39,* 1–50.

Gordon, P. 1985. Level-ordering in lexical development. *Cognition, 21,* 73–93.

Görlach, M. 1991. *Introduction to Early Modern English.* New York: Cambridge University Press.

Greenberg, J. H., Ferguson, C. A., & Moravcsik, E. A. (Eds.). 1978. *Universals of human language.* 4 vols. Stanford, CA: Stanford University Press.

Gropen, J., Pinker, S., Hollander, M., Goldberg, R., & Wilson, R. 1989.

The learnability and acquisition of the dative alternation in English. *Language, 65,* 203–257.

Gross, M., Say, T., Kleingers, M., Clahsen, H., & Münte, T. F. 1998. Human brain potentials to violations in morphologically complex Italian words. *Neuroscience Letters, 241,* 83–86.

Grossman, M., Carvell, S., Stern, M. B., Gollomp, S., & Hurtig, H. I. 1992. Sentence comprehension in Parkinson's disease: The role of attention and memory. *Brain and Language, 42,* 347–384.

Growden, J., & Corkin, S. 1986. Cognitive impairments in Parkinson's Disease. In M. Yahr & K. Bergmann (Eds.), *Advances in neurology.* Vol. 45. New York: Raven Press.

Haber, L. R. 1975. The Muzzy Theory. In *Proceedings of the Eleventh Regional Meeting of the Chicago Linguistics Society.* Chicago: Chicago Linguistics Society, University of Chicago Press.

Haber, L. R. 1976. Leaped and leapt: A theoretical account of linguistic variation. *Foundations of Language, 14,* 211–238.

Hadley, R. F. 1994a. Systematicity in connectionist language learning. *Mind and Language, 9,* 247–272.

Hadley, R. F. 1994b. Systematicity revisited: Reply to Christiansen and Chater and Niklasson and Van Gelder. *Mind and Language, 9,* 431–444.

Hadley, R. F., & Hayward, M. 1994. Strong semantic systematicity from unsupervised connectionist learning. Technical Report CSS-IS TR94–02, School of Computing Science, Simon Fraser University, Burnaby, BC.

Hagoort, P., Brown, C., & Swaab, T. 1996. Lexical semantic event-related potential effects in patients with left hemisphere lesions and aphasia, and patients with right hemisphere lesions without aphasia. *Brain, 119,* 627–649.

Halford, G. S., Wilson, W. H., Gray, B., & Phillips, S. 1997. A neural net model for mapping hierarchically structured analogs. In *Proceedings of the Fourth Conference of the Australasian Cognitive Science Society.* Department of Psychology, University of Newcastle, Australia.

Halle, M. 1973. Prolegomena to a theory of word formation. *Linguistic Inquiry, 4,* 3–16.

Halle, M., & Marantz, A. 1993. Distributed morphology and the pieces of inflection. In K. Hale & S. J. Keyser (Eds.), *The view from Building 20: Essays in honor of Sylvain Bromberger.* Cambridge, MA: MIT Press.

Halle, M., & Mohanan, K. P. 1985. Segmental phonology of modern English. *Linguistic Inquiry, 16*, 57–116.

Hare, M., & Elman, J. 1992. A connectionist account of English inflectional morphology: Evidence from language change. In *Proceedings of the Fourteenth Annual Conference of the Cognitive Science Society.* Mahwah, NJ: Erlbaum.

Hare, M., Elman, J., & Daugherty, K. 1995. Default generalization in connectionist networks. *Language and Cognitive Processes, 10*, 601–630.

Harris, C. L. 1992. Understanding English past-tense formation: The shared meaning hypothesis. In *Proceedings of the Fourteenth Annual Conference of the Cognitive Science Society.* Mahwah, NJ: Erlbaum.

Hartley, D. 1775/1973. *Hartley's theory of the human mind.* New York: AMS Press.

Hawkins, J. (Ed.). 1988. *Explaining language universals.* Cambridge, MA: Blackwell.

Heine, B., Claudi, U., & Hunnemeyer, F. 1991. *Grammaticalization: A conceptual framework.* Chicago: University of Chicago Press.

Hinton, G. E., McClelland, J. L., & Rumelhart, D. E. 1986. Distributed representations. In D. Rumelhart, J. McClelland, and the PDP Research Group, *Parallel distributed processing: Explorations in the microstructure of cognition.* Vol. 1. *Foundations.* Cambridge, MA: MIT Press.

Hobbes, T. 1651/1957. *Leviathan.* New York: Oxford University Press.

Hoeffner, J. 1992. Are rules a thing of the past? The acquisition of verbal morphology by an attractor network. In *Proceedings of the Fourteenth Annual Conference of the Cognitive Science Society.* Mahwah, NJ: Erlbaum.

Hogg, R. M. 1988. Snuck: The development of irregular preterite forms. In G. Nixon & J. Honey (Eds.), *An historic tongue: Studies in English linguistics in memory of Barbara Strang.* London: Routledge.

Humboldt, W. von. 1836/1972. *Linguistic variability and intellectual development.* G. C. Buck & F. Raven, Trans. Philadelphia: University of Pennsylvania Press.

Hume, D. 1748/1955. *Enquiry concerning human understanding.* Indianapolis: Bobbs-Merrill.

Hummel, J. E., & Biederman, I. 1992. Dynamic binding in a neural network for shape recognition. *Psychological Review, 99*, 480–517.

Hummel, J. E., & Holyoak, K. J. 1997. Distributed representations of structure: A theory of analogical access and mapping. *Psychological Review, 104,* 427–466.

Hymes, D. 1964. *Language in culture and society: A reader in linguistics and anthropology.* New York: Harper and Row.

Illes, J. 1989. Neurolinguistic features of spontaneous language production dissociate three forms of neurodegenerative disease: Alzheimer's, Huntington's, and Parkinson's. *Brain and Language, 37,* 628–642.

Indefrey, P., Brown, C., Hagoort, P., Herzog, H., Sach, M., & Seitz, R. J. 1997. A PET study of cerebral activation patterns induced by verb inflection. *NeuroImage, 5,* S548.

Jackendoff, R. 1975. Morphological and semantic regularities in the lexicon. *Language, 51,* 639–671.

Jackendoff, R. 1983. *Semantics and cognition.* Cambridge, MA: MIT Press.

Jackendoff, R. 1997. *The architecture of the language faculty.* Cambridge, MA: MIT Press.

Jaeger, J. J., Lockwood, A. H., Kemmerer, D. L., Van Valin, R. D., Murphy, B. W., & Khalak, H. G. 1996. A positron emission tomography study of regular and irregular verb morphology in English. *Language, 72,* 451–497.

Jakobson, R. 1971. Quest for the essence of language. In *Roman Jakobson: Selected writings II: Word and Language.* The Hague: Mouton.

James, W. 1890/1950. *The principles of psychology.* New York: Dover.

Janda, R. D. 1991. Frequency, markedness, and morphological change: On predicting the spread of noun-plural -*s* in Modern High German and West Germanic. In Y. No & M. Libucha (Eds.), *ESCOL '90.* Ithaca, NY: Cornell Linguistics Circle Publications.

Jespersen, O. 1938/1982. *Growth and structure of the English language.* Chicago: University of Chicago Press.

Johnson, K. 1986. Fragmentation of strong verb ablaut in Old English. *Ohio State University Working Papers in Linguistics, 34,* 108–122.

Karmiloff-Smith, A., Grant, J., & Berthoud, I. 1993. Within-domain dissociations in Williams syndrome: A window on the normal mind. Paper presented at the Biennial Meeting of the Society for Research in Child Development, New Orleans, March.

Karmiloff-Smith, A., Grant, J., Berthoud, I., Davies, M., Howlin, P., & Udwin, O. 1997. Language and Williams syndrome: How intact is 'intact'? *Child Development, 68,* 2, 246–262.

Karmiloff-Smith, A., Klima, E. S., Bellugi, U., Grant, J., & Baron-Cohen, S. 1995. Is there a social module? Language, face processing, and Theory of Mind in individuals with Williams syndrome. *Journal of Cognitive Neuroscience, 7,* 196–208.

Karttunen, L., Kaplan, R. M., & Zaenen, A. 1992. Two-level morphology with composition. *Proceedings of Coling-92,* Nantes, France.

Karttunen, L., Koskenniemi, K., & Kaplan, R. M. 1987. A compiler for two-level phonological rules. In M. Dalrymple (Ed.), *Tools for morphological analysis.* Stanford, CA: Center for the Study of Language and Information.

Katz, L., Rexer, K., & Lukatela, G. 1991. The processing of inflected words. *Psychological Research, 53,* 25–32.

Kelly, M. H. 1992. Darwin and psychological theories of classification. *Evolution and Cognition, 2,* 79–97.

Kemmerer, D. 1999. Impaired comprehension of raising-to-subject constructions in Parkinson's disease. *Brain and Language, 66,* 311–328.

Kemmerer, D., Dapretto, M., & Bookheimer, S. Y. 1999. An fMRI study of regular and irregular inflectional morphology in English. Unpublished manuscript, Department of Neurology, University of Iowa.

Kemper, T. 1994. Neuroanatomical and neuropathological changes during aging and dementia. In M. Albert & J. Knoefel (Eds.), *Clinical neurology of aging.* New York: Oxford University Press.

Kempley, S. T., & Morton, J. 1982. The effects of priming with regularly and irregularly related words in auditory word recognition. *British Journal of Psychology, 73,* 441–454.

Keyser, S. J. & O'Neil, W. 1985. *Rule generalization and optionality in language change.* Dordrecht: Foris.

Kiefer, F. 1985. The possessive in Hungarian: A problem for natural morphology. *Acta Linguistica Academiae Scientiarum Hungaricae, 35,* 85–116.

Kim, J. J. 1999. The semantics hypothesis of grammatical structure: A response to Shirai. Unpublished manuscript, Department of Psychology, San Francisco State University.

Kim, J. J., Marcus, G. F., Pinker, S., Hollander, M., & Coppola, M. 1994.

Sensitivity of children's inflection to morphological structure. *Journal of Child Language, 21*, 173–209.

Kim, J. J., Pinker, S., Prince, A., & Prasada, S. 1991. Why no mere mortal has ever flown out to center field. *Cognitive Science, 15*, 173–218.

Kiparsky, P. 1982. Lexical phonology and morphology. In I. S. Yang (Ed.), *Linguistics in the morning calm*. Seoul: Hansin.

Köpcke, K.-M. 1988. Schemas in German plural formation. *Lingua, 74*, 303–335.

Kruschke, J. K. 1992. ALCOVE: An example-based connectionist model of category learning. *Psychological Review, 99*, 22–44.

Kuczaj, S. A. 1976. *-ing, -s, & -ed:* A study of the acquisition of certain verb inflections. Doctoral dissertation, Department of Psychology, University of Minnesota.

Kuczaj, S. A. 1977. The acquisition of regular and irregular past tense forms. *Journal of Verbal Learning and Verbal Behavior, 16*, 589–600.

Kuczaj, S. A. 1978. Children's judgments of grammatical and ungrammatical irregular past tense verbs. *Child Development, 49*, 319–326.

Kuryłowicz, J. 1964. *The inflectional categories of Indo-European.* Heidelberg: C. Winter.

Lachter, J., & Bever, T. G. 1988. The relation between linguistic structure and associative theories of language learning – A constructive critique of some connectionist learning models. *Cognition, 28*, 195–247. Reprinted in S. Pinker & J. Mehler (Eds.), *Connections and symbols*. Cambridge, MA: MIT Press.

Lahiri, A., & Marslen-Wilson, W. 1991. The mental representation of lexical form: A phonological approach to the recognition lexicon. *Cognition, 38*, 245–294.

Lakoff, G. 1987. *Women, fire, and dangerous things: What categories reveal about the mind.* Chicago: University of Chicago Press.

Laudanna, A., & Burani, C. 1995. Distributional properties of derivational affixes: Implications for processing. In L. B. Feldman (Ed.), *Morphological aspects of language processing*. Mahwah, NJ: Erlbaum.

Lederer, R. 1990. *Crazy English.* New York: Pocket Books.

Lenneberg, E. 1964. The capacity for language acquisition. In J. A. Fodor & J. J. Katz (Eds.), *The structure of language: Readings in the philosophy of language*. Englewood Cliffs, NJ: Prentice-Hall.

Leonard, L. B. 1998. *Children with Specific Language Impairment.* Cambridge, MA: MIT Press.

Levelt W. J., Praamstra, P., Meyer, A. S., Helenius, P., & Salmelin, R. 1998. An MEG study of picture naming. *Journal of Cognitive Neuroscience, 10,* 553–67.

Levin, S. R. 1964. A reclassification of the Old English strong verbs. *Language, 40,* 156–161.

Lieber, R. 1980. *On the organization of the lexicon.* Doctoral dissertation, Department of Linguistics and Philosophy, MIT, Cambridge, MA. Distributed by the Indiana University Linguistics Club.

Lieber, R. 1992. *Deconstructing morphology: Word formation in syntactic theory.* Chicago: University of Chicago Press.

Lieberman, P., Kako, E., Friedman, J., Tajchman, G., Feldman, L. S., & Jiminez, E. B. 1992. Speech production, syntax comprehension, and cognitive deficits in Parkinson's disease. *Brain and Language, 43,* 169–189.

Ling, C., & Marinov, M. 1993. Answering the connectionist challenge: A symbolic model of learning the past tenses of English verbs. *Cognition, 49,* 235–290.

Luria, A. R. 1968. *The mind of a mnemonist: A little book about a vast memory.* New York: Basic Books.

MacWhinney, B. 1978. Processing a first language: the acquisition of morphophonology. *Monographs of the Society for Research in Child Development, 43.*

MacWhinney, B., & Leinbach, J. 1991. Implementations are not conceptualizations: Revising the verb learning model. *Cognition, 40,* 121–157.

MacWhinney, B., & Snow, C. E. 1985. The Child Language Data Exchange System. *Journal of Child Language, 12,* 271–296.

MacWhinney, B., & Snow, C. E. 1990. The Child Language Data Exchange System: An update. *Journal of Child Language, 17,* 457–472.

Marchand, H. 1969. *The categories and types of present-day English word-formation.* 2d ed. Munich: C. H. Beck.

Marchman, V. 1993. Constraints on plasticity in a connectionist model of the English past tense. *Journal of Cognitive Neuroscience, 5,* 215–234.

Marcus, G. F. 1993. Negative evidence in language acquisition. *Cognition, 46,* 53–85.

Marcus, G. F. 1995a. The acquisition of inflection in children and multilayered connectionist networks. *Cognition, 56*, 271–279.

Marcus, G. F. 1995b. Children's overregularization of English plurals: A quantitative analysis. *Journal of Child Language, 22*, 447–459.

Marcus, G. F. 1997. Review of 'Exercises in Rethinking Innateness.' *Trends in Cognitive Sciences, 1*, 318–319.

Marcus, G. F. 1998. Rethinking eliminative connectionism. *Cognitive Psychology, 37*, 243–282.

Marcus, G. F. 2000. Two kinds of representations. In E. Deitrich & A. Markman (Eds.), *Cognitive dynamics: Conceptual and representational change in humans and machines.* Mahwah, NJ: Erlbaum.

Marcus, G. F. in press. Rethinking eliminative connectionism. *Cognitive Psychology.*

Marcus, G. F. 2000. *The algebraic mind: integrating connectionism and cognitive science.* Cambridge, MA: MIT Press.

Marcus, G. F., Brinkmann, U., Clahsen, H., Wiese, R., & Pinker, S. 1995. German inflection: The exception that proves the rule. *Cognitive Psychology, 29*, 189–256.

Marcus, G. F., Pinker, S., & Larkey, L. 1995. Do overregularizations come from a grammatical reorganization? Paper presented at the Twentieth Annual Boston University Conference on Language Development, Boston, November 3–5.

Marcus, G. F., Pinker, S., Ullman, M., Hollander, M., Rosen, T. J., & Xu, F. 1992. Overregularization in language acquisition. *Monographs of the Society for Research in Child Development, 57*.

Marin, O., Saffran, E. M., & Schwartz, M. F. 1976. Dissociations of language in aphasia: Implications for normal function. In S. R. Harnad, H. S. Steklis, & J. Lancaster (Eds.), *Origin and evolution of language and speech. Annals of the New York Academy of Sciences.* Vol. 280. New York: New York Academy of Sciences.

Marslen-Wilson, W. D. 1987. Functional parallelism in spoken word recognition. *Cognition, 25*, 71–102.

Marslen-Wilson, W. D. (Ed.). 1989. *Lexical representation and process.* Cambridge, MA: MIT Press.

Marslen-Wilson, W. D. 1998. Dissociating types of lexical computation? Paper presented at the Department of Brain & Cognitive Sciences, MIT, April 17.

Marslen-Wilson, W. D., Hare, M., & Older, L. 1995. Priming and blocking in the mental lexicon: The English past tense. Paper presented at the Meeting of the Experimental Society, London, January.

Marslen-Wilson, W. D., & Tyler, L. K. 1997. Dissociating types of mental computation. *Nature, 387*, 592–594.

Marslen-Wilson, W. D., & Tyler, L. K. 1998. Rules, representations, and the English past tense. *Trends in Cognitive Science, 2*, 428–435.

Martin, J. H., Brust, J. C. M., & Hilal, S. 1991. Imaging the living brain. In E. R. Kandel, J. H. Schwartz, & T. M. Jessel (Eds.), *Principles of neural science.* 3d ed. Norwalk, CT: Appleton & Lange.

McCarthy, J., & Prince, A. 1990. Foot and word in prosodic morphology: The Arabic broken plural. *Natural Language and Linguistic Theory, 8*, 209–283.

McCarthy, J., & Prince, A. in press. *Prosodic morphology.* Cambridge, MA: MIT Press.

McClelland, J. L., & Rumelhart, D. E. 1985. Distributed memory and the representation of general and specific information. *Journal of Experimental Psychology: General, 114*, 159–188.

McClelland, J. L., Rumelhart, D. E., & the PDP Research Group. 1986. *Parallel distributed processing: Explorations in the microstructure of cognition.* Vol. 2. *Psychological and biological models.* Cambridge, MA: MIT Press.

McCloskey, M., & Glucksberg, S. 1978. Natural categories: Well-defined or fuzzy sets? *Memory and Cognition, 6*, 462–472.

McGuinness, D. 1997. *Why our children can't read.* New York: Free Press.

Mencken, H. L. 1919/1986. *The American language.* (One-volume abridgement of the 4th ed. and the two supplements. R. I. McDavid, Jr., Ed.) New York: Knopf.

Mencken, H. L. 1936. *The American language.* New York: Knopf.

Middleton, F., & Strick, P. 1994. Anatomical evidence for cerebellar and basal ganglia involvement in higher cognitive function. *Science, 226*, 458–461.

Miller, G. A. 1967. The psycholinguists. In G. A. Miller, *The psychology of communication.* London: Penguin Books.

Miller, G. A. 1991. *The science of words.* New York: W. H. Freeman.

Minsky, M., & Papert, S. 1988. Epilogue: The new connectionism. In *Perceptrons: Expanded edition.* Cambridge, MA: MIT Press.

Mishkin, M., Malamut, B., & Bachevalier, J. 1984. Memories and habits: Two neural systems. In G. Lynch, J. McGaugh, & N.

Weinberger (Eds.), *Neurobiology of learning and memory.* New York: Guilford Press.

Mithun, M. 1988. Lexical categories and the evolution of number marking. In M. Hammond & M. Noonan (Eds.), *Theoretical morphology: Approaches in modern linguistics.* New York: Academic Press.

Moravcsik, E. 1975. Borrowed verbs. *Wiener Linguistische Gazette, 5,* 3–31.

Morgan, J. L., Bonamo, K., & Travis, L. L. 1995. Negative evidence on negative evidence. *Developmental Psychology, 31,* 180–197.

Morgan, J. L., & Travis, L. L. 1989. Limits on negative information on language learning. *Journal of Child Language, 16,* 531–552.

Mozer, M. 1991. *The perception of multiple objects: A connectionist approach.* Cambridge, MA: MIT Press.

Münte, T. F., Say, T., Clahsen, H., Schiltz, K., & Kutas, M. 1998. Decomposition of morphologically complex words in English: Evidence from event-related brain potentials. *Cognitive Brain Research, 7,* 241–253.

Murdoch, B. E., Chenery, H. J., Wilks, V., & Boyle, R. S. 1987. Language disorders in dementia of the Alzheimer type. *Brain and Language, 31,* 122–137.

Murray, T. E. 1998. More on *drug/dragged* and *snuck/sneaked*: Evidence from the American Midwest. *Journal of English Linguistics, 26,* 209–221.

Myers, J. 1998. Rules vs. analogy in Mandarin classifier selection. Paper presented at the Sixth International Symposium on Chinese Languages and Linguistics, July 14–16, Academia Sinica, Taipei, Taiwan. Unpublished manuscript, Graduate Institute of Linguistics, National Chung Cheng University, Taiwan.

Myers, S. 1987. Vowel shortening in English. *Natural Language and Linguistic Theory, 5,* 485–518.

Nakisa, R. C., & Hahn, U. 1996. Where defaults don't help: The case of the German plural system. In G. W. Cottrell (Ed.), *Proceedings of the Eighteenth Annual Conference of the Cognitive Science Society.* Mahwah, NJ: Erlbaum.

Napps, Shirley E. 1989. Morphemic relationships in the lexicon: Are they distinct from semantic and formal relationships? *Memory and Cognition, 17(6),* 729–739.

Newell, A. 1990. *Unified theories of cognition.* Cambridge, MA: Harvard University Press.

Newman, A. H., Neville, H., & Ullman, M. T. 1998. Neural processing of inflectional morphology: An event-related potential study of English past tense. Paper presented at the Fifth Annual Meeting of the Cognitive Neuroscience Society, San Francisco.

Newman, A., Izvorski, R., Davis, L., Neville, H., & Ullman, M. T. 1999. Distinct electro-physiological patterns in the processing of regular and irregular verbs. Paper presented at the Sixth Annual Meeting of the Cognitive Neuroscience Society, Washington.

Nosofsky, R. M., Palmeri, T. J., & McKinley, S. C. 1994. Rule-plus-exception model of classification learning. *Psychological Review, 101,* 53–79.

Oetting, J. B., & Horohov, J. E. 1997. Past tense marking by children with and without Specific Language Impairment. *Journal of Speech, Language, and Hearing Research, 40,* 62–74.

Oetting, J. B., & Rice, M. 1993. Plural acquisition in children with Specific Language Impairment. *Journal of Speech and Hearing Research, 36,* 1236–1248.

Omar, M. K. 1973. *The acquisition of Egyptian Arabic as a native language.* The Hague: Mouton.

Orsolini, M., & Marslen-Wilson, W. D. 1997. Universals in morphological representation: Evidence from Italian. *Language and Cognitive Processes, 12,* 1–47.

Papanicolaou, A. C., Simos, P. G., & Basile, L. F. H. 1998. Applications of magnetoencephalography to neurolinguistic research. In B. Stemmer & H. A. Whitaker (Eds.), *Handbook of neurolinguistics.* San Diego: Academic Press.

Patterson, K. E., Seidenberg, M. S., & McClelland, J. L. 1989. Connections and disconnections: Acquired dyslexia in a computational model of reading processes. In R. G. M. Morris (Ed.), *Parallel Distributed Processing: Implications for psychology and neurobiology.* New York: Oxford University Press.

Penke, M., Weyerts, H., Gross, M., Zander, E., Münte, T. F., & Clahsen, H. 1997. How the brain processes complex words: An ERP study of German verb inflection. *Cognitive Brain Research, 6,* 37–52.

Pinker, S. 1984. *Language learnability and language development.* Cambridge, MA: Harvard University Press. Reprinted with a new introduction, 1996.

Pinker, S. 1989. *Learnability and cognition: The acquisition of argument structure.* Cambridge, MA: MIT Press.

Pinker, S. 1994. *The language instinct.* New York: HarperCollins.

Pinker, S. 1997. *How the mind words.* New York: Norton.

Pinker, S., & Birdsong, D. 1979. Speakers' sensitivity to rules of frozen word order. *Journal of Verbal Learning and Verbal Behavior, 18,* 497–508.

Pinker, S., & Mehler, J. (Eds.). 1988. *Connections and symbols.* Cambridge, MA: MIT Press.

Pinker, S., & Prince, A. 1988. On language and connectionism: Analysis of a Parallel Distributed Processing model of language acquisition. *Cognition, 28,* 73–193. Reprinted in S. Pinker & J. Mehler (Eds.), *Connections and symbols.* Cambridge, MA: MIT Press.

Pinker, S., & Prince, A. 1994. Regular and irregular morphology and the psychological status of rules of grammar. In S. D. Lima, R. L. Corrigan, & G. K. Iverson (Eds.), *The reality of linguistic rules.* Philadelphia: Benjamins.

Pinker, S., & Prince, A. 1996. The nature of human concepts: Evidence from an unusual source. *Communication and Cognition, 29,* 307–361. Reprinted in P. Van Loocke (Ed.), *The nature, representation and evolution of concepts.* London: Routledge, 1999. Reprinted also in R. Jackendoff, P. Bloom, & K. Wynn (Eds.), *Language, logic, and concepts: Essays in memory of John Macnamara.* Cambridge, MA: MIT Press, 1999.

Plaut, D. C., McClelland, J. L., Seidenberg, M. S., & Patterson, K. 1996. Understanding normal and impaired word reading: Computational principles in quasi-regular domains. *Psychological Review, 103,* 56–115.

Plunkett, K., & Marchman, V. 1991. U-shaped learning and frequency effects in a multi-layered perceptron: Implications for child language acquisition. *Cognition, 38,* 43–102.

Plunkett, K., & Marchman, V. 1993. From rote learning to system building. *Cognition, 48,* 21–69.

Plunkett, K., & Nakisa, R. 1997. A connectionist model of the Arab plural system. *Language and Cognitive Processes, 12,* 807–836.

Pollack, J. B. 1990. Recursive distributed representations. *Artificial Intelligence, 46,* 77–105.

Posner, M. I., & Raichle, M. E. 1994. *Images of mind.* New York: W. H. Freeman.

Prasada, S., & Pinker, S. 1993. Generalizations of regular and irregular morphological patterns. *Language and Cognitive Processes, 8,* 1–56.

Prasada, S., Pinker, S., & Snyder, W. 1990. Some evidence that irregular forms are retrieved from memory but regular forms are rule generated. Paper presented at the Thirty-first Annual Meeting of the Psychonomic Society, New Orleans, November 16–18.

Prince, A., & Pinker, S. 1989. Wickelphone ambiguity. *Cognition, 30,* 189–190.

Pyles, T., & Algeo, J. 1982. *The origins and development of the English language.* 3d ed. New York: Harcourt Brace Jovanovich.

Quine, W. V. O. 1969. Natural kinds. In W. V. O. Quine, *Ontological relativity and other essays.* New York: Columbia University Press.

Quinlan, P. 1991. *Connectionism and psychology: A psychological perspective on new connectionist research.* Chicago: University of Chicago Press.

Quirk, R. 1970. Aspect and variant inflexion in English verbs. *Language, 46,* 300–311.

Quirk, R., Greenbaum, S., Leech, G., & Svartvik, J. 1985. *A comprehensive grammar of the English language. New York: Longman.*

Raymond, E. S. (Ed.). 1991. *The new hacker's dictionary.* Cambridge, MA: MIT Press.

Rayner, K., & Pollatsek, A. 1994. *The psychology of reading.* Mahwah, NJ: Erlbaum.

Reiner, A., Albin, R., Anderson, K., D'amato, C., Penney, J., & Young, A. 1988. Differential loss of striatal projection neurons in Huntington's Disease. *Proceedings of the National Academy of Science USA, 85,* 5733–5777.

Renfrew, C. 1987. *Archaeology and language: The puzzle of Indo-European origins.* New York: Cambridge University Press.

Rey, G. 1983. Concepts and stereotypes. *Cognition, 15,* 237–262.

Rhee, J., Pinker, S., & Ullman, M. T. 1999. A magnetoencephalographic study of English past tense production. Paper presented at the Sixth Annual Meeting of the Cognitive Neuroscience Society, Washington.

Ridley, M. 1986. *Evolution and classification: The reformation of cladism.* New York: Longman.

Ritchie, G. D., Russell, G. J., Black, A. W., & Pulman, S. G. 1991. *Computational morphology: Practical mechanisms for the English lexicon.* Cambridge, MA: MIT Press.

Roberts, T. P. L., Poeppel, D., & Rowley, H. A. 1998. Magnetoencephalography and magnetic source imaging. *Neuropsychiatry, Neuropsychology, and Behavioral Neurology, 11,* 49–64.

Rosch, E. 1978. Principles of categorization. In E. Rosch & B. B. Lloyd (Eds.), *Cognition and categorization.* Mahwah, NJ: Erlbaum.

Rosch, E. 1988. Coherences and categorization: A historical view. In F. Kessel (Ed.), *The development of language and of language researchers: Papers presented to Roger Brown.* Mahwah, NJ: Erlbaum.

Rosenblatt, J., & Mitchison, T. J. 1998. Actin, cofilin, and cognition. *Nature, 393,* 739–740.

Rossen, M., Klima, E. S., Bellugi, U., Bihrle, A., & Jones, W. 1996. Interaction between language and cognition: Evidence from Williams syndrome. In J. H. Beitchman, N. J. Cohen, M. M. Konstantareas, & R. Tannock (Eds.), *Language, learning and behavior disorders: Developmental, biological, and clinical perspectives.* New York: Cambridge University Press.

Rosten, L. C., & Ross, L. Q. 1989. *The education of H*Y*M*A*N K*A*P*L*A*N.* New York: Harcourt Brace.

Rumelhart, D. E., & McClelland, J. L. 1986. On learning the past tenses of English verbs. In J. L. McClelland, D. E. Rumelhart, & the PDP Research Group, *Parallel distributed processing: Explorations in the microstructure of cognition.* Vol. 2. *Psychological and biological models.* Cambridge, MA: Bradford Books/MIT Press.

Rumelhart, D. E., McClelland, J. L., and the PDP Research Group. 1986. *Parallel distributed processing: Explorations in the microstructure of cognition.* Vol. 1. *Foundations.* Cambridge, MA: MIT Press.

Rumelhart, D. E., Smolensky, P., McClelland, J. L., & Hinton, G. E. 1986. Schemata and sequential thought processes in PDP models. In J. L. McClelland, D. E. Rumelhart, and the PDP Research Group, *Parallel distributed processing: Explorations in the microstructure of cognition.* Vol. 2. *Psychological and biological models.* Cambridge, MA: Bradford Books/MIT Press.

Sampson, G. 1987. A turning point in linguistics. *Times Literary Supplement,* June 12, 1987, 643.

Saussure, F. de. 1916/1959. *Course in general linguistics.* New York: McGraw-Hill.

Say, T. 1998. Regular and irregular inflection in the mental lexicon: Evidence from Italian. Unpublished manuscript, Department of Language and Linguistics, University of Essex.

Schneider, W. 1987. Connectionism: Is it a paradigm shift for psychology? *Behavior Research Methods, Instruments, and Computers, 19,* 73–83.

Schreuder, R., & Baayen, R. H. 1995. Modeling morphological processing. In L. Feldman (Ed.), *Morphological aspects of language processing.* Mahwah, NJ: Erlbaum.

Seidenberg, M. S. 1992. Connectionism without tears. In S. Davis (Ed.), *Connectionism: Theory and practice.* New York: Oxford University Press.

Seidenberg, M. S., & Bruck, M. 1990. Consistency effects in the generation of past tense morphology. Paper presented at the Thirty-First Annual Meeting of the Psychonomic Society, New Orleans, November 16–18.

Selkirk, E. O. 1982. *The syntax of words.* Cambridge, MA: MIT Press.

Sellar, W. C., & Yeatman, R. J. 1930/1970. *1066 and all that.* London: Methuen & Co.

Senghas, A., Kim, J. J., & Pinker, S. 1999. Plurals-inside-compounds: Morphological constraints and their implications for acquisition. Unpublished manuscript, Department of Brain & Cognitive Sciences, MIT.

Senghas, A., Kim, J. J., Pinker, S., & Collins, C. 1991. Plurals-inside-compounds: Morphological constraints and their implications for acquisition. Paper presented at the Sixteenth Annual Boston University Conference on Language Development, October 18–20.

Sereno, J. A., & Jongman, A. 1997. Processing of English inflectional morphology. *Memory and Cognition, 25,* 425–437.

Seuss, Dr. 1987. *The tough coughs as he ploughs the dough: Early writings and cartoons by Dr Seuss.* New York: William Morrow.

Shaoul, C. 1993. Regularization in the inflection of French nouns. Unpublished manuscript, Department of Brain & Cognitive Sciences, MIT.

Shastri, L. 1999. Advances in SHRUTI: A neurally motivated model of relational knowledge representation and rapid inference using temporal synchrony. *Applied Intelligence.*

Shastri, L., & Ajjanagadde, V. 1993. From simple associations to systematic reasoning: A connectionist representation of rules, variables, and dynamic bindings using temporal synchrony. *Behavioral and Brain Sciences, 16,* 417–494.

Shenkman, K. D. 1994. Structure sensitivity and language processing in adult learners of English. Doctoral dissertation, Department of Psychology, University of Rochester.

Shepard, R. N. 1987. Toward a universal law of generalization for psychological science. *Science, 237*, 1317–1323.

Shirai, Y. 1997. Is regularization determined by semantics, or grammar, or both? Comments on Kim, Marcus, Pinker, Hollander & Coppola 1994. *Journal of Child Language, 24*, 495–501.

Shopen, T. (Ed.). 1985. *Language typology and syntactic description.* 3 vols. New York: Cambridge University Press.

Sloman, S. A. 1996. The empirical case for two systems of reasoning. *Psychological Bulletin, 119*, 3–22.

Smith, E. E., Langston, C., & Nisbett, R. 1992. The case for rules in reasoning. *Cognitive Science, 16*, 1–40.

Smith, E. E., & Medin, D. L. 1981. *Categories and concepts.* Cambridge, MA: Harvard University Press.

Smith, E. E., Medin, D. L., and Rips, L. J. 1984. A psychological approach to concepts: Comments on Rey's 'Concepts and Stereotypes.' *Cognition, 17*, 265–274.

Smolensky, P. 1988. On the proper treatment of connectionism. *Behavioral and Brain Sciences, 11*, 1–74.

Smolensky, P. 1990. Tensor product variable binding and the representation of symbolic structures in connectionist systems. *Artificial Intelligence, 46*, 159–216.

Smolensky, P. 1995. Reply: Constituent structure and explanation in an integrated connectionist/symbolic cognitive architecture. In C. MacDonald & G. MacDonald (Eds.), *Connectionism: Debates on psychological explanations.* Vol. 2. Cambridge, MA: Blackwell.

Spencer, A. 1991. *Morphological theory.* Cambridge, MA: Blackwell.

Spieler, D. H., & Balota, D. A. 1997. Bringing computational models of word naming down to the item level. *Psychological Science, 8*, 411–416.

Sproat, R. 1992. *Morphology and computation.* Cambridge, MA: MIT Press.

Squire, L. R., Knowlton, B., & Musen, G. 1993. The structure and organization of memory. *Annual Review of Psychology, 44*, 453–495.

Stanners, R. F., Neiser, J. J., Hernon, W. P., & Hall, R. 1979. Memory representation for morphologically related words. *Journal of Verbal Learning and Verbal Behavior, 18*, 399–412.

Staten, V. 1992. *Ol' Diz: A biography of Dizzy Dean.* New York: Harper Collins.

Stemberger, J. P. 1983. Inflectional malapropisms: Form-based errors in English morphology. *Linguistics, 21*, 573–602.

Stemberger, J. P. 1985. An interactive activation model of language production. In A. Ellis (Ed.), *Progress in the psychology of language.* Vol. 1. Mahwah, NJ: Erlbaum.

Stemberger, J. P. 1989. The acquisition of morphology: Analysis of a symbolic model of language acquisition. Unpublished manuscript, Department of Linguistics, University of Minnesota.

Stemberger, J. P., & MacWhinney, B. 1986a. Frequency and the lexical storage of regularly inflected forms. *Memory and Cognition, 14*, 17–26.

Stemberger, J. P., & MacWhinney, B. 1986b. Form-oriented inflectional errors in language processing. *Cognitive Psychology, 18*, 329–54.

Stemberger, J. P., & MacWhinney, B. 1988. Are inflected forms stored in the lexicon? In M. Hammond & M. Noonan (Eds.), *Theoretical morphology: Approaches in modern linguistics.* New York: Academic Press.

Stevens, K., & Keyser, S. J. 1989. Primary features and their enhancement in consonants. *Language, 65*, 81–106.

Stevens, T., & Karmiloff-Smith, A. 1997. Word learning in a special population: Do individuals with Williams syndrome obey lexical constraints? *Journal of Child Language, 24*, 737–765.

Strauss, S. (Ed.). 1982. *U-shaped behavioral growth.* New York: Academic Press.

Stromswold, K. 1994. What a mute child tells us about language acquisition. Unpublished manuscript, Center for Cognitive Science, Rutgers University.

Stromswold, K. 1998. Genetics of spoken language disorders. *Human Biology, 70*, 297–324.

Swadesh, M. 1964. Linguistic overview. In J. Jennings & E. Norbeck (Eds.), *Prehistoric man in the new world.* Chicago: University of Chicago Press.

Taft, M. 1994. Interactive-activation as a framework for understanding morphological processing. *Language and Cognitive Processes, 9*, 271–294.

Tanz, C. 1971. Sound symbolism in words relating to proximity and distance. *Language and Speech, 14*, 266–276.

Tenenbaum, J. 1999. A Bayesian framework for concept learning.

Doctoral dissertation, Department of Brain & Cognitive Sciences, MIT.

Teuber, H.-L. 1955. Physiological psychology. *Annual Review of Psychology, 9,* 267–296.

Tiersma, P. M. 1982. Local and general markedness. *Language, 58,* 832–849.

Tooby, J., and DeVore, I. 1987. The reconstruction of hominid evolution through strategic modeling. In W. G. Kinzey (Ed.), *The evolution of human behavior: Primate models.* Albany: SUNY Press.

Tramo, M. J., Loftus, W. C., Thomas, C. E., Green, R. L., Mott, L. A., & Gazzaniga, M. S. 1995. Surface area of human cerebral cortex and its gross morphological subdivisions: In vivo measurements in monozygotic twins suggest differential hemispheric effects of genetic factors. *Journal of Cognitive Neuroscience, 7,* 267–91.

Twain, M. 1880/1979. The awful German language. Reprinted in *The Unabridged Mark Twain.* Philadelphia: Running Press.

Tyler, L. K., Karmiloff-Smith, A., Voice, J. K., Stevens, T., Grant, J., Udwin, U., Davies, M., & Howlin, P. 1997. Do individuals with Williams syndrome have bizarre semantics? Evidence for lexical organization using an on-line task. *Cortex, 33,* 515–527.

Ullman, M. T. 1993. The computation of inflectional morphology. Doctoral dissertation, Department of Brain & Cognitive Sciences, MIT.

Ullman, M. T. 1999. Acceptability ratings of regular and irregular past-tense forms: Evidence for a dual-system model of language from word frequency and phonological neighborhood effects. *Language and Cognitive Processes, 14,* 47–67.

Ullman, M., Bergida, R., & O'Craven, K. M. 1997. Distinct fMRI activation patterns for regular and irregular past tense. *NeuroImage, 5,* S549.

Ullman, M., Corkin, S., Coppola, M., Hickok, G., Growdon, J. H., Koroshetz, W. J., & Pinker, S. 1997. A neural dissociation within language: Evidence that the mental dictionary is part of declarative memory, and that grammatical rules are processed by the procedural system. *Journal of Cognitive Neuroscience, 9,* 289–299.

Ullman, M. T., Corkin, S., Pinker, S., Coppola, M., Locascio, J., & Growdon, J. H. 1993. Neural modularity in language: Evidence from Alzheimer's and Parkinson's diseases. Paper presented at the

Twenty-third Annual Meeting of the Society for Neuroscience, Washington, D.C.

Ullman, M. T., & Gopnik, M. 2000. The production of inflectional morphology in hereditary specific language impairment. *Applied Psycholinguistics.*

Ullman, M. T., & Pinker, S. 1990. Why do some verbs not have a single past tense form? Paper presented at the Fifteenth Annual Boston University Conference on Language Development, October 19–21.

van Dam, J. 1940. *Handbuch der deutschen sprache. Zweiter Band: Wortlehre.* Groningen: J. B. Wolter's Uitgevers-Maatschappij N. V.

van der Lely, H. K. J. 1997. Language and cognitive development in a grammatical SLI boy: Modularity and innateness. *Journal of Neurolinguistics, 10,* 75–107.

van der Lely, H. K. J., & Christian, V. 1998. Lexical word formation in specifically language impaired children. Unpublished manuscript, Department of Psychology, Birkbeck College, University of London.

van der Lely, H. K. J., Rosen, S., & McClelland, A. 1998. Evidence for a grammar-specific deficit in children. *Current Biology, 8,* 1253–1258.

van der Lely, H. K. J., & Stollwerck, L. 1996. A grammatical specific language impairment in children: An autosomal dominant inheritance? *Brain and Language, 52,* 484–504.

van der Lely, H. K. J., & Ullman, M. T. 1996. The computation and representation of past tense morphology in specifically language impaired and normally developing children. In A. Stringfellow, D. Cahana-Amitay, E. Hughes, & A. Zukowsiki (Eds.), *Proceedings of the Twentieth Annual Boston University Conference on Language Development.* Vol. 2. Somerville, MA: Cascadilla Press.

van der Lely, H. K. J., & Ullman, M. T. 1999. Past tense morphology in specifically language impaired and normally developing children. Unpublished manuscript, Department of Psychology, Birkbeck College, University of London.

Vargha-Khadem, F., Watkins, K., Alcock, K., Fletcher, P., & Passingham, R. 1995. Praxic and nonverbal cognitive deficits in a large family with a genetically transmitted speech and language disorder. *Proceedings of the National Academy of Sciences USA, 92,* 930–933.

Warnow, T. 1997. Mathematical approaches to comparative linguistics. *Proceedings of the National Academy of Sciences, 94,* 6585–6590.

Warnow, T., Ringe, D., & Taylor, A. 1996. Reconstructing the evolutionary history of natural languages. *Proceedings of the Seventh Annual ACM/SIAM Symposium on Discrete Algorithms.* New York: Association of Computing Machinery / Philadelphia: Society for Industrial and Applied Mathematics.

Watkins, C. 1992. The Indo-European origin of English. In *The American Heritage Dictionary of the English language.* 3d ed. Boston: Houghton Mifflin.

Weng, Z., & Sokal, R. R. 1995. Origins of Indo-Europeans and the spread of agriculture in Europe: Comparison of lexicostatistical and genetic evidence. *Human Biology, 67,* 577–594.

Westcott, R. 1970. Types of vowel alternation in English. *Word, 26,* 309–343.

Westermann, G., & Goebel, R. 1995. Connectionist rules of language. In J. D. Moore & J. F. Lehman (Eds.), *Proceedings of the Seventeenth Annual Conference of the Cognitive Science Society.* Mahwah, NJ: Erlbaum.

Weyerts, H., Penke, M., Dohrn, U., Clahsen, H., & Münte, T. F. 1997. Brain potentials indicate differences between regular and irregular German noun plurals. *NeuroReport, 8,* 957–962.

Whaley, L. 1997. *An introduction to typology: The unity and diversity of language.* Thousand Oaks, CA: Sage.

Wickelgren, W. 1969. Context-sensitive coding, associative memory, and serial order in (speech) behavior. *Psychological Review, 76,* 1–15.

Wiese, R. 1996. *The phonology of German.* New York: Oxford University Press.

Williams, E. 1981. On the notions 'lexically related' and 'head of a word.' *Linguistic Inquiry, 12,* 245–274.

Williams, J. 1975. *Origins of the English language: A social and linguistic history.* New York: Free Press.

Wilson, E. O. 1998. *Consilience: The unity of knowledge.* New York: Knopf.

Wise, S., Murray, E., & Gerfen, C. 1996. The frontal cortex-basal ganglia system in primates. *Critical Reviews in Neurobiology, 10,* 317–356.

Wittgenstein, L. 1953. *Philosophical investigations.* New York: Macmillan.

Woodworth, N. 1991. Sound symbolism in proximal and distal forms. *Linguistics, 29,* 273–299.

Wunderlich, D. 1992. A minimalist analysis of German verb morphology.

Arbeiten des Sonderforschungsberereichs 282: Theorie des Lexikons. Vol 21. Heinrich-Heine-Universität: Düsseldorf.

Xu, F., & Pinker, S. 1995. Weird past tense forms. *Journal of Child Language, 22,* 531–556.

Young, A., & Penney, J. 1993. Biochemical and functional organization of the basal ganglia. In J. Jankovic & E. Tolosa (Eds.), *Parkinson's Disease and movement disorders.* Baltimore: Williams & Wilkins.

Zwicky, A. 1970. A double regularity in the acquisition of English verb morphology. *Papers in Linguistics, 3,* 411–418.

Zwicky, A. M. 1975. Settling on an underlying form: The English inflectional endings. In D. Cohen & J. Wirth (Eds.), *Testing linguistic hypotheses.* Washington: Hemisphere.

INDEX